The University of Georgia Humanities Center Series on Science and the Humanities

Betty Jean Craige, Series Editor

This series makes available to a broad audience of scholars and interested lay readers original works of scholarship in such areas as the history and philosophy of science, the relationship of aesthetic expression to scientific expression, the ethics of science and technology, environmental ethics, the social consequences of scientific research and knowledge, and the study of Western and non-Western conceptions of nature.

Thomas Henry Huxley's Place in Science and Letters

Thomas Henry Huxley's Place in Science and Letters

CENTENARY ESSAYS

Edited by Alan P. Barr

THE UNIVERSITY OF GEORGIA PRESS

ATHENS & LONDON

Set in Monotype Bulmer by G&S Typesetters, Inc.
Printed and bound by Braun-Brumfield Inc.
The paper in this book meets the guidelines for
permanence and durability of the Committee on
Production Guidelines for Book Longevity of the
Council on Library Resources.

Printed in the United States of America

01 00 99 98 97 C 5 4 3 2 1

LIBRARY OF CONGRESS CATALOGING IN PUBLICATION DATA

Thomas Henry Huxley's place in science and letters :
centenary essays / edited by Alan P. Barr.
p. cm.
Includes bibliographical references.
ISBN 0-8203-1865-5 (alk. paper)
1. Science—History—19th century. 2. Huxley, Thomas Henry,
1825–1895—Contributions in science. I. Barr, Alan P.
Q125.T427 1997
574′.092—dc20 96-13705

BRITISH LIBRARY CATALOGING IN PUBLICATION DATA AVAILABLE

For Dace—a joyous evolution

CONTENTS

ACKNOWLEDGMENTS

I WOULD LIKE FIRST TO EXPRESS MY APPRECIATION TO MY colleagues around the world as well as across disciplines who have been so patient and forthcoming in contributing their essays to this volume. The inevitable delays, the complexities of the process—from solicitation to final editing—and the frustrations of communication did not, I am delighted to say, take their toll. Often I was astonished at the grace and speed with which my requests and queries were satisfied.

I am also grateful for all the gracious assistance I received from Anne Barrett, the archivist at Imperial College, London—from guiding me through the materials to securing reproducible photographs. I am likewise indebted to the Wellcome Institute Library, London, for permission to use the picture "Huxley Illustrating a Gorilla."

My own university, Indiana University Northwest, has been most generous in its support of this research through Summer Research Grants—for which I am thankful.

Finally, I express my gratitude for the wonderful encouragement and assistance I have received at all levels from the University of Georgia Press—from Betty Jean Craige's early enthusiasm to Malcom Call's efficient follow-through to Kelly Caudle's expeditious supervision and to Anne R. Gibbons's superb editing.

Centenary publications can, in fact, be appropriately celebratory.

ABBREVIATIONS

Throughout these essays, the following abbreviations have been used:

CE — *Collected Essays.* 9 vols. 1894. N.Y.: Greenwood Press, 1968.

Diary — *T. H. Huxley's Diary of the Voyage of the HMS* Rattlesnake. Ed. Julian Huxley. London: Chatto & Windus, 1935.

HP — Huxley Papers. Imperial College of Science, Technology, and Medicine Archives, London.

Life — Leonard Huxley. *Life and Letters of Thomas H. Huxley.* 2 vols. N.Y.: Appleton, 1901.

SM — *The Scientific Memoirs of Thomas Henry Huxley.* Ed. Michael Foster and E. Ray Lankester. 4 vols. plus supplement. London: Macmillan, 1898–1903.

INTRODUCTION

On the occasion of his death in 1895, the elegies to Thomas Henry Huxley resonated with praises of his sturdy forthrightness and integrity, his scientific accomplishments and influence, and his stature as a sage who had ventured into cultural and religious criticism. To celebrate his hundredth birthday, *Nature,* a journal that Huxley had been instrumental in founding and nurturing, issued a supplement devoted to recollections of him. Because his effect as a scientist was still so recent (a number of his students were contributors), many of the essays dealt with Huxley as a scientist (in biology, anthropology, botany, or paleontology) and as a science educator; his relationship to evolution was of course also very visible.

Seventy years later, one hundred years after his death, the critical landscape has in part changed and in part remained strangely familiar. Huxley's specific contributions as a scientist have faded from prominence—perhaps not surprisingly. Robert Reid's essay is the only one in this volume that sets out explicitly to discuss his scientific accomplishments, importantly reminding us that his impress on biology was in fact significant and that biologists continue to explore issues he presented. Generally, there is far greater interest in the more humanistic dimensions of Huxley: the philosophical implications of his science, the exact color of his evolutionary convictions, the dynamics of his controversies with the church and sundry authorities, and his political and social views. His campaign as an advocate for science and as a science educator has always been recognized and commented on; Cyril Bibby alone has filled half a shelf on the subject. Huxley's prose style and rhetoric, whether written or extemporaneous, have likewise been consistently appreciated and often analyzed.

A century after his death, Huxley remains a formidable and compelling figure. He combined his individual genius, sparkle, and capacity with the Victorian drive to produce, to succeed, and to be socially responsible.

Whatever other reservations he may have had about Francis Bacon, he assuredly shared his unreserve about feeling all the world's knowledge to be his province. His energy and range of interests and competencies remain striking: science (biology, paleontology, geology, zoology, anatomy, physiology, and botany), anthropology, education, languages (German, Italian, French, Latin, and Greek), philosophy, literature, politics, philology, and even theology and biblical studies. He has the further attraction of always seeming to write and speak clearly, effectively, and authoritatively—in a way that conveys honesty and intelligence.

For over thirty years Huxley was at the center of the British scientific establishment, for the research he did, the issues he debated, the education reforms that he affected, and the policies he helped set. From the time he entered the lists of the British Association meeting at Oxford in 1860 as Darwin's "bulldog," he was never far from the center of controversy and policy making. As a founding editor of *Nature*, as an early mover in the X-Club, as president of the Royal Society—and the list of offices and positions becomes extensive—Professor Huxley was remarkably able to make his views, particularly his championing of science education, heard.

Many of the issues that Huxley engaged are still pertinent, sometimes even in the same form. He may have felt confident that by the 1870s there could be no serious intellectual disagreement left about the validity of evolution; today we have colleges of Creation Science and school boards wanting to give creationism "equal time" in science curricula. Even if the truth of evolution is not a seriously debatable question among scientists today, there is no shortage of discussion about its details. Contemporary biologists may be far more comfortable with the mechanism of natural selection than Huxley ever was, but the other difference he had with Darwin—over saltation—has certainly not disappeared, as Gould and Eldredge argue persuasively for "punctuated equilibria."

Huxley campaigned relentlessly to free science from what he called ecclesiasticism and to make public education secular. He demanded of intellectual inquiries that they be free and open (disinterested), uncolored and uncoerced by any preferred conclusions. We now face the prospect of an amendment allowing school prayer, behind which is the sentiment that the United States is, after all, a Christian nation and should tailor its education to Christian principles. The terms of the disagreement can be dismayingly similar, as we hear certain political voices opting for right thinking and then eerily recall Huxley's inveighing against Dr. Wace for suggesting that it ought to be "an unpleasant thing for a man to have to say plainly that he does not believe in Jesus Christ." Subordinating truth to desire has not lost its appeal, which makes Huxley's concept of agnosticism refreshing in its rigor and appropriateness.

This zeal for truth (that Reed finds at the basis of Huxley's thinking about ethics and that others in this collection observe) led me in my essay to concentrate on what this term, which he used so often and became so associated with, meant for Huxley. Truth, honesty, veracity, and the like punctuate his own writings (formal as well as informal) and the responses to him. I tried to do more than just see him as another product of a high Victorian moral position—by looking at the views he inherited, including the ones that do not usefully apply, how the demand for truth works rhetorically, and how it abets his scientific, anticlerical, professional, and educational campaigns—issues that Michael Ruse likewise visits.

As a principal player in the world of nineteenth-century science, Huxley was very aware of its power to affect the world, our views of it, and our lives. But this was no simple or trivial development. Once science began to achieve real power, the question of its relation to ethics became a serious and engaging one. With seeming inevitability, Huxley's later writings turn increasingly to questions of ethics, culminating in "Ethics and Evolution" and its "Prolegomena" (of 1893 and 1894). These are the treatises that John Reed is primarily drawn to in his examination of the scientist-philosopher and the question of morality. Driving Huxley's ethics was a commitment to view the world with honesty and reason and a trust that this kind of "veracity of thought" and resoluteness in facing things as they are is possible (perhaps, as Reed suggests, a trust that might strike our more jaded sensibilities as naive).

Galileo's telescope and Copernicus's renovated cosmology may have incommoded traditional beliefs, but such upsets were comparatively rare before the nineteenth century. Now, from the monumentally aged rocks of Lyell to the Darwinian reconstruction of biological history to such transforming innovations as electricity and steam, the old world order was indeed under siege. That humans might have evolved along with other apelike mammals, that evolutionary science dispensed with the need for "a creation" (and perhaps, by implication, a creator) thrust the party of science smack into the middle of the animated Victorian religious controversies.

More than ever before in history, science seemed to be irrevocably at odds with theology, which, willy-nilly, made it a part of the energetic discussion of ethics.

Huxley's considerable part in the confrontation between the Christian moral tradition he inherited and the ethics of the skeptical scientific outlook he cultivated is Reed's concern. He elaborates the difficult paradox of "an independent being warring against that which gave it its being"—the image that Huxley spells out in "Evolution and Ethics" of the need for ethical humanity to combat amoral nature (a combat all the more complicated by the questions evolution raises, including whether an ethical sense has itself evolved). These issues are very much alive, as we recall the recent furor over the idea of sociobiology; we apparently prefer that nature—at least the human part of it—not be quite so indifferent to morality, not quite so much the product of a cosmic-genetic crapshoot.

David Knight's review of Huxley's philosophy of science, where he had "to explore the nature of scientific thinking and practice itself," discovers this kind of truthfulness or agnostic freedom of inquiry to be fundamental. He is cognizant of Huxley's awareness of the scientific traditions he inherited and of almost consciously having to shape his own philosophy. Knight describes Huxley's method as a scientist as more empirical than theoretical, a view that Michael Ruse shares. Nor can Knight resist the temptation (any more than in similar contexts Robert Reid and John Reed can) to imagine Huxley's practical, empirical method of science transposed into a modern setting of uncertainty principles and quantum theory, the highly theoretical world of physics.

One of the considerable differences between the content of this collection and the previous criticisms of Huxley is the attention paid to Huxley's early career—before the publication of Darwin's *Origin.* Prior to the publication by Julian Huxley (in 1935) of his grandfather's *Diary of the Voyage of HMS* Rattlesnake, the material, as Joel Schwartz notes, was difficult to obtain. Even since, the temptation has been (understandably) to focus on the better-known and more colorful mature scientist and controversialist. Schwartz's essay usefully reminds us of the inherent interest of the *Rattlesnake* years and of its significance for his subsequent doings. In showing how Huxley, as a careful observer, moved from recapitulation theory and the mutability of species, skeptical of the transitions required of evolution and disinclined to theorize, to an increasingly less-hesitant Darwinian, Schwartz returns us to the familiar question of Huxley's relation to Darwinism.

Sherrie Lyons, Peter Bowler, Michael Ruse, and Mario di Gregorio all return to this question—one that continues to be at the center of modern discussions of Huxley—and offer revisions of what it means to see Huxley as an evolutionist, as a Darwinian. When and to what extent did he endorse Darwin's views? How long and how deeply did he remain skeptical about the mechanism of natural selection, and how serious was their difference over saltation? Lyons concentrates on *Man's Place in Nature* as evidence of Huxley's commitment to evolution. She sees him, as "*the* premier advocate of science in the nineteenth century," as having, in *Man's Place,* seized upon the centrality of the question of human ancestry. She sees him striving, as an evolutionist, to convince his audiences that the similarities between humans and apes did not diminish the dignity of either. The questions of origins and of ethics are distinct, culminating in "Ethics and Evolution," and recalling us to the subject of John Reed's essay.

Peter Bowler makes it clear from the onset that Huxley's "conversion" to Darwinism is still an open subject. Huxley's innovative work with horse and crocodilian fossils, especially the transitions he asserted between vertebrate forms, spoke for his conversion. Still, Bowler prefers to retain his term "pseudo-Darwinian" as an accurate description of Huxley. Though he showed little interest in the actual process of adaptation and might admit the possibility of "internally programmed trends" driving evolution, Huxley did see the present "as the product of a historical process governed by the interaction between the possibilities of evolution and the limitations of the environment," and he rejected any preprogrammed goals or mysterious developmental forces behind evolution.

Michael Ruse is less uncertain or qualifying. He wants to be "absolutely clear" about Huxley's commitment to the truth of evolution—even though he was not an early comer to the fold and was cautious about the evidence for natural selection. What intrigues Ruse is the paradox of his justifiable reputation as a Darwinist and his exclusion of evolution from his own teaching. He finds the resolution in Huxley's suspicion of evolution as *science* (rather than as a new religion, of which he was a prophet) and in his own scientific roots in the German tradition of "types"—a subject that Mario di Gregorio pursues.

Di Gregorio is persuasive that the ongoing discussion of Huxley's relationship to Darwin has to take into account Huxley's strong connections with German science. It was, he suggests, the biology of Karl Ernst von Baer, Carl Gegenbaur, and especially Ernst Haeckel that determined Huxley's "evolutionism" more than Darwin's (which differs from Ruse's image of Spencer as the primary influence). Considering Huxley's initial training in types, Haeckel's *Generelle Morphologie* would have been more congenial to him—with its accommodation of *Descendenztheorie* and its independence of natural selection and gradualism—than strict Darwinism. The entire back-and-forth relationship between Huxley and Haeckel emerges as one that invites further exploration.

Although Lyons, Bowler, Ruse, and di Gregorio all make use of Huxley's actual science in their discussions, Robert Reid concentrates on the significant and lasting value of the science itself. As a zoologist, Reid finds Huxley's biology exciting and vital. Like Schwartz, Reid is interested in Huxley's early work, locating his *Rattlesnake* biology within the tradition of Cuvier and Owen—but even so, with "an evolutionistic leaning." He is impressed with Huxley's diligence and his impatience, right from the start, with careless or questionably honest work. Like Lyons, he appreciates *Man's Place in Nature* as a still-useful text, specifically as a historical introduction to anthropology. His sense is that it was in paleontology that Huxley's science was most original. Joining a chorus that precedes this collection, Reid emphasizes how important in the development of biology Huxley's teaching, teaching advocacy and reforms, and textbooks were. "His influence," Reid says, "on the teaching of comparative anatomy and physiology is still felt."

If the scientific traditions and organizations figured into Huxley's fashioning of himself as a professional, so, too, from a different perspective, did his marriage. As a person who was able to express in letters and in an essay such as "Emancipation—Black and White" attitudes toward women and their education that were liberal for his society, Huxley still had to forge a domestic arrangement for himself, an arrangement that would inevitably be influenced by that society. Moreover, as Paul White discusses, as a scientist in a world that had not yet found a place for and come to respect scientists, Huxley's marriage had an extra weight to carry. Issues of masculinity and femininity were not so comfortably clear-cut for the Victorian scientist, who worked at home, thereby participating in the "feminine sphere." The questions multiply and fascinate: how a scientific career was connected to the place of women (particularly those married to scientists); how the issue of religion loomed over such households; how much a working professional husband located within the house intensified the separation of spheres or broke them down—or both. The whole relationship of Henrietta and Thomas Huxley, as White makes clear, is fascinating—and up to now inadequately examined.

In discussing Huxley's marriage, White comments on the scientific quest that brought Huxley to Australia, where he met Henrietta, and on how in such journeys the notions of progress (in this case scientific) and imperialism (British) dovetailed. That imperialism, which intersected with Huxley's career so early, is what Pat Brantlinger unravels. Brantlinger would look beneath the commonplaces of Huxley's Victorian liberal political opinions to "situate Huxley's general scientific and sociological thinking in the context of late-Victorian imperialism." It is a world, we should remember, where war and empire-building were the norm and where—less obviously—the mastery of nature's laws paralleled political-imperial mastery. In an unexpected twist, one that is much harder on Huxley's ethics than John Reed's treatment, the ethical process for Huxley even leads to imperialism, as the more civilized ("fittest") groups prevail—often with the help of the workers in science. Huxley may have been one of the milder and more humane of those whose views converged upon Social Darwinism, but Brantlinger finds him very much a believer in the power of scientific knowledge and the effective translation of that power into colonization. It is fitting that Huxley's career began as a scientific surveyor in the service of the Royal Navy—and that Brantlinger's argument is in a volume alongside Schwartz's differently focused discussion of the *Rattlesnake* voyage.

From the time he was a child, Huxley was evidently interested in drawing and adept at it. An important part of what he brought back from his navy voyage was his drawings and watercolors—of the specimens he had examined and of the terrain, people, and structures he had seen. Peers and students of his often commented on his ability to illustrate; Schwartz, Ruse, and White remark on it in passing. But George Bodmer seems to be the first to have studied it in the context of Victorian cultural training, printing and drawing techniques, and Huxley's own writing and teaching. Illustration enabled Huxley to express his own visual experiences as his world became wider and more intricate; through drawing he could convey this vision and the arguments it implied—in texts and in the lecture hall. Bodmer shows how Huxley, a competent drawer, used his art to illustrate and enhance scientific arguments.

Even Huxley's late career, which has received considerable attention and analysis, turns out to be amenable to new retrospections. Bruce Sommerville and Michael Shortland use the example of Huxley's homage to Voltaire, "The Method of Zadig," to propose the unexpected conclusion that behind their very different surfaces and reputations, Huxley the vigorous rationalist and agnostic and H. G. Wells the romantic fantasist actually proceeded by similar methods. Their triumphant protagonists, Zadig and the Time Traveller, both work in ways that are essentially scientific, and both progress (moving back-and-forth between hypotheses and observations) toward closer approximations to "scientific truth." Not only does their essay return this volume to the subject of truth, it also reaches again into the issue of the amorality of nature and evolution.

Huxley's final controversy, one that was unresolved at his death, was with A. J. Balfour. The usual image passed on of Huxley is that of a victorious controversialist; from his early tangles with Owen and Wilberforce to his late wranglings with Wace and Gladstone, his opponents did not fare well. Lightman turns to Huxley's rarely discussed two-part review of Balfour's *Foundations of Belief* (the second half was unrevised at Huxley's death) to question this conventional wisdom. He finds the touted rationalist at least matched in this last inning. By 1895 a more conservative wave was evident, and scientific naturalism was no longer quite so well received; Balfour was part of an intellectual spirit more respectful of tradition and tolerant of authority arguments than Huxley was.

My hope in soliciting these new commentaries on Thomas Henry Huxley is that they will contribute toward a more general appreciation of his stature as a thinker, writer, and scientist of the century that gave birth to ours. In his own right he is a striking figure; in the history of science (as a scientist, a historian of science, and a science teacher) his eminence is, I think, unarguable. His almost equal conversance with the world of humane letters stands as sharp instructor to those complacently on one side or the other of the "two cultures." Most of all, he is challenging and intellectually rewarding. He makes us aware of how much another age can say to ours.

"Common Sense Clarified"

Thomas Henry Huxley's Faith in Truth

ALAN P. BARR

SURELY ONE OF THE MOST STRIKING CHARACTERISTICS OF Thomas Henry Huxley's prose is its pervasive concern with truth and the related notions of sincerity, veracity, and integrity. In 1903 Henrietta Huxley compiled a collection of 385 aphorisms and reflections excerpted from her husband's writings. Twenty-nine of these centrally speak of truth and honesty. Three-quarters of a century later, James Paradis used as the epigraph for his study of Huxley a quote from "The Progress of Science: 1837–1887": "Nothing great in science has ever been done by men, whatever their powers, in whom the divine afflatus of the truth-seeker was wanting" (*CE* 1:56). Neither Mrs. Huxley nor Professor Paradis was specifically concerned with this issue of honesty; nevertheless, their selections, in this sense randomly chosen, reflect Huxley's preoccupation with integrity. Testimonials by contributors to the 1925 supplement to *Nature* on the centenary of Huxley's birth similarly emphasize this trait. Not surprisingly, so do the recollections written by Leonard, Julian, and Aldous Huxley.

Huxley's concern with truth is apparent in his style, tone, and rhetoric, as well as in his more overt choices of diction and subjects. It was an integral part of his behavior and judgment as a scientist, teacher, essayist, and social critic. His desire to be faithful to truth and his declared passion for truthfulness often colored his responses to people (and any ensuing controversies), his philosophical expressions, and his scientific proceedings. No motif punctuates Huxley's prose as characteristically as the demand for honesty.

Diogenes' lamp and Bacon's cynical shading of it did not, of course, pass exclusively to Huxley's podium. The Romantics' compulsion to be honest and the zealous earnestness of their Victorian inheritors contributed emphatically to the cultural climate of the nineteenth century. Rejecting what they saw as the artificial posturing and diction of the Augustan age, poets such as Wordsworth determined to speak simply (unguardedly) to ordinary people, and others like Shelley and Keats had limitless faith in the power of truth (and its equivalent, beauty) to perform like unacknowledged legislators working to ameliorate society's inequities. If Wordsworth suggested a tone of guileless sincerity in *The Prelude,* this manner defined the confessional writers from de Quincey and Carlyle to Ruskin and Oliphant—notably in Newman's *Apologia Pro Vita Sua,* where it provides the most persuasive justification for his conversion to Roman Catholicism.[1]

Although Huxley was by no means a suprahuman paragon of honesty, neither was he simply another manifestation or extension of a Victorian obsession. The attestations of his integrity by his contemporaries are numerous and impressive. O. C. Marsh, upon Huxley's death, declared: "His honesty, in the broadest sense of the word, was the dominant feature of the man. His love of truth for its own sake, wherever it might lead him, was one of the strongest elements in his character, and this resulted not only in his well-known intellectual honesty, but also in his hatred of the opposite, wherever found" (183). Leslie Stephen, in 1900, recalled that "Huxley's most prominent intellectual quality was his fidelity to fact" (908).

In the 1925 supplement to *Nature,* it was Leonard who was strongest on the subject of his father's integrity. After attributing Huxley's influence on his family to his sincerity and truthfulness, in word and deed, he concluded:

> I think that of all forms of immorality—and naturally he avoided that unscholarly euphemism which delicately restricts the word to the least delicate breaches of the moral code—he hated most the lie, dishonesty of word or act. Veracity he felt and knew to be the very foundation not only of intellectual but also of moral and social life. Firmly and inevitably he broke off relations with people whom he found he could not trust, no matter how close their former association, or how powerful their influence in the world where he moved. Indeed, against a lively talker who argued that truth was no virtue in itself, but must be upheld for expediency's sake only, he declared himself to be "almost a fanatic for the sanctity of truth." ("Home Memories" 699)

In trying to analyze his youthful feelings toward his father, Leonard speculated, "it was this living intensity of the passion for veracity which was at the bottom of the sense of awe that crept, as I have said, into our regard" (700).

The memorial volume also finds his protégé E. Ray Lankester referring to Huxley as "the unflinching champion of veracity" and quoting the master's private journal from 1856 on his "toleration for everything but lying" (702, 704). Likewise, E. B. Poulton cited Huxley's description of himself as "almost a fanatic for the sanctity of truth" and caring only about his reputation for honesty (705). Finally, Stephen Paget observed how,

> Like all prophets, [Huxley] had his share of resolute fighting for the truth. To him a man's principal duty was to hold and defend the truth, and to let nothing in life come between him and it. We had in Huxley a perfect example of zeal in preaching: there never was a more faithful servant of the truth.
>
> All his life, Huxley cared only to tell the truth, the whole truth, and nothing but the truth: he could not bring himself to any acceptance of half-truths, compromises, superstitions, and all sorts of guess-work. This was the secret of his love of fighting: so often as he came across a half-truth he had to hit it. . . . It was part and parcel of Huxley's preaching of the national need for truthfulness at all costs. (749)

As perceived by himself and his family and from the perspective of his colleagues and the generation succeeding him, Huxley was associated preeminently with the virtue of integrity. To the modern reader (me, for example), the fascination with truth in his prose is striking. This is not to accept at face value and perpetuate this identification of Huxley with truth. Subsequent commentators have certainly expressed reservations about the extent of his veracity. Rather, I am interested in what forms this familiar Victorian concern took with Huxley, how he used the term rhetorically as well as substantively, how he seems to have viewed it philosophically, and even how he may have wavered and stumbled.

Truth and its substantial cognates—integrity, veracity, honesty, and even virtue and morality—which occupy so pronounced a place in Huxley's discourse, have been perennial favorites of philosophers. In attempting to unravel a more precise appreciation of what it meant to him and how he used it, good sense, rigor, and academic virtue all urge a preliminary glance at the daunting library of philosophical theorizings about truth—even if only to dismiss these considerations as largely peripheral.

Philosophers, particularly ethicists, commonly recognize four principal theories of truth (the correspondence, coherence, performative, and pragmatic theories) and two ancillary or derivative approaches (through semantics and redundancy theory).[2] The various modern articulators, from F. H. Bradley, William James, and G. E. Moore to Bertrand Russell and from Alfred Tarski to P. F. Strawson, and their often elaborate expositions, however, are all dedicated to testing a statement's truth: whether it corresponds to our experience of reality, coheres within a larger system of meaningful assertions, is validated in practice or semantically. Generally, these are issues that Huxley ignored or presumed obvious. Rarely does he entertain the philosopher's or the semanticist's querulous concern with the nature of truth or meaning. Much more to his point are the questions of intention and practice: Did the person intend to tell the truth; is he or she being honest? Given a fact or truth, Huxley asks, do we acknowledge, report, and deal with it truthfully, rather than distorting or suppressing it.

If the relevance of truth theory to Huxley's usage is at best indirect, ethics, with its focus on sincerity, integrity, and virtue, offers another possibility. But as moral philosophy tended to evolve since the 1940s from an examination of truth statements and criteria of meaning to the study of ethical language, it seems equally remote from the concerns and linguistic habits of Huxley. Just as the intricately elaborated theories of truth set off—by contrast—a Victorian sage enviably confident that truth was real and discernible, so too do the expositions of people like R. M. Hare and Alasdair MacIntire set in relief that same Victorian equably assured of the moral assertions he invokes. Huxley would presumably have found appealing MacIntire's image of the cohesively integrated virtuous person situated within a cultural tradition—an ethical paradigm MacIntire locates at least as far back as Aristotle. He would, however, have wondered why a scholarly monograph was needed to argue that it was preferable to the emotive criterion of virtue, where what feels good becomes the basis of a hedonistic morality.

Because Huxley so much participated in what Lionel Trilling identifies as the reputation the English, especially in the nineteenth century, had for rectitude and sincerity, he would have been puzzled by and impatient with Trilling's very modern reflection *Sincerity and Authenticity*. The issues these lectures raise as they trace the history of sincerity from Rousseau and Diderot to the modern, post-Freudian world of angst would seem academic and contrived to a mind confident of its ability to recognize sincerity and to behave sincerely. Twentieth-century existentialism, with its decrying of inauthenticity, no matter how alluring, and the understandings of R. D. Laing and David Cooper of psychosis as a legitimate rebellion against an inauthentic world (both positions are discussed by Trilling) would appear evasive and dissembling from the perspective of Victorian moral eminence.

One recent study that does begin to address the issues surrounding integrity in a way that would be pertinent to Huxley is Mark Halfon's *Integrity: A Philosophical Inquiry*. Halfon explores different moral conundrums that can arise: What situations and actions demonstrate integrity, how compromise impinges, and the roles of adversity and consistency in defining our moral beings. What particularly delighted and amused me, as someone concerned with discussing Huxley and truth, was one of the examples Halfon conjures to make his point about intellectual integrity and its demands:

> Any person who has intellectual integrity is ordinarily committed to seeking the truth. . . . Let us imagine a nineteenth-century scientist who is alleged to have intellectual integrity and also believes that Darwinism is false. That the scientist believes Darwinism is false does not by itself betoken a lack of integrity. A relevant factor, though, is the extent to which the scientist has considered all of the relevant evidence that is available. If none of the available evidence weighs in favor of Darwinism, then the scientist's integrity can remain intact. If the available evidence provides support for Darwinism, then the scientist—as a person of intellectual integrity—must either refute the evidence or reevaluate and perhaps abandon his disbelief in Darwinism. If this occurs, the scientist's integrity can remain untarnished. The possibility arises, however, that the scientist, when confronted with evidence supporting Darwinism, chooses a different path. The scientist may elect to ignore the data. At this point a loss of integrity may result, since the scientist has compromised the commitment to seek the truth. If a person of intellectual integrity has pledged to pursue the truth, then all relevant and available evidence—conflicting or otherwise—must be acknowledged and examined. The failure to do so will ordinarily result in a loss of integrity. (33)

That Halfon next imagines a fundamentalist Christian who is committed to the defense of Christian principles but is confronted with recalcitrant evidence strongly suggests that he in fact did have some of Huxley's antagonists in mind—not only the scientist Richard Owen but also Bishop "Soapy" Sam Wilberforce and perhaps even Gladstone. Still, Huxley would not have sympathized with the complex, nuanced give-and-take of the ensuing discussion.

Especially when placed in the context of philosophical enquiries, Huxley emerges as appreciating very little ambiguity in the idea of truth and even less ambivalence about truthfulness. But the ways in which truth operated for Huxley were various and often interestingly linked. His frequent identification of scientific facticity with truth made it easy for him to see the scientist as the truth-seeker. This, in turn, projected onto the newly minted image of the scientist a crucially important moral patina. Paradis begins his chapter "The New Victorian" with the observation that

> If Mary Shelley presented the nineteenth century with the figure of the obsessed man of science, and Charles Dickens and George Eliot created images of the intense young researcher and physician, it was Huxley who formulated for the cultural imagination the ideal scientist as moral figure, prophet of order, and relentless seeker after truth. Selecting almost at random from Huxley's essays, one finds the general outlines and characteristics of the new man of science in the frequent references to what the scientist is, what he is concerned with, and how he carries out his quest for truth, in the voice and assumptions of the essays themselves. (11)

Huxley frequently waged his crusade for scientific truth with language and emphases more usually associated with moral and theological diatribes. He demanded of scientists the kind of integrity he often castigated clergymen for lacking, and he was not forgiving when it had been circumvented. Even Huxley's quasi-philosophical advocacy of agnosticism is a part of his image of truth-seeker, being alike appropriate to scientific, religious, or ethical inquiries.

If, as Huxley was fond of reporting, he owed to his youthful reading of Carlyle a disdain for sham and hypocrisy, he was soon able to parley this into a more encompassing vision.[3] As Paradis suggests, "Fusing his idea of the scientist with the stern moral vision that defined many of Carlyle's great men, he dramatized himself as both scientist and Carlylean truth seeker" (50). Throughout his career Huxley's conception of the scientist was of someone who sought the truths of nature—the only really reliable, interesting, and useful truths. This was a primary part of the allure that science held for him. As early as May 1851, having just been elected a Fellow of the Royal Society and groping to explain his puzzling devotion to science, he proclaimed to Henrietta: "A man who chooses a life of science chooses not a life of poverty, but, so far as I can see, a life of *nothing,* and the art of living upon nothing at all has yet to be discovered. You will naturally think, then, 'Why persevere in so hopeless a course?' At present I cannot help myself. For my own credit, for the sake of gratifying those who have hitherto helped me on—nay, for the sake of truth and science itself, I must work out fairly and fully complete what I have begun" (*Life* 1:74–75).

Nine years later, as a more secure and visible professional, Huxley anticipated, in "The Origin of Species," a time when the truths of science would replace hypotheses such as special creation and the whole history of such supernatural pleadings would yield to an understanding of "the natural order of phenomena" (*CE* 2:58). Shortly thereafter, in his major single volume, *Man's Place in Nature,* Huxley announced rather tersely and unequivocally, "Science has fulfilled her function when she has ascertained and enunciated truth" (7:151). In his 1866 address to the British Association (the year before the lecture at Oxford in which Matthew Arnold famously dismissed coal as mere "machinery"—*Culture and Anarchy*), Huxley "felt sure that at the present time, the important question for England was not the duration of her coal, but the due comprehension of the truths of science, and the labours of her scientific men" (*Life* 1:298–99). Discussing universities in "Actual and Ideal," Huxley pointedly adapts a quote from John Stuart Mill to insist upon this connection between science and truth: "the workers in Science . . . inspire no skepticism about the reality of truth or indifference to its pursuit. The noblest enthusiasm, both for the search after truth and for applying it to its highest uses, pervaded those writers" (*CE* 3:211).

This linkage is one that was coterminous with Huxley's professional life. In one of the last things he wrote, the 1894 preface to volume 6 of *Collected Essays,* he spoke of the intimate dependence of "natural knowledge" (science as well as philosophy; the volume includes his monographs on Hume and Berkeley) on truth:

> The development of exact natural knowledge in all its vast range, from physics to history and criticism, is the consequence of working out, in this province, of the resolution to "take nothing for truth without clear knowledge

> that it is such"; to consider all beliefs open to criticism [terms that are appropriate to skepticism as well as to his own formulation of agnosticism]. . . . It is that spirit which works and will work "without haste and without rest," gathering harvest after harvest of truth into its barns and devouring error with unquenchable fire. (ix–x)

As the figure or operative pursuing the labyrinth of nature, the scientist becomes for Huxley a philosopher in search of truth. Not only must he demonstrate "the divine afflatus of the truth seeker," but he also needs the associated qualities of rationality, disinterestedness, justness, and even self-abnegation. Often, Huxley's conception of the scientist assumes exalted, quasi-mythic status. Because he thought it might be "most useful," he attempted, in writing about Descartes, to give "some vision of this Extra-christian world [science], as it appears to a person who lives a good deal in it" and to try to show "by what methods the dwellers therein [scientists] try to distinguish truth from falsehood, in regard to some of the deepest and most difficult problems that beset humanity, 'in order to be clear about their actions, and to walk surefootedly in this life,' as Descartes says" (*CE* 1:195). Much of Huxley's campaign against theologians involved his reappropriating the language of religion from them for the scientists.

Science is not simply another profession or vocation. The image of the scientist, in Huxley's conception, is very much connected with morality, progress, and the larger life. Paradis indicates how "the struggle of the scientist, linked as it was to both truth and adversity, became symbolic for Huxley in a personal as well as a professional sense. It became part of the nobler life" (28–29). Many of the judgments Huxley rendered on other scientists were heavily tinted by how honest he thought them. He was critical of Francis Bacon for not being ethically true to the comprehensive quest for knowledge that he had taken for his patrimony, selling "that birthright for the mess of pottage of Court favour and professional success" (*CE* 1:47). Huxley's general distrust of Richard Owen's professional demeanor was capped by his contempt for the anatomist's dishonesty. The ultimate violation of any scientific code was to distort, suppress, or lie about evidence. This Owen did, for example, in the debatc over the hippocampus minor. Huxley's own behavior in openly acknowledging his error over Bathybius haeckelii provides a telling contrast.[4]

His positive response to Joseph Priestly similarly reflected the high value he placed upon veracity. He praised Priestly for being one of those "to whom the satisfaction of throwing down a triumphant fallacy is as great as that which attends the discovery of a new truth," concluding his appreciation of his forerunner's scientific achievement by emphasizing his contribution to the store of truth: "To all eternity, the sum of truth and right will have been increased by [the means of men like Priestly]; to all eternity, falsehood and injustice will be the weaker because they have lived" (*CE* 3: 13, 37).

In the middle of his career, in his essay on Descartes, Huxley made it clear that his estimation of this early modern was strongly tied to his intellectual integrity. He condensed Descartes's moral injunction into "learn what is true, in order to do what is right" and then called it the "summing up of the whole duty of man." One of "Descartes' great claims to our reverence as a spiritual ancestor [is] that at three-and-twenty, he saw clearly that this [search after knowledge which leads to action] was his duty, and acted up to his conviction" (*CE* 1:168–69). Sections of the *Discourse* provided a particularly congenial anticipation of Huxley's notion of agnosticism and of his ethic: "The central propositions of the whole 'Discourse' are these. There is a path which leads to truth so surely, that any one who will follow it must needs reach the goal, whether his capacity be great or small. And there is one guiding rule by which a man may always find this path, and keep himself from straying when he has found it. This golden rule is—give unqualified assent to no propositions but those the truth of which is so clear and distinct that they cannot be doubted" (1:169).

This is strikingly parallel to Huxley's description of agnosticism, a parallel that becomes even stronger as he continues: "Thus prepared to go on living while he doubted, Descartes proceeded to face his doubts like a man. One thing was clear to him, he would not lie to himself—would, under no penalties, say, 'I am sure' of that which he was not sure" (*CE* 1:171). Exactly opposite is the kind of figure Dr. Wace cuts in Huxley's essays on agnosticism, a person who feels it ought to be unpleasant to deny belief in Jesus Christ or the miracle of the Gadarene swine. "A thousand times, no!" Huxley remonstrates, "It ought *not* to be unpleasant to say that which one honestly believes or disbelieves." At the core of what Huxley erects as his agnostic platform is a plank that joins it to the quest for truth. An agnostic undertakes not to assert the truth of anything about which he or she is uncertain, which closely echoes what Huxley discovered in *The Discourse.* This "honesty of word or deed" he called "that most valuable of all qualities" (5:241), despite Wace's perverse determination to call it "infidelity" (5:307–8). Even in the comparatively mundane setting of the classroom, Huxley reported how he should have been "utterly ashamed of myself if, when I stood up as an instructor of others, I had not taken every pains to assure myself of the truth of that which I was about to say" (5:93).

Huxley's comments on Descartes also reflect the strong, professional connection he urged between truth and morality. It is not just a matter of another version of the commandment against bearing false witness. Truth, particularly scientific truth, is characteristically equated to fact or evidence, and being factually or scientifically right merges with being ethically right. Knowledge, wisdom, and morality, in effect, similarly begin with truth. In discussing such provocative doctrines as, for example, whether humans are a part of the evolutionary continuum of conscious automata, Huxley suggests that the "only question which any wise man can ask himself, and which any honest man will ask himself, is whether a doctrine is true or false" (*CE* 1:244). To the difficult and interesting problem of the Synoptic Gospels, he would, very much in the spirit of the then contemporary higher criticism, invoke the standard of historical truth: "And what is historical truth but that of which the evidence bears strict scientific investigation?" (5:270). Always, truth is to be supported by evidence and can survive objective scrutiny. Facts, whether historical, epistemological, or biological, are, as Huxley termed the facts of organic nature, "wonderful truths" (2:11). The truth or falsehood of a proposition, whether it be about Jesus or about the boundaries between species, "is a matter of evidence" not a personal desideratum (8:205). Speaking to a group of working men in 1863, Huxley described how scientists do not pledge themselves to creeds or beliefs, but instead follow that which is factual (that is, true)—and useful (2:469).

Scientific truths are useful because they directly contribute to progress, as Huxley explained through the metaphor of a chess game: "It is a very plain and elementary truth, that the life, the fortune, and the happiness of every one of us, and, more or less, of those who are connected with us, do depend upon our knowing something of the rules of a game infinitely more difficult and complicated than chess. . . . The chess-board is the world, the pieces are the phenomena of the universe, the rules of the game are what we call the laws of Nature" (*CE* 3:82). Because our well-being is so intimately connected with our competence as players, part of being liberally educated is having one's mind "stored with a knowledge of the great and fundamental truths of Nature and of the laws of her operations" (3:86). That person who neglects the truths of nature risks becoming her shuttlecock.

It was but a relatively short step for Huxley from the practical utility of scientific knowledge to its ethical utility. His provocative Romanes lecture "Evolution and Ethics" accomplished this step. Since "the cosmic process has no sort of relation to moral ends," he would have "us understand, once for all, that the ethical progress of society depends, not on imitating [it], still less in running away from it, but in combating it" (*CE* 9:83). In moral terms, nature is a fickle wanton. Only by study can we hope to channel it—not by abjectly relying upon the justice and benevolence of some divine engineer. His meliorism, however, envisioned "no limit to the extent to which intelligence and will . . . may modify the conditions of existence" (9:85).

With its insistence upon careful observation, honest reporting, and intellectual openness, science inculcates morality. At an ideal university, "the very air [the student] breathes should be charged with that enthusiasm for truth, that fanaticism for veracity . . . for veracity is the heart of morality" (*CE* 3:205). Such a university, needless to say, would emphasize science in its curriculum. Huxley, borrowing support from John Stuart Mill's inaugural address at St. Andrew's, trumpeted "the value of Science as knowledge and discipline. . . . 'In cultivating, therefore,' science as an essential ingredient in education, 'we are all the while laying an admirable foundation for ethical and philosophical culture'" (3:211–12). As Cyril Bibby, who has written extensively on Huxley's views of education, commented: "Outweighing all other arguments for science teaching in Huxley's mind was his belief that it could lead to what in his view was at the very center of morality, more honest thinking" (14). Huxley's many years of involvement with education boards and his commitment to public lecturing attest to the high value he placed on science education—for its specific content, but even more for its ethical instruction. For him, "The most zealous of popular lecturers can aim at nothing more than the awakening of a sympathy for abstract truth, in those who do not really follow his arguments; and of a desire to know more and better in the few who do" (*CE* 8:vii). More important than the provisional hypotheses and the specific truths of science are its intellectual forthrightness and its openness. An indispensable companion to truthfulness was reason: "The scientific spirit is of more value than its products, and irrationally held truths may be more harmful than reasoned errors" (2:229). It is also worth recognizing in passing that in no sense did Huxley imagine that only scientific education facilitated the teaching of honesty. One of the primary reasons he advocated including drawing in the school curriculum was that it fostered accuracy of observation, a necessary component of truthfulness.[5]

The ethical component of science, anchored in its devotion to truth, gave it—for Huxley—a much surer claim to moral authority than theology. The intellectual duplicity or chicanery of many theologians only intensified his ire. He confidently asserted in 1886 that "the moral sense and the really scientific method of seeking truth are once more predominating over false science." With the "spread of true scientific culture," he anticipated "a constant elevation of the standard of veracity" and that theology "will cease to have any relation to ethics" (*CE* 4:371–72).

Much of Huxley's argument with the clergy was essentially over ethics. As he states unambiguously in an 1893 preface to *Collected Essays,* "The truth is that the pretension to infallibility, by whomsoever made, has done endless mischief" (4:ix). Such a pretension violated his demand for a scientific, open search for truth and thus drew Huxley's especial rancor. It is the dangerously misguided cleric who

> asserts that it is morally wrong not to believe certain propositions, whatever the results of a strict scientific investigation of the evidence of these propositions. He tells us "that religious error is, in itself, of an immoral nature." He declares that he has prejudged certain conclusions, and looks upon those who show cause for arrest of judgment as emissaries of Satan. It necessarily follows that, for him, the attainment of faith, not the ascertainment of truth, is the highest aim of mental life. And, on careful analysis of the nature of this faith, it will too often be found to be, not the mystic process of unity with the Divine, understood by the religious enthusiast; but that which the candid simplicity of a Sunday scholar once defined it to be. "Faith," said this unconscious plagiarist of Tertullian, "is the power of saying you believe things which are incredible." (*CE* 5:313–14)

Such a position, so anathema to the formulation of a rigorous agnosticism, is as unenlightened as it is retrograde. This is why, by contrast, Huxley applauded Descartes's admonition to "learn what is true, in order to do what is right." He disdained any complacency before "the east wind of authority" (*CE* 1:168), in any profession, declaring that "Ecclesiasticism in science is only unfaithfulness to truth" (2:149).

The philosophical heinousness of ecclesiasticism is that it compromises our intellectual and moral obligation to attend open-mindedly to the world as it actually is. It perverts the skeptical and scientific spirit of inquiry, a spirit solely concerned with the truth or falsity of a doctrine. The history of theology (or of ecclesiasticism of any stripe) is one of ongoing antagonism to this attitude. It presumes to know "truth" by fiat or by faith, to value its declared truths above truthfulness, above the search for truth. Huxley's specific battles with the figures of orthodoxy were only, as he implied, the last chapter in an unfortunate but ancient war. "Who shall number," he chastened, "the patient and earnest seekers after truth, from the days of Galileo until now, whose lives have been embittered and their good name blasted by the mistaken zeal of Bibliolaters?" (*CE* 2:52).

From Huxley's earliest joust, as the heady upstart assaulting the parapets of Oxford and its bishop, to the public controversies in his sixties with Gladstone, Dr. Wace, and the Duke of Argyll, one thread that persists is his rage at the misappropriation of authority he saw in their biased pronouncements. When Gladstone insisted upon a Miltonic cosmogony, obdurately blind to scientific evidence, Huxley dismissed it simply as "not true," his ultimate criticism (*CE* 4:156). He went on, in "The Interpreters of Genesis and the Interpreters of Nature," to explain that "The antagonism of science is not to religion, but to the heathen survivals and the bad philosophy under which religion herself is often well-nigh crushed." He thought one of the "most beneficent functions" of true science was to relieve "men from the burden of false science which is imposed upon them in the name of religion" (4:162–63). Understanding orthodoxy as an older, rigid, false image of science not only enabled Huxley to present modern (true) science as rational and preferable but also made it clear why the "Lights of the Church" should be so strenuously hostile to the newer movement:

> This declaration of war to the knife against secular science, even in its most elementary form; this rejection, without a moment's hesitation, of any and all evidence which conflicts with theological dogma—is the only position which

> is logically reconcilable with the axioms of orthodoxy. If the Gospels truly report that which an incarnation of the God of Truth communicated to the world, then it surely is absurd to attend to any other evidence touching matters about which he made any clear statement, or the truth of which is distinctly implied by his words. If the exact historical truth of the Gospels is an axiom of Christianity, it is as just and right for a Christian to say, Let us "close our ears against suggestions" of scientific critics, as it is for the man of science to refuse to waste his time upon circle-squarers and flat-earth fanatics. (*CE* 4:230)

If Scriptural religion is an outmoded, fraudulent, or exploded "science," it is so because its standard of truth is hopelessly incompatible with a more historically sound, reasoned, verifiable view—and therefore untenable, though perhaps internally consistent. Happily, the contest is uneven. Everything, even time, Huxley assured, "is powerless against truth" (*CE* 1:255). "Extinguished theologians lie about the cradle of every science . . . and history records that whenever science and orthodoxy have been fairly opposed, the latter has been forced to retire from the lists, bleeding and crushed if not annihilated; scotched if not slain" (2:52).

Huxley demanded of thoughtful people, especially those in public offices, what his friend Matthew Arnold so pointedly demanded of criticism: disinterestedness. Lamentably, the voices of ecclesiasticism have betrayed that trust by narrowly, biasedly "knowing" in advance of the facts and arguments what the truths of evolution or of creation or even of Scripture were. Sometimes their evasions of what the best evidence seemed to indicate were smugly deliberate (thus, Huxley's thrust at an unctuous Oxford bishop who misuses his powers); sometimes they had the witlessness of zeal. Either way, it was "unfaithfulness to truth"—and retrogressive, anything but disinterested. Very different is what Mario di Gregorio describes as "Huxley's ideal of neutral, objective science," with its painstaking search for truth, where we "must remain detached and unemotional in the face of scientific observations; only thus will we be able to reach genuinely scientific conclusions" (143).[6]

In his writings it seems clear that Huxley saw himself as in harmony with this image of the honest and disinterestedly objective scientist. At twenty-eight, despondent about the difficulties of making his way in science in London, and separated by half a world from Henrietta, he could still write to her: "I am full of faults, but I am real and true and the whole devotion of an earnest soul cannot be overprized" (*Life* 1:91). In notes that Huxley made to himself on New Year's Eve 1856, as he awaited the birth of his first child, Noel, this sentiment assumed a distinctly Carlylean timbre. He undertook "To smite all humbugs, however big; to give a nobler tone to science; to set an example of abstinence from petty personal controversies, and of *toleration for everything but lying;* to be indifferent as to whether the work is recognized as mine or not, so long as it is done—are these my aims? 1860 will show" (1:162; italics mine). One of the things 1860 would show was how successful his projected three-year plan of study had been.

What 1860 did produce was one of the century's most remarkable letters. Noel had just died of scarlet fever, leaving the Huxleys devastated. Charles Kingsley wrote to offer consolation from his own Christian view of immortality. Huxley's long reply to him (six pages in *Life*) is sincere, poignant, and unyielding in its insistence upon honesty, no matter how wrenching the circumstances, no matter how tempting the consolation of immortality might seem. "Oh devil! truth is better than much profit. I have searched over the grounds of my belief, and if wife and child and name and fame were all to be lost to me one after the other as the penalty, still I will not lie. . . . The longer I live, the more obvious it is to me that the most sacred act of a man's life is to say and to feel, 'I believe such and such to be true'" (1:233–34).

It was useless, he explained to Kingsley, to say that he should accept the doctrine of immortality because the aspirations of humanity and perhaps his own (which he doubted) urged it. This was only asking him in grand words to believe a thing because he liked it. "Science," he countered, "has taught me the opposite lesson" (*Life* 1:235). Even at this emotional nadir of his life, he could not help what hard names people might call him; "One thing people shall not call me with justice," he proclaimed, "and that is—a liar. As you say of yourself, I too feel and lack courage; but if ever the occasion arises when I am bound to speak, I will not shame my boy" (1:238).

The next decade found a still less tentative scientist and public speaker able to confide to his wife: "But to say truth I am not greatly concerned about my reputation except that of being entirely honest and straightforward, and that reputation I think and hope I have" (*Life* 1:427). The brief autobiography that Huxley wrote in 1890 includes as an important part of its summary statement a eulogy to veracity (a eulogy that implies his own personal veracity):

> If I may speak of the objects I have had more or less definitely in view since I began the ascent of my hillock, they are briefly these: To promote the increase of natural knowledge and to forward the application of scientific methods of investigation to all the problems of life to the best of my ability, in the conviction which has grown with my growth and strengthened with my strength, that there is no alleviation for the sufferings of mankind except veracity of thought and of action, and the resolute facing of the world as it is when the garment of make-believe by which pious hands have hidden its uglier features is stripped off. (*CE* 1:16)

Huxley's style and rhetoric, which derived from his own certainty, contributed to this aura of honesty. His tendency to refer to the "truths of science" or of nature and even to personify Truth is apparent to the most casual reader. Julian Huxley recognizes that "it would not be unfair to say he occasionally made abstractions of truth and morality, that, in other words, he tended to deify them" (146).

More striking are Huxley's frequent rhetorical uses of "truth." Often it functions as a simple intensifier, not actually contributing to the meaning. For example, he wrote, "In truth, the intermediate step," "In truth, the new philosophy deserves," "In sober truth, to those who," and "But to say truth, although," rather than beginning directly with the next phrase, which, in each case, introduces the substance of his sentence (*CE* 4:334, 1:52, 9:195, 2:469). Huxley's fondness for the term "truth" led him to use it where others might have employed another, perhaps more economical expression. He spoke of "modern scientifically ascertained truths" where we might speak of theories or results; of estimates as "approximations to the truth" rather than simply approximate; and of its being "probably true that" instead of probable (4:191, 4:251, 1:309). His inclination was to use true (or truth or fact) where others might use correct (or accurate). "Surely, Aristotle was much nearer the truth in this matter" (1:307) may be smoother than simply saying, "Aristotle was more correct"; it also has the effect of building on Huxley's numerous repetitions and advancing his campaign for truth and for himself and other scientists as truth-seekers. Such examples abound. Whether this resorting to rhetorical intensifiers can be interpreted as a nervous mannerism masking a fear of not being believed, on the surface it projects a preoccupation with integrity. In a 1902 essay James R. Routh decided "the most marked thing about his style is its clearness and sincerity . . . a frankness and an earnestness which come from the bottom of the soul and carry conviction—a conviction of the absolute truthfulness of the man" (394). To his grandfather's declaration that "veracity is the heart of morality," Aldous noted, "It was also the heart of his literary style" (62).

Huxley's view of himself was clearly that of an intrepid straight-dealer; the witnesses to his professional and personal life attest to his uprightness; and his own writings and actions undeniably convey his concern with truth. These in no way, however, necessarily represent a closed and unimpeachable portrait. He talked a great deal about the virtue of integrity—including his own—and those who knew him generally concurred. Commencing, however, in the early years of the new century there is a note of doubt sounded. Paul Elmer More, in his strong visceral preference for a more romantic humanism than the unabashed scientism of Huxley, was suspicious of Huxley's passion for truth. He rather curiously saw this champion of science as "one of the cunning artificers lost to letters, an essayist who, if he had devoted his faculties to the more permanent aspect of truth, might have taken a place among the great masters of literature" (209). More concluded that even if "in one sense his strongest intellectual trait was, as his son says, 'an uncompromising passion for truth,' yet in the sum of his thinking he was one of the master sophists of the age"—because "his ends are not always the same, and . . . in the total effect of his works there lies concealed an insoluble ambiguity" (210–11). From his traditional, Christian rampart, Robert Shafer complained that Huxley's sense of papal infallibility "implies a temper very different from that of the seeker after truth. At times it seems, indeed, as if he felt so confident that the truth was already within him that he could venture boldly to play fast and loose with it" (154). A somewhat more urbane but equally insubstantial and general reservation about Huxley's celebrated honesty was that of A. O. J. Cockshut. In Cockshut's survey of "the scientific sage's" moral zeal, he described him as "angry, humorous, unbalanced in his emotion, very honest, and a little too conscious of it, clear in his thoughts and impatient of subtleties" (87). But this Victorian fashion of sincerity, which Cockshut found suspect, of course embraced a good many besides Huxley. The fashion, "begun perhaps by

Carlyle, for boasting continually of one's honesty" rankled Cockshut. He objected, "They were all very quick to defend what no one questioned. Sometimes it seems as if it was not so much their honesty they were really asserting as their masculinity" (88).

Among modern critics, Walter Houghton and Evelleen Richards have, from different perspectives, questioningly examined Huxley's "reputation for candour and sincerity," his position as "something like an apostle of veracity" (Houghton 161). Houghton finds that "when Huxley turned from straight exposition to generalizations about the values of science or religion, usually their relative values, and to cognate passages on scientists and clergymen, he was by no means reluctant to bolster a shaky or a biased argument, or even to support a good one, with the extra-emotional influence of rhetorical methods" (161). In this sense, he was like every other Victorian, other than "the saint of rationalism, J. S. Mill," neither saint nor sinner, not vicious nor unscrupulous, and usually kept straight by his love of humor even more than by his passion for veracity (Houghton 173). Houghton concludes:

> If Huxley's style reveals a bias, a failure to recognize both sides of an issue, a play of exaggeration and insinuation calculated to impose on unwary readers and critics, this is not because he was simply a slick politician with a party platform to defend. His anti-religious rhetoric had its source in a firm belief that the term "'Nature' covers the totality of that which is" [from *CE* 5:39], and that the religious view of man and the universe was not only an illusion, but that in the hands of its official upholders it was an obstacle to the advancement of scientific truth. (173–75)

It is as if the minor rhetorical bendings and stretchings pale before the larger truths of science and nature.

Evelleen Richards's challenges are more specific. Her essay about Huxley and woman's place in science argues that there was a disturbing gap between Huxley's articulated support of women's rights and some of his dealings with individual women pursuing those rights. While he paid lip service in such essays as "Emancipation—Black and White" to democratic sentiments in favor of equality of opportunity, in practice he agreed to exclude women from participation in scientific debates and from medical schools. Richards's evidence is powerful,[1] and we are left with the image of Huxley as far less thorough an advocate for women and far less consistent than we might have wished. (He is again especially unsaintly when seen alongside Mill.) Some of the incidents Richards cites (e. g., his refusal to intercede when Sophia Jex-Blake was denied admission to medical school) are disconcertingly at odds with his professions.

Still, how profoundly this compromises Huxley's concern with integrity—at least his philosophical commitment to it—is arguable. Like his willingness to bend rhetoric, his failures here may represent an all-too-human and common capacity to rationalize. Even for a scientist and self-proclaimed zealot for truth, "truth" can assume a disarmingly obliging malleability. Like Keats's beauty, it is often in the eye of the beholder or, more accurately in this instance, unapparent to the blinking or diverted eye. Huxley shared the common Victorian view that women were by and large less intellectually capable than men; he also genuinely felt that was all the more reason not to handicap them further by denying them equal access to education.[7]

Perhaps more interesting than Huxley's fall from liberal philosophical grace is what that fall illustrates. Even in the case of someone so overtly concerned with honesty, when a particularly compelling concern arose, he could overlook certain inconvenient contradictions. Most of us choose the areas where we will be especially insistent upon integrity, and my sense is that Huxley was far more zealous about integrity in questions of "hard" facts, of science and nature reporting, than he was in matters of rhetoric or of social or political debate. I can well imagine his contending that the advancement and professionalizing of science took precedence over the risky, in his eyes, admission of "ill-prepared" women to the debates. That he stumbled and fell should not obscure what remained a pervasive concern for him. He may just have been a far more impressive egalitarian in his essays than in his professional doings.[8]

What emerges, then, in Huxley is a recognizably Victorian partisan of truth, quite capable of unwittingly twisting and tampering with it (never seriously in doubt about his own veracity), able because of this conviction to transmute the moldy garment of clericism into the seedpods of natural science. He had the pulpiteer's flair for the rhetoric of sincerity and the scientist's respect for real, disinterested truth-seeking. His own compulsions to champion right-thinking causes and produce noteworthy results on the one hand made him occasionally unaware of his insensitivities and inconsistencies, but on the other left those around him thoroughly persuaded of his impeccable integrity. These compulsions also left him one of the most effective figures in the history of nineteenth-century science. If, as Paradis suggested, it was the image of the "ideal scientist as moral figure, prophet of order, and relentless seeker after truth" that Huxley was most successful in establishing, this was a powerful image and a transforming achievement. Huxley was an extraordinarily effective advocate for science and science education. A good deal of his success was the result of his reputation for honesty, including his image as an honest speaker with a strong, clear prose style. As an "honest speaker," Huxley was an incomparably successful and effective purveyor of science, in large measure because he thought science akin to truth and that truth in the long run was "only common sense clarified" (*CE* 3:282).

NOTES

1. See George P. Landow, ed., *Approaches to Victorian Autobiography* (Athens: Ohio UP, 1979), for an indication of the range of Victorian autobiographies and their fascination with honesty.

2. For a useful introduction to these theories of truth see the *Encyclopedia of Philosophy,* which provides bibliographies leading to such figures as Russell, James, and Tarski.

3. See William Irvine for a full discussion of the elder sage's influence on Huxley. As late as December 2, 1890, Huxley wrote to Leonard: "This world is not a very lovely place, but down at the bottom, as old Carlyle preached, veracity does really lie, and will show itself if people won't be impatient" (*Life* 2:289).

4. For a discussion of the hippocampus minor argument, see Mario di Gregorio (134–39). For a recounting of the "discovery" of the Bathybius haeckelii and an incidental contrast of Huxley's behavior with E. H. Haeckel, his German coworker and co-Darwinian, see Nicolaas Rupke.

5. Huxley explains in some detail the advantages someone "endowed with the artistic sense of form" would have over the "mere osteologist . . . if he lacked the artistic visualizing faculty" in discerning and conveying biological class relationships. He prizes "the artist's power of visualizing the result of his mental processes, of embodying the facts of resemblance in a visible 'type'" (*SM* 4:664–65).

6. Sheridan Gilley and Ann Loades specifically attribute Huxley's antagonism to Comte to his "interestedly" perverting the workings of science: "Comte gave Huxley greatest cause for offense by arguing that science should serve the end of

perfecting a new social order; Huxley thought that this ideal would annihilate science as the disinterested pursuit of truth" (298).

7. Huxley writes in "Emancipation—Black and White": "Granting the alleged defects of women, is it not somewhat absurd to sanction and maintain a system of education which would seem to have been specially contrived to exaggerate all these defects?" He concludes the essay charging man with the duty to see "that injustice is not added to inequality" (*CE* 3:71, 75).

8. For a good discussion of Huxley's involvement in the movement to professionalize science, see Frank Turner.

See Alan Barr for a discussion of the often injudicious response of nineteenth-century ethnologists and biological scientists to the women's movement. Huxley was very much toward the sane and temperate end of that spectrum of scientific voices.

REFERENCES

Barr, Alan P. "Evolutionary Science and the Woman Question." *Victorian Literature and Culture* 20 (1992): 25–54.

Bibby, Cyril. "Science as an Instrument of Culture: An Examination of the Views of T. H. Huxley." *Researches and Studies* U of Leeds Institute of Education 15 (1957): 7–23.

Cockshut, A. O. J. *The Unbelievers: English Agnostic Thought, 1840–1890.* N.Y.: New York UP, 1966.

di Gregorio, Mario A. *T. H. Huxley's Place in Natural Science.* New Haven: Yale UP, 1984.

Gilley, Sheridan, and Ann Loades. "Thomas Henry Huxley: The War between Science and Religion." *Journal of Religion* 61 (1981): 285–308.

Halfon, Mark S. *Integrity: A Philosophical Inquiry.* Philadelphia: Temple UP, 1989.

Houghton, Walter. "The Rhetoric of T. H. Huxley." *University of Toronto Quarterly* 18 (1949): 159–75.

Huxley, Aldous. *The Olive Tree and Other Essays.* London: Chatto & Windus, 1936.

Huxley, Henrietta A., ed. *Aphorisms and Reflections from the Works of Thomas Henry Huxley.* London: Macmillan, 1907.

Huxley, Julian. *Essays in Popular Culture.* London: Chatto & Windus, 1925.

Huxley, Leonard. "Home Memories." *Nature* 115 (1925): 698–702.

———. *Life and Letters of Thomas H. Huxley.* 2 vols. N.Y.: Appleton, 1901.

Huxley, Thomas Henry. *Collected Essays.* 9 vols. 1894. N.Y.: Greenwood Press, 1968.

———. *The Scientific Memoirs of Thomas Henry Huxley*. Ed. Michael Foster and E. Ray Lankester. 4 vols. plus supplement. London: Macmillan, 1898–1903.

Irvine, William. "Carlyle and T. H. Huxley." *Victorian Literature: Modern Essays in Criticism*. Ed. Austin Wright. N.Y.: Oxford UP, 1961. 193–207.

Lankester, E. Ray. "Huxley." *Nature* 115 (1925): 702–4.

MacIntire, Alasdair. *After Virtue: A Study in Moral Theory*. Notre Dame: U of Notre Dame P, 1981.

Marsh, Othniel C. "Thomas Henry Huxley." *American Journal of Science* 3d series 50 (1895): 177–83.

More, Paul Elmer. *The Drift of Romanticism*. Shelburne Essays. 8th series. Boston: Houghton-Mifflin, 1913.

Paget, Stephen. "Truth and Righteousness." *Nature* 115 (1925): 748–50.

Paradis, James G. *T. H. Huxley: Man's Place in Nature*. Lincoln: U of Nebraska P, 1978.

Poulton, E. B. "Thomas Henry Huxley." *Nature* 115 (1925): 704–8.

Richards, Evelleen. "Huxley and Woman's Place in Science: The 'Woman Question' and the Control of Victorian Anthropology." *History, Humanity, and Evolution: Essays for John C. Greene*. Ed. James R. Moore. Cambridge: Cambridge UP, 1989. 253–84.

Routh, James R., Jr. "Huxley as a Literary Man." *Century* 63 (1902): 392–98.

Rupke, Nicolaas A. "Bathybius Haeckelii and the Psychology of Scientific Discovery." *Studies in History and Philosophy of Science* 7 (1976): 53–62.

Shafer, Robert. *Christianity and Naturalism: Essays in Criticism*. 2d series. New Haven: Yale UP, 1926.

Stephen, Leslie. "Thomas Henry Huxley." *Nineteenth Century* 48 (1900): 905–18.

Trilling, Lionel. *Sincerity and Authenticity*. 1970 Norton Lectures. Cambridge: Harvard UP, 1972.

Turner, Frank M. "The Victorian Conflict between Science and Religion: A Professional Dimension." *Isis* 69 (1978): 356–76.

Thomas Henry Huxley and the Question of Morality

JOHN R. REED

IN A LATE PIECE ENTITLED SIMPLY "AUTOBIOGRAPHY," THOMAS Henry Huxley wrote that "there is no alleviation for the sufferings of mankind except veracity of thought and of action, and the resolute facing of the world as it is when the garment of make-believe by which pious hands have hidden its uglier features is stripped off" (*CE* 1:16). This determination to face the world as it is resembles his friendly antagonist Matthew Arnold's similar desire for a critical disinterestedness that sees things as they really are. Both men, in trying to root their substitutes for Christian faith in an accurate perception of "the world," proposed, whether they directly asserted it or not, a new method of moral judgment. Both men dismissed the authority of religion as a source of moral authority, but in transferring that authority to the realm of human intelligence, they had to employ skillful rhetorical techniques in the service of persuasion. Arnold's moral assertions have enjoyed a good deal of commentary, but Huxley has, at least in our day, been less scrutinized for his moral position. In this essay I examine the rhetorical methods by which Huxley constructed his scheme of human morality.

To begin with, Huxley's philosophical position concerning morality changed over time. James G. Paradis summarizes Huxley's development from an idealistic youthful Romanticism to a determinism that lodged man's only hope in revolt against nature (3). Huxley began by thinking that seeing nature steadily and whole would assist humankind in its attempt to alleviate the human condition; he ended by believing that such an enterprise was possible only by opposing nature. Removing the veil from nature was at first commensurate with moral progress. This metaphor of disclosure, which Huxley very likely appropriated from Carlyle, whose writings had early fascinated him, is itself a dangerous rhetorical tool. Huxley accused "pious hands" of hiding the "uglier features" of the world. When he himself removed its merciful garments, he eventually found he had little more to offer his audience than a clearer view of the world's ugly and even hostile attributes.

Although Huxley was himself willing to live with the ugliness thus disclosed, he had to perform some interesting intellectual contortions to explain how human beings, themselves inextricably part of nature, could develop a morality to transcend a nature once thought motherly but now become a monster. One essential step was to explain the nature of necessity and humankind's capacity to alter its own destiny, always a difficult argument to assemble. In his study of Hume, Huxley criticizes Kant's treatment of necessity. Kant, Huxley says, raises volition to the level of the noumenon above the phenomenal world. "Hence volition is uncaused, so far as it belongs to the noumenon; but, necessary, so far as it takes effect in the phenomenal world" (*CE* 6:226). By contrast, Hume is utterly practical and pragmatic in his notion of necessity, basing his philosophy on the absoluteness of the law of cause and effect. This necessary association of cause and effect, Huxley argues, is essential to moral responsibility, not destructive of it, as some opponents of necessity argue. If men did not believe that actions necessarily followed from will and will from motives, they could not assign blame, for all actions would be random and unrelated to what preceded or followed them (6:221–23). Yet if all acts are determined, how can the will be free to alter the pattern of cause and effect? Huxley labeled himself an agnostic but, by the time he came to write "Evolution and Ethics," he seems as much a Gnostic, since he felt obliged to divide the world into a malign force associated with the natural world and a benign power confined to human consciousness.[1] In fact, he sounds a good deal like the Kant he earlier criticized for removing will into the realm of the noumenous.

Like Hume, Huxley based human morality on the concept of responsibility. You cannot blame persons for actions they cannot control, only for actions they intend. Like many other speculators on this issue, Huxley allows a lapse in his necessitarian argument, for he does not consider that the intention itself, though conscious, is nonetheless determined—today we might say genetically programmed—and therefore beyond the range of blame. Human society might require punishment, discipline, and control, but it does not require a theory of intention to justify such forms of restraint. The debate over free will and necessity was central to philosophical and general intellectual discourse in the nineteenth century, though it remained unresolved then as it does to this day.[2] Huxley's proffered resolution to the dilemma is not one of the strongest.

There have been many challenges to Huxley's view about necessity and free will; Robert Shafer, writing in 1929, offers an early, simple one. Huxley, says Shafer, claims that humans are conscious automata endowed with free will, but nonetheless part of an unbroken continuity. Shafer resists the claim that automata can be reconciled with free will. "The existence of a single volitional agent is an unaccountable anomaly in Huxley's unitary, natural order" (143). Albert Ashforth inadvertently calls attention to a disparity in Huxley's assertions about human control. At one point he cites Huxley's denial that he is a materialist, since that position would require "an unswerving doctrine of necessity," but in another place, Ashforth quotes Huxley's remark that he believes in absolute determinism (66–67; 98). Though consistent in his overall approach, Huxley was not always entirely consistent in his means of expression. For most disputants in the nineteenth century, necessity and determinism were synonymous. This does not seem to have been the case for Huxley. Huxley is similarly ambiguous about morality, an ambiguity not necessarily attributable to differences between his early and mature outlooks on life. For example, in a letter to W. Platt Ball in October 1890, Huxley writes: "Of moral purpose I see no trace in Nature. That is an article of exclusively human manufacture—and very much to our credit" (*Life* 2:285). In a letter to W. P. Clayton in November 1892, Huxley describes the moral sense as "a very complex affair—dependent in part upon associations of pleasure and pain, approbation and disapprobation formed by education in early youth, but in part also on an innate sense of moral beauty and ugliness (how originated need not be discussed), which is possessed by some people in great strength, while some are totally devoid of it" (2:324).[3] He likens this innate gift to the capacity for appreciating music. But if the moral sense is partly innate, the question that Huxley says need not be discussed becomes strikingly important. How, in fact, did the moral sense come to be innate? If it evolved in man over time, then it is something begotten by nature and the cosmic process. If it appeared suddenly and spontaneously, it has violated the law of cause and effect, an unacceptable condition in Huxley's thinking.

It seems apparent that while he wished to place all experience on the dissecting table for dispassionate examination, Huxley nonetheless wished to preserve some quality peculiar to humankind that sets it apart from the rest of nature. This quality might be described as spirit. As early as "On the Physical Basis of Life," Huxley cautioned against utter materialism and asserted his fundamental faith in skepticism, praising Hume for stipulating the limits of philosophical inquiry. In this essay, Huxley demonstrates that all life, including human life, is physical. He acknowledges that this assertion will disturb those who fear such an outlook threatens to extinguish spirit in favor of matter; they "are alarmed lest man's moral nature be debased by the increase of wisdom." But Huxley asks, What do we know of "matter" or of "spirit," and replies that "matter and spirit are but names for the imaginary substrata of groups of natural phaenomena" (*CE* 1:160). He declares that "the materialistic position that there is nothing in the world but matter, force, and necessity, is as utterly devoid of justification as the most baseless of theological dogmas" (1:162), and concludes that the fundamental doctrines of materialism and spiritualism lie outside the limits of philosophical inquiry.

Huxley's final position in "On the Physical Basis of Life" shifts the ground of discussion from a description of how all of life shares a common rudimentary character to the means of this description. Having presented a body of scientific information, he discusses the *communication* of that information. If "matter" and "spirit" are only names for natural phenomena, all such terms are simply representations of the natural world. Humankind extends its knowledge through the manipulation of symbols, and nature itself may be seen as an accumulation of symbols to be interpreted and reinterpreted, as Carlyle argues in *Sartor Resartus.* But Huxley is concerned here with the symbols that do the interpreting. He is concerned with appropriate discourse: "If we find that the ascertainment of the order of nature is facilitated by using one terminology, or one set of symbols, rather than another, it is our clear duty to use the former; and no harm can accrue, so long as we bear in mind, that we are dealing merely with terms and symbols"

(*CE* 1:164). Huxley's position here is disingenuous, since he promptly declares materialist terminology superior for scientific investigation. But he also wishes to leave open the possibility of any discourse that orders existence according to a novel scheme and purpose.

Although Huxley himself calls attention to the fundamental importance of choosing one discourse over another for a specific interpretive task, he nonetheless dismisses problems arising from dependence upon a materialist discourse by announcing that we are "dealing merely with terms and symbols" (*CE* 1:164). But terms and symbols are not insignificant issues in interpretation and expression—they are everything and shape the nature of that which we perceive through their use. A peculiar illustration of the complications involved in choosing an appropriate discourse is provided in a book contemporary with Huxley's mature period. In *Natural Law in the Spirit World,* Henry Drummond states that his argument "is not that the Spiritual Laws are analogous to the Natural Laws, but that *they are the same Laws*" (11). Drummond refers to Carlyle's view that matter is the emblem of the spirit and then states his own position: "Nature is not a mere image or emblem of the spiritual. It is a working model of the Spiritual. In the Spiritual World the same wheels revolve—but without the iron" (27).

Drummond bases his argument on the Law of Continuity, which he calls the law governing laws requiring that what operates now will continue to operate. He says that science has proved there is no spontaneous generation in matter, and we can therefore assume there is no spontaneous generation of spirit. Hence man did not develop spirit out of his moral nature; it was a gift of divinity. Drummond explores his thesis in chapters on biogenesis, degeneration, growth, death, mortification, and so forth. The whole work, adapting scientific terminology to a spiritualist purpose, reads like a parody of Huxley's approach. Drummond argues that natural laws extend into the spiritual world, insisting that there is no reason to suppose they would not. The weakness of Drummond's position is that he begs the question of there being a spiritual world in the first place—something he never attempts to prove.

It is interesting to notice how those whose beliefs remained rooted in Christianity nonetheless, like Drummond, sought to absorb science into their vision of existence. Huxley pronounced scientific method the best means for discovering truth. James Hinton, in *Man and His Dwelling Place,* writes approvingly that science "brings man face to face with nature, and makes him know *himself*" (xxxi). He, too, had faith in scientific method as a means of revealing truth. Just as Huxley urges the disinterested role of the scientist, Hinton asserts, "Self-abnegation is the law of knowing" (26). "Phenomena," he remarks, "are appearances" (8). But if Huxley and Hinton agree on these rudimentary issues, Hinton departs from anything Huxley called scientific method by assuming, like Drummond, that behind physical phenomena is a spiritual reality. Hinton offers some hard-headed, practical explanations about the material world in *Life in Nature,* but his conclusion is still that life in nature is spiritual life. For Hinton the laws of nature themselves teach morality. "All strife and failure, all subjection, baffling, wrong;—these are nutrition, they are the instruments of life, the prophecies of its perfect ends" (209). Nature's laws reconcile us to life by revealing its higher purpose. "Of death, and life raised up from death; of life bestowed by death, and perfected through it; of sacrifice, which is the law of being and the root of joy; of these things Nature speaks to us. She points us to her Maker, in Him who gave his Son" (220).

This language about nature that Hinton heard so clearly was not audible to Huxley. He could accept neither the rhetoric nor the a priori assumptions of Hinton's position, which, against Huxley's stated view, located morality precisely in nature. Without Hinton's certainty about the divine source of morality, Huxley had to provide a credible history of human morality that excluded both divinity and nature. In attempting this history, Huxley had recourse to rhetorical devices as questionable in their way as Hinton's use of metaphor and analogy.

In "Evolution and Theology" Huxley stressed the fact that morality precedes theology, a position he reasserted in his prologue to "Controverted Questions," where he argued that morality began with the establishment of human society (*CE* 5:1–58). Because primitive humans had little understanding about the natural basis for social rules, they attributed their power to a supernatural source. In this argument, Huxley clearly endorses Darwin's idea that the moral sense is a consequence of human social instinct.[4] Huxley was not comfortable with Herbert Spencer's bold extension of evolutionary ideas into social realms. At one point in "Administrative Nihilism," he expresses this uneasiness over Spencer's analogy of body and government, saying that the brain constantly interferes with the functions of the body in a way that Spencer would not want his analogous government to operate (*CE* 1:251–89). The analogy is inoperable, Huxley declares, and then substitutes an analogy of his own. This rhetorical gesture is highly risky. Having just observed that analogies are suspect, Huxley invites a similar criticism of his trope (*CE* 1:270–74). Ed Block applauds Huxley's analogizing abilities particularly in the scientific essays, but makes no mention of this bolder and more contentious use of analogy (366). Huxley's analogy is more detailed than Spencer's, but is similarly concerned with body and society.

> Every society, great or small, resembles such a complex molecule, in which the atoms are represented by men, possessed of all those multifarious attractions and repulsions which are manifested in their desires and volitions, the unlimited power of satisfying which, we call freedom. The social molecule exists in virtue of the renunciation of more or less of this freedom by every individual. It is decomposed, when the attraction of desire leads to the resumption of that freedom, the suppression of which is essential to the existence of the social molecule. And the great problem of that social chemistry we call politics, is to discover what desires of mankind may be gratified, and what must be suppressed, if the highly complex compound, society, is to avoid decomposition. (*CE* 1:275)

Granted, Huxley is writing with some degree of humor, and answering an analogy with another analogy is as much playful as it is argumentative. Still, the analogy creates difficulties. If morality arose out of human social instincts and through the conscious choices of individual participants in society, to what extent can this all-important consciousness be likened to the chemical activities of a molecule? Huxley slips in the terms "desires and volitions" but does not clearly explain how these can be equated with chemical attractions and repulsions. It is the old dilemma of an automaton being able to make free choices.

The point of "Administrative Nihilism" is to argue for a form of government that encourages cooperation and self-denial, but that also supports individual liberty. "If individuality has no play," Huxley argues, "society does not advance; if individuality breaks out of all bounds, society perishes" (*CE* 1:277). This is a balanced view shared by many of Huxley's contemporaries, among them Walter Bagehot, who presented a similar view in "Physics and Politics" (1872). Near the end of his career Huxley reasserted this position in "Government: Anarchy or Regimentation," where he argued against the two extremes of excessive freedom for individuals within society and tyrannical regulation of the community (*CE* 1:383–430).

Huxley's position is a sensible one and echoes John Stuart Mill's ideas about the relationship of the citizen to the state. Other contemporaries carried this basic line of thought further, reflecting a familiar topos of Victorian thinking—self-renunciation. William Kingdon Clifford, for example, in "On the Scientific Basis of Morals," responds to the utilitarian emphasis on society's obligation to provide the greatest good for the greatest number.

> Next, the end of Ethic is not the greatest happiness of the greatest number. Your happiness is of no use to the community, except in so far as it tends to make you a more efficient citizen—that is to say, happiness is not to be desired for its own sake, but for the sake of something else. If any end is pointed to, it is the end of increased efficiency in each man's special work, as well as in the social functions which are common to all. A man must strive to be a better citizen, a better workman, a better son, husband, or father." (2:95)

Piety, he concludes, is not altruism, but the forgetting of the self in the service of the community. This self-forgetting, which may be equated with morality, he declares after Darwin in "Body and Mind," is an acquired characteristic: "the only right motive to right action is to be found in the social instincts which have been bred into mankind by hundreds of generations of social life" (2:49).

If morality is the product of socialization in humankind, how did the process begin and how does it operate? Many of Huxley's contemporaries offered theories, as did he himself. One such theory that commanded a good deal of respect was presented by Edward B. Tylor in *Primitive Culture.*[5] In Tylor's scheme of ethnography, culture develops through recognizable stages, beginning with savage or primitive and arriving at the civilized society of western Europe. Contemporary savage or "lower" races represent primitive societies in their comparison with the "higher nations" (1:24). Tylor declares, "From an ideal point of view, civilization may be looked upon as the general improvement of mankind by higher organization of the individual and of society, to the end of promoting at once man's goodness, power, and happiness" (1:27). By taking Western culture as his model of the most civilized society, Tylor has already begged an important question, nor does he give much attention to the increased suffering that a more complex society can occasion—something that our contemporary students of culture, especially since Hiroshima, have had to consider. But Tylor is more concerned with the development of what might generally be called morality. Much of his study explores the gradual movement from an animistic toward a materialistic outlook. This division, he says, is greater than that between any two religions. Unlike Huxley, Tylor purposely avoids discussion of the truths of religion, though he admits that "the doctrines here examined bear not only on the development but the actual truth of religious systems" (2:358). Like Huxley and other interested writers of the time, Tylor regards morality as a social construct only crudely present in savage society.

> Savage animism is almost devoid of that ethical element which to the educated modern mind is the very mainspring of practical religion. Not, as I have said, that morality is absent from the life of the lower races. Without a code of morals, the very existence of the rudest tribe would be impossible; and indeed the moral standards of even savage races are to no small extent well-defined and praiseworthy. But these ethical laws stand on their own ground of tradition and public opinion, comparatively independent of the animistic beliefs and rites which exist beside them. The lower animism is not immoral, it is unmoral. (2:360)

Tylor is confident that a scientific examination of human moral sentiment and usage will produce an unprejudiced picture of the "long and intricate world-history of right and wrong" (2:449).

Tylor is confident of the eventual results of his investigations into the history of morality because he is equally confident about human reason and its powers to apply rigorous method to the examination of nature and of humankind. Huxley also was a determined advocate of reason and located its best expression in scientific method—the chief means of unveiling truth. Yet he also realized that Darwin's theory of natural selection and the popularized notion of evolution were two different things. He could not project the modest successes that Tylor anticipated from the rational investigation of the human past; far less could he propose the human accomplishments trumpeted in William Winwood Reade's *Martyrdom of Man.* Reade took Darwin's findings, especially in *The Origin of Species,* as proof that man had experienced, and would continue to experience, a gradual upward evolution. This evolution was, and would continue to be, intellectual and moral.

Huxley had democratic, even socialist, aspirations for human society, but he had little hope that humankind could achieve the kind of harmony he was capable of imagining. Reade had little doubt about humankind's power. "For, as we become more and more enlightened, we perceive more and more clearly that it is with the whole human population as it was with the primeval clan: the welfare of every individual is dependent on the welfare of the community, and the welfare of the community depends on the welfare of every individual" (531). Love will conquer fear and ultimately unite the human race. "The world will become a heavenly Commune to which men will bring the inmost treasures of their hearts, in which they will reserve for themselves not even a hope, not even the shadow of a joy, but will give up all for all mankind. With one faith, with one desire, they will labor together in the Sacred Cause—the extinction of disease, the extinction of sin, the perfection of genius, the perfection of love, the invention of immortality, the exploration of the infinite, the conquest of creation" (537).

This is heady stuff with as much of religious fervor, though Reade abjured religion, as a zealous Christian might exhibit. Indeed, Reade here sounds very like Henry Drummond in *The Ascent of Man,* who declares that Huxley had the right idea (in "Evolution and Ethics") but the wrong method in urging humankind to turn away from the struggle of life, for the true struggle leading to the welfare of humankind is the struggle for others (23). Drummond opposes the social Darwinism he detects in the writings of Benjamin Kidd and rejects the struggle for life as the supposed supreme fact disclosed by biological study. Instead, he argues that altruism is that supreme fact and is a part of nature, not of religion (56). "The path of progress and the path of Altruism are one. Evolution is nothing but the Involution of Love, the revelation of Infinite Spirit, the Eternal Life returning to Itself. Even the great shadow of Egoism which darkens the past is revealed as shadow only because we are compelled to read it by the higher light which has come. In the very act of judging it to be shadow, we assume and vindicate the light" (36).

Drummond's confidence is premised on the idea that humans are the final goal of life. Huxley had no such comforting assumption for his theorizing, realizing that all species may emerge and pass away and that there is no *story* in the evolutionary process, only a process. More than once he even complained about the dangerous tendency to refer to laws of nature as though they were analogous to rules; he recognized such "laws" to be simple descriptions of processes. Moreover, Huxley was quite sure that humankind had scarcely achieved the plateau of excellence suggested by Drummond and Reade in their different ways. In "The Struggle for Existence in Human Society," he declares that Englishmen are not living according to their own ethical standard, as the internecine struggles with commercial rivals so clearly indicate. The most hopeful sign of real progress, he adds, is in the attention being given to educating the whole community (*CE* 9:213).

Here, then, is Huxley's situation. On one side are writers who resist Darwinian thought altogether, rejecting man's intimate connection with the natural world and founding morality in divine edict. On the other are those who are willing to engage Darwinian concepts, but who are also tempted to bend those concepts to their own ideologies, whether Christian or materialist. Huxley's dilemma was that he fully accepted the theory of natural selection and the consequent idea of evolution, but did not credit them with teleological qualities. Yet if there is no inherent direction or purpose in the evolutionary process, where does morality fit in? Huxley asserted morality was a construction of human society that gradually developed into what was known as morality, or ethics, in his own day. But he also said that morality was, in most humans, partially innate, that is, if man is a part of nature and a part of him has a built-in sense of morality, then morality itself must be an inherent part of nature. As Ashforth comments, Huxley, in the "Prolegomena" to "Evolution and Ethics" had to concede that ethical nature is a part of cosmic nature, though more often than not its operations and objectives are opposed to cosmic nature (152). Ashforth summarizes Huxley's position: "However, since the ethical process runs counter to the cosmic process, man must realize that the standards (ethical and moral codes) he invokes to govern society are not absolute nor can they be justified by reference to cosmic nature" (153).

A central difficulty with the whole issue of human morality as a product of the cosmic process that operates against the cosmic process is the apparently absurd paradox that nature would create what would oppose natural process. But a quick glance around the world today might indicate something that was less apparent in Huxley's time—that much more than the moral impulse in man opposes natural processes, particularly if we limit those natural processes to planet Earth. Much of what humans can do seems to be opposed to natural process, not merely those "positive" activities that we associate with morality and the control of "savage" instincts in humankind.

That Huxley was aware of the anomalous role of morality in his own philosophy is apparent in his tendency to employ rhetorical devices as a means of persuasion. Many authors easily turned to metaphor and analogy to give their arguments force, and Huxley was a sharp critic of such rhetorical embellishments. Nevertheless, he often made use of such devices himself.[6] His use of the Cinderella metaphor in "Science and Morals" and of the Jack-and-the-Beanstalk analogy in the "Prolegomena" to "Evolution and Ethics" are two striking instances where a forceful figure supports, with all its childhood, intuitive associations and its narrative energy, a far from secure argument.

W. S. Lilly, to whom Huxley is responding in his essay, fears that if science replaces Christianity all hope for morality will be lost. But Huxley argues that Christianity itself inherited much from Paganism and Judaism. Alluding to his Cinderella conceit, but also echoing a long-standing metaphor in his writing, Huxley contends, "if morality has survived the stripping off of several sets of clothes which have been found to fit badly, why should it not be able to get on very well in the light and handy garments which science is ready to provide?" (145). The "sinful sisters" Theology and Philosophy should have known better than poor Cinderella Science, Huxley argues. Instead, by bickering over matters of which they know nothing because they are unknowable, they have been the cause of modern disbelief. Cinderella, by contrast, is content to serve simple material needs, knowing her own ignorance.

> But in her garret she has fairy visions out of the ken of the pair of shrews who are quarrelling down stairs. She sees the border which pervades the seeming disorder of the world; the great drama of evolution with its full share of pity and terror, but also with abundant goodness and beauty, unrolls itself before her eyes; and she learns, in her heart of hearts, the lesson, that the foundation of morality is to have done, once and for all, with lying; to give up pretending to believe that for which there is no evidence, and repeating unintelligible propositions about things beyond the possibilities of knowledge.
>
> She knows that the safety of morality lies neither in the adoption of this or that philosophical speculation, or this or that theological creed, but in a real

> and living belief in that fixed order of nature which sends social disorganisation upon the track of immorality, as surely as it sends physical disease after physical trespasses. And of that firm and lively faith it is her high mission to be the priestess. (*CE* 9:146)

Here Huxley lays claim to science as the priestess of a new faith based on truth that will replace the old religion of ignorance. Matthew Arnold similarly wished to replace the tattered *Aberglaube* of Christianity with the trim new vesture of culture. But, as Raymond Williams indicates in *Culture and Society,* because it is a process, culture has no absolute ground and therefore is no sound basis for a faith (127). Science, too, is a process and though its purpose may be the revelation of truth, "truth" itself is not an absolute but varies with the perceiver.

Huxley trusted profoundly in the operation of reason. His philosophical tradition was that of Enlightenment thinking. Jürgen Habermas comments that Enlightenment philosophers believed that "objective science, universal morality and law, and autonomous art" would shape human nature and lead civilization toward perfection (9). Huxley was unable to confirm this notion of perfectibility, but it survived in him early as a possibility, and later as a "fairy vision." Today, we may look back and consider Huxley's trust in reason and truth naive. Thanks to Thomas Kuhn we can appreciate how each era has its own scientific paradigm for constructing its picture of existence. Thanks to Werner Heisenberg we can realize that the very act of observation alters what is being observed. Thanks to quantum physics we must remain open-minded about the nature of a "truth" that accommodates two competing theories of energy. Scientists today may have the same aims Huxley had, but they must be more circumspect about the possibility of isolating truth and of doing so objectively. Science requires imagination and intuition as much as it requires reason. Often it is imagination that provides the innovative model according to which reason will pursue its purpose of disrobing nature. In many ways Huxley understood that, and his tendency to use metaphors, analogies, and other similitudes, shows it.[7] A few examples of Huxley's use of these devices will show how dangerously slippery they can be.

"Science and Morals" ends with a Cinderella metaphor. "Evolution and Ethics" begins with an elaborate Jack-and-the-Beanstalk reference. Like Jack, who discovered a like-yet-unlike world in the aerial sphere of the beanstalk, Huxley wants to lead his audience into a strange new world by means of a bean. The cycle of the bean's life, he demonstrates, is the cycle of all life—begetting, growth, and decay. Jack is the scientist who has climbed his beanstalk to explore the cosmic process. Huxley goes on to his central point that human beings, while learning about the cosmic process have also learned that nature is amoral, in fact, a "powerful enemy" who will last as long as the world does (*CE* 9:85). Humankind is no longer a child. "We are grown men, and must play the man" (9:86). Huxley quotes the ending of Tennyson's "Ulysses" ("To strive, to seek, to find, and not to yield") to reinforce this call to mature effort. Scientists and other humans are grown men now, not boys like adventurous Jack. They are at war with a tenacious enemy, not merely climbing a beanstalk. But lurking in that fairy tale is another picture of Jack the Scientist. He climbs into an unforeseen world, despoils it, and in fleeing from his own rash act brings the avenue to that strange world crashing down behind him. Science, as we today realize, can be a formidable enemy of nature, capable of vast destruction. But this aspect of science rarely, if ever, seems to have troubled Huxley. One might say that for him science itself was morality.

In *Man's Place in Nature,* his first truly dramatic appearance before the public, Huxley offered the shocking theory that humankind was descended from the lower animals. He knew that his audience was likely to resist this proposition, and he urged them to be open-minded. He acknowledged that it takes an originality beyond most people to answer the large questions underlying all others, answers that are themselves likely to be supplanted in time. Yet he insisted that the effort must be made. Huxley illuminates the history of human intellect with the "well-worn metaphor" of the caterpillar's metamorphosis into the butterfly. "But the comparison may be more just as well as more novel, if for its former term we take the mental progress of the race. History shows that the human mind, fed by constant accessions of knowledge, periodically grows too large for its theoretical coverings, and bursts them asunder to appear in new habiliments, as the feeding and growing grub, at intervals, casts its too narrow skin and assumes another, itself but temporary. Truly the imago state of Man seems to be terribly distant, but every moult is a step gained, and of such there have been many" (*CE* 7:79). Huxley declared it the duty of every good citizen to assist in the removal of the old skin. In using this familiar metaphor from nature, Huxley emphasized man's intimate connection with the natural world; however, the "surface" or "integument" being removed is not something material, but an abstract concept, a proposition that cannot truly be verified. Moreover, Huxley's metaphor implies a teleology that he would later deplore—that "moultings" of successive philosophies are steps toward a grand intellectual liberty resembling the elegant emergence of the butterfly. The history of human thought could scarcely be said to consist of a steady moulting of erroneous creeds in the direction of enlightenment. In many ways the history of human intelligence can be viewed in Darwinian terms, with various species of belief evolving slowly, adapting, or dying out with no direction and no purpose. The very argument Huxley's book was making suggested a less-splendid narrative than that figured in his metaphor.

One more example may clarify this point. Determined to make his theory acceptable to his audience, Huxley said no one needed to be offended by the notion that humankind's ancestors were animals. Is the greatness of a poet, philosopher, or artist diminished because he is "the direct descendant of some naked and bestial savage," he asked (*CE* 7:154). Then he turned to a dramatic metaphor to serve his purpose. When a traveler views the awesome splendor of Alpine scenery, "he may be excused if, at first, he refuses to believe the geologist who tells him that these glorious masses are, after all, the hardened mud of primeval seas, or the cooled slag of subterranean furnaces—of one substance with the dullest clay, but raised by inward forces to that place of proud and seemingly inaccessible glory" (7:155). In like manner, humankind, emerging from lower forms of being, has become a noble pinnacle in the natural world. Huxley was here harnessing the Romantic topos of the illustrious mountain peak for his own purpose. But hidden in this metaphor is an ominous alternative reading. Mountains do not necessarily signify nobility and may inspire fear rather than wonder. More important, if mountains stand above surrounding terrains, they are less productive of growth, less congenial to life in general, their chief service being to store water in the form of ice for a season that it might refresh the lower regions. In this scenario, mountains are barren excrescences, not sublime elevations.

Inherent in any metaphor or analogy is the capacity for more than one reading, since the fit between the things likened must always be loose or there is no need for the metaphor in the first place. Thus, metaphors are always open to interpretation in a way that more unadorned expository discourse is not. Huxley was fully aware of the dangerous tendency of metaphors employed in argumentation to betray those who used them, observing that "telling metaphors showed a curious capacity for being turned to account by the other side."[8] What Huxley did not want to admit is that the concept of morality functioned in his thinking as a species of metaphor. It was that power, partly innate, partly the result of social accommodation, that made for righteousness. But "morality" here only means the agreed-upon code of a given society, tribe, or civilization. It has no abiding residence. It is the image for that which members of a community mean when they say "good." But "good" might mean killing your daughters or eating your neighbors.

For Huxley, morality ultimately meant the control of the State of Nature. It was based upon the fundamental notion of restraint and discipline. He was fully alert to the irony that humankind, although part of the cosmic process, had to oppose itself to that process. For some interpreters, this contradiction is absurd. Moreover, Huxley understood that the human race, though extraordinary, was not immortal. He admitted how short-lived its tenure of control might be in his conclusion to the "Prolegomena" to "Evolution and Ethics": "That which lies before the human race is a constant struggle to maintain and improve, in opposition to the State of Nature, the State of Art of an organized polity; in which, and by which, man may develop a worthy civilization, capable of maintaining and constantly improving itself, until the evolution of our globe shall have entered so far upon its downward course that the cosmic process resumes its sway; and, once more, the State of Nature prevails over the surface of our planet" (*CE* 9:44–45).

In a sense, what William Irvine wrote of Huxley some years ago is ironically true. "In all but doctrine," he declared, "Huxley remained a staunch Victorian Christian throughout his life" (11). He could not accept the moral implications of the Darwinian theory that he endorsed. But the rhetoric he used often hid behind its optimistic metaphors and analogies a more sinister likelihood. Ultimately, his need to believe in a moral order led him to construct one of the most fantastic, if ancient, metaphors of all—the metaphor of an independent being warring against that which gives it its being.

Throughout his career he had done battle against those who argued for a Divine Being and for human institutions that transmitted the authority of that Being. In the later years of his life, he came to feel the need to rebel as well against the power that, in his youth, he had lauded. Great Nature had itself become the enemy. Human morality had become more valuable than that entity out of which it arose. Humankind had become a doomed Savior teaching the lessons of renunciation, charity, and love.

It is interesting to speculate on how Huxley's concept of morality might have been altered had he been permitted a view of the contemporary world and the role of science in it. Science has allowed the human race to create greater and greater organizational structures for the convenience of human beings. If viewed from a strictly physical position, this power has its grave consequences, since the second law of thermodynamics declares that forms of energy tend to change from greater to lesser degrees of order, and thus in any organizing process we put more energy in than we retrieve from it. The most obvious manifestation of this notion is that humankind may create wonders by depleting most sources of energy on the planet. But beyond such a rudimentary example is the larger concept that all of humankind's capacity to create through the use of intellect actually contributes to the more rapid disordering of material existence. Put another way, it means that to fight against the State of Nature is to increase, not decrease, the progress toward dissolution. But set against this scenario of scientific disaster is another possibility not available to Huxley. It involves the human capacity to alter the genetic structure of all living things. The discovery of DNA and science's increasing ability to massage instead of assault nature through genetic engineering, offers another script in which humankind again locates itself within nature, now as a working partner rather than as a subordinate. But in both of these scenarios morality is still a central concern. Few today would try to maintain Huxley's conjunction of skeptical materialism (and despite his denials, that is what it must be called) and innate morality, yet many feel the need for an ethical or moral code to contain not only the unruly impulses of nature but the disciplined contributions of science as well. Huxley felt that the endeavor of science—the search for truth—was an expression of morality. What he could scarcely be expected to anticipate was that science itself would become the focus of moral and ethical concern and that the disciplined handmaiden of morality would require a morality based upon the very laws of nature she studied to control her own excesses.

NOTES

1. See Ed Block Jr. for a discussion of Huxley's views on volition, especially as related to the mind-brain controversy.

J. Vernon Jensen sees a similar opposition of good and evil underlying Huxley's rhetorical strategy: "The basic nature of Huxley's rhetorical contributions was the visualization of a dramatic conflict between the forces of good and the forces of evil, the former being science and the latter being theological orthodoxy" (126).

2. See John R. Reed's *Victorian Will* (Athens: Ohio UP, 1989) for an extended discussion of this subject.

3. As James Paradis notes, Huxley was following Darwin's lead, for Darwin argued, against Mill's position in *Utilitarianism,* that the moral sense was innate rather than acquired (Paradis and Williams 27).

4. James G. Paradis's account of Huxley's view of instinct in humans suggests that he endorsed some form of individualism, though Huxley spoke out time and again against Individualism as a philosophical position. "Darwin believed," says Paradis, "that man's moral sense, the conscience, resulted from the sense of unfulfilled social instinct an individual might experience upon giving in to some antisocial passion or impulse. Peer pressure was thus a matter of instinct, a collective expression of moral intuition, enjoining the individual to follow his social instinct. To Huxley, instinct seemed primarily antisocial, and he was inclined to see peer pressure as the threat of superior group force over the instincts of the individual. Society was at war with instinct, preempting primitive man's prerogatives and 'desire for unlimited self-gratification' " (Paradis 149).

5. Paradis discusses Huxley's relationship to Victorian anthropology (Paradis and Williams 18–21).

6. A substantial body of criticism discusses metaphor in Darwin's work, from Stanley Edgar Hyman's *Tangled Bank* (New York: Atheneum, 1962) to James Elie Adam's "Woman Red in Tooth and Claw: Nature and the Feminine in Tennyson and Darwin," *Victorian Studies* 33.1 (1989), 7–27. Huxley's rhetorical techniques have had less attention, but he has certainly not been ignored. Walter Houghton's "Rhetoric of T. H. Huxley," *University of Toronto Quarterly* 18 (1949), 159–75, is an early example, and Charles S. Blinderman's "Semantic Aspects of T. H. Huxley's Literary Style," *Journal of Communication* 12 (1962), 171–78, a somewhat more

recent one. Huxley has been the subject of literary analysis at least since Aldous Huxley's "T. H. Huxley as a Literary Man," collected in *The Olive Tree and Other Essays* (London: Chatto & Windus, 1936), but there is room for a good deal of further study of Huxley's rhetoric and style.

7. Ed Block Jr. demonstrates how Huxley's rhetoric combined a wide range of techniques, including the injunctive, mimetic, and hortatory modes (373). I have concentrated on more limited and specific devices.

8. Quoted in Jensen (167).

REFERENCES

Ashworth, Albert. *Thomas Henry Huxley.* N.Y.: Twayne, 1969.

Block, Ed, Jr. "T. H. Huxley's Rhetoric and the Popularization of Victorian Scientific Ideas: 1854–1874." *Victorian Studies* 29 (1986), 363–86.

Clifford, William Kingdon. *Lectures and Essays.* 2 vols. Ed. Leslie Stephen and Sir Frederick Pollock. London: Macmillan, 1901.

Drummond, Henry. *The Ascent of Man.* N.Y.: James Pott, 1895.

———. *Natural Law in the Spirit World.* N.Y.: James Pott, 1887.

Habermas, Jürgen. "Modernity—An Incomplete Project." *The Anti-Aesthetic: Essays on Postmodern Culture.* Ed. Hal Foster. Port Townsend, Wash.: Bay Press, 1983. 3–15.

Hinton, James. *Life in Nature.* London: Smith Elder, 1862.

———. *Man and His Dwelling Place. An Essay toward the Interpretation of Nature.* 1857. N.Y.: Appleton, 1872.

Huxley, Leonard. *Life and Letters of Thomas H. Huxley.* 2 vols. N.Y.: Appleton, 1901.

Huxley, Thomas Henry. *Collected Essays.* 9 vols. 1894. N.Y.: Greenwood Press, 1968.

Irvine, William. *Apes, Angels, and Victorians: Darwin, Huxley, and Evolution.* N.Y.: McGraw-Hill, 1972.

Jensen, J. Vernon. *Thomas Henry Huxley: Communicating for Science.* Newark: U of Delaware P, 1991.

Paradis, James G. *T. H. Huxley: Man's Place in Nature.* Lincoln: U of Nebraska P, 1978.

Paradis, James G., and George C. Williams. *Evolution and Ethics: T. H. Huxley's "Evolution and Ethics," with New Essays on Its Victorian and Sociobiological Context.* Ed. Paradis and Williams. Princeton: Princeton UP, 1989.

Reade, William Winwood. *The Martyrdom of Man.* 1872. N.Y.: Truth Seeker, n.d.
Shafer, Robert. *Christianity and Naturalism: Essays in Criticism Second Series.* New Haven: Yale UP, 1926.
Tylor, Edward B. *Primitive Culture: Researches into the Development of Mythology, Philosophy, Religion, Language, Art, and Custom.* 2 vols. N.Y.: Henry Holt, 1889.
Williams, Raymond. *Culture and Society, 1780–1950.* N.Y.: Harper & Row, 1966.

Thomas Henry Huxley and Philosophy of Science

DAVID KNIGHT

HUXLEY WAS A BOLD, ACCESSIBLE, AND ABOVE ALL CONTROVERSIAL writer, at his best defending a friend or attacking an enemy—a David in constant search of Goliaths, if we may use the kind of biblical imagery in which he delighted. Like Aristotle, another keen student of living organisms, Huxley developed his positions in argument with others, living or dead. Unlike Aristotle, he is not much cited in philosophical writings today. Charles Darwin was "Philos" to his shipmates on HMS *Beagle*, and like him and like Michael Faraday, Huxley saw himself as a natural philosopher. The new word "scientist" had been coined in his childhood, by analogy with "artist," but it was not popular in the nineteenth century; it implied narrowness, whereas Huxley and others were determined to develop and promote a worldview, like the ancient Ionians.

We do not, by contrast, find in Darwin or in Faraday much explicit discussion of what we would call philosophy of science; they both had new ways of looking at the world, involving natural selection and fields of force, and had to promote these to win acceptance among the scientific community of their day. But they did not have to explore the nature of scientific thinking and practice itself, in the way that Huxley did in his various conflicts—in which he was at first the outsider challenging the Establishment, like Galileo breaking into the patronage system, and then the champion of underdogs when he had himself rather surprisingly risen to the pinnacle of the scientific world. While interested in, and interesting on, philosophy, Huxley may best be seen not as a philosopher, but as a sage like Carlyle or Ruskin.

Huxley's writing on philosophy of science is thus mostly found in his more popular writings, especially where he described scientific method. This was something that often came up in the context of teaching science in Huxley's generation. It seemed that the method of science might be separated from the mass of detail involved in learning any particular science, and that this method should be of very wide applicability, and hence great educational importance. Huxley's views (like those of most nineteenth-century Britons) were what we might broadly call Baconian: for him, science was an open-minded search for truth, in contrast to dogmatic tradition, especially that represented by the Roman Catholic Church—the great bugbear for John Tyndall and others of Huxley's circle, as well as for British men of science more generally. It is in controversy then, and on education, that we shall meet Huxley philosophizing: but we shall also want to see how far his method is actually exemplified in his research, because he was a practitioner generalizing from his own experience and that of those he admired. His greatest pupil, Michael Foster, wrote that after 1860 Huxley "to a large extent deserted scientific research and forsook the joys which it might bring to himself, in order that he might secure for others that full freedom of inquiry which is the necessary condition for the advance of natural knowledge" (qtd. in Lodge 36). This "full freedom of inquiry," which became for Huxley agnosticism, is thus the first key to his thinking.

Huxley was, like Bacon, particularly good at aphorisms (what we might call soundbites); his widow edited a little book of his "reflections." One announces that, "Agnosticism, in fact, is not a creed but a method." It was not just for intellectuals, but for everyone. Henrietta Huxley reminds us also of the working men, "whose cause my husband so ardently espoused" (*Aphorisms and Reflections* 35, vi). Agnosticism for Huxley was as old as Socrates, and was the basis of science, while science was more a creed to live by than something to philosophize about (see Lightman 14). It would thus empower working men or schoolchildren. It was not, like religions, mysterious; its pursuit did not require an elite priesthood; as common sense, it was democratic and accessible: all of us constantly generalize and reason back to causes. In the natural history sciences common sense seemed a plausible guide, but when Huxley confronted mathematical physicists he found himself out of his depth. We can best examine his philosophy of science by seeing him interact with the ideas of Auguste Comte, René Descartes, David Hume, and A. J. Balfour: and then, because his philosophizing acquired authority from his scientific position, we shall look at his practice—what he did as well as what he said.

Darwin, like others of his generation, had been much impressed by Sir John Herschel's *Preliminary Discourse* of 1830, which set out a sophisticated Baconianism, with examples from the physical sciences. Herschel, the son of a great astronomer, was a most distinguished all-round man of science who had just failed to be elected (as a Reform candidate) President of the Royal Society. One might have expected that Huxley would also have taken up Herschel, but he does not seem to use him much. His references are to John Stuart Mill, whose *System of Logic* (1843) owed much to Herschel, and then back to David Hume. Mill's great adversary was William Whewell, Master of Trinity College, Cambridge (see Fisch; Fisch & Schaffer; Yeo), whose vision of science had strongly Kantian elements. For Whewell, the intuitive leap of reason that meant getting the right end of the stick, the appropriate fundamental idea, different in different sciences, was crucial—any science consisted of facts ordered by theory. Huxley had Kantian interests too, notably in epistemology and ethics, but on science his pronouncements put him squarely in Mill's empiricist camp; thus in 1854 he explained that "Science is not, as many would seem to suppose, a modification of the black art, suited to the tastes of the nineteenth century, and flourishing mainly in consequence of the decay of the Inquisition. Science is, I believe, nothing but *trained and organized common sense,* differing from the latter only as a veteran may differ from a raw recruit" (*CE* 3:45–46). He was opposed to any idea of a hierarchy of sciences, in which physicists would look down indulgently upon those engaged in the less exacting discipline of the more descriptive sciences.

Reaction to Comte

In his fascinating new study of Richard Owen, Nicolaas Rupke describes Huxley as a positivist, indeed as one of "the three musketeers of positivism," with Herbert Spencer and G. H. Lewes (Rupke 205). This is surely mistaken, because even in the early 1850s Huxley felt it necessary to distance himself from Auguste Comte, whose *Cours de Philosophie Positive* had begun to appear in 1830. An English version was prepared in 1853 by Harriet Martineau, a friend of Charles Darwin's brother, Erasmus, and positivism became very popular in the nineteenth century. Comte's notion that knowledge, in the race and the individual alike, begins theologically, passes into a metaphysical phase, and then (with time and education) into its final, positive, and scientific stage, was taken up into a kind of scientism. This provoked counterattacks; just as Charles Darwin needed to distance himself from "gutter evolutionists" (see Desmond and Moore), so Huxley kept his agnosticism distinct from vulgar atheism and its connections with immorality, as exemplified in the seamy life of John Chapman of the *Westminster Review* (Haight 81). Huxley had to demonstrate, as he did in his life and in his writing, that agnosticism was the route to truth, responsibility, and respectability. Edward James Mortimer Collins (1827–76), a schoolmaster turned professional writer, in his poem "The Positivists" provides a good clue to the dangers Huxley's reputation faced.

Life and the Universe show Spontaneity;
Down with ridiculous notions of Deity!
  Churches and creeds are all lost in the mists;
  Truth must be sought with the Positivists.

Wise are their teachers beyond all comparison,
Comte, Huxley, Tyndall, Mill, Morley, and Harrison;
  Who will adventure to enter the lists,
  With such a squadron of Positivists?

Social arrangements are awful miscarriages;
Cause of all crime is our system of marriages;
  Poets with sonnets, and lovers with trysts,
  Kindle the ire of the Positivists.

Husbands and wives should be all one community,
Exquisite freedom with absolute unity;
  Wedding rings worse are than manacled wrists,—
  Such is the creed of the Positivists.

There was an APE in the days that were earlier;
Centuries passed and his hair became curlier;
  Centuries more gave a thumb to his wrist,—
  Then he was MAN—and a Positivist.

If you are pious, (mild form of insanity,)
Bow down and worship the mass of humanity.
Other religions are buried in mists;
We're our own gods, say the Positivists.
(qtd. in Watson 240)

These were not propositions to which Huxley wished to assent, and just as he dissented vigorously from the evolutionary doctrines set out in *Vestiges* (1844), so he attacked Comte; those who think rather like us are often most provoking. In particular, he saw in Comte (as in *Vestiges*) poor science: dogma, inaccuracies, inconsistencies, misinterpretations, and hasty generalizations. As Huxley remarked bluntly: "M. Comte, as his manner is, contradicts himself two pages further on, but that will hardly relieve him from the responsibility of such a paragraph as the above" (*CE* 3:49n). This was with reference to Comte's remark that the biological sciences were concerned with observation rather than experiment (and thus probably ranked below physics); its context was Huxley's urging, in 1854, the educational value of the natural history sciences.

In 1863, in a famous series of lectures to working men, Huxley argued again for the unity of sciences, believing that the man of science thought just as the man of sense would do: and he explicitly referred to Mill the empiricist "those who wish to study fully the doctrines of which I have endeavoured to give some rough-and-ready illustrations" (*CE* 2:363, 376n). There was nothing mysterious in "Baconian philosophy" or scientific method. Everyone could appreciate it, and perhaps join in.

By 1863 Huxley had already emerged as Darwin's bulldog, notably at the British Association for the Advancement of Science in 1860. Defending a wide-ranging theory like evolution by natural selection might seem to involve abandoning inductivism, and Huxley's reputation in science was as a formidable expert, with no time for genial amateurs who might, like Bishop Wilberforce, propose "common sense" ideas at scientific meetings (see Barton). But although he accepted the need for hypotheses in science, Huxley never found Darwin's deductive pattern of thought congenial. His empiricism thus looked to outsiders like positivism, and in 1868 in a lecture in Edinburgh he returned to the task of distancing himself from Comte. Not long before, the Archbishop of York had in the same city identified Huxley's agnostic "New Philosophy" with positivism. Huxley would have none of it:

"In so far as my study of what specially characterises the Positive Philosophy has led me, I find therein little or nothing of any scientific value, and a great deal which is as thoroughly antagonistic to the very essence of science as anything in ultramontane Catholicism. In fact, M. Comte's philosophy in practice might be compendiously described as Catholicism minus Christianity" (*CE* 1:156). This statement provoked an outcry, though Comte's "Religion of Humanity" might seem to us to fit this designation rather well. Huxley was pleased with it, and in 1869 set about defending himself against a critic, Richard Congreve, translator of Comte's *Catechism* where the calendar and doctrines of positivism are set out.

Whether against his old adversary Richard Owen the anatomist *(Man's Place)* or later against St. George Mivart (*CE* 2:125–50). Huxley was prepared to absorb his opponents' learning in order to defeat them on their own ground. Now, he picked out Comte's failures to see what was best in the science of his own day: his approval of phrenology (the science of the bumps) and dismissal of psychology, his description of Georges Cuvier as "brilliant but superficial" (*Lay Sermons* 155). Next, he attacked the whole notion of the "three states" as being neither self-consistent nor empirically true; "the positive state has more or less co-existed with the theological, from the dawn of human intelligence" (158). Anyone who had watched a child would know that Comte's schema was false; and the word "positive" was anyway "in every way objectionable" (161n). He believed that the "last and greatest of all speculative problems" was "Does human nature possess any free, volitional, or truly anthropomorphic element, or is it only the cunningest of all Nature's clocks? Some, among whom I count myself, think that the battle will for ever remain a drawn one, and that, for all practical purposes, this result is as good as anthropomorphism winning the day" (164). In short, we shall never reach Comte's positive stage.

Comte's taxonomy of the sciences was also faulty for Huxley, who emphasized unity rather than hierarchy but who also saw how scientific progress inevitably undermines any such arrangement. He pointed to the "dogmatism and narrowness" with which Comte discussed doctrines he disliked, and his "meddling systematization and regulation" (*Lay Sermons* 170). Comte was a prescriptive philosopher of science, laying down rules for good science, whereas Huxley was a man of science hoping to generalize from his experience and that of others to produce a descriptive philosophy based upon excellent practice. Comte, moreover, hoped to found a spiritual power exercising enormous influence over temporal affairs; this, to Huxley, smacked of Roman Catholicism. Whatever Collins may have supposed in writing his poem, Comteans like Congreve and Frederic Harrison in Britain knew that Huxley was not really one of them.

France and Germany

For over two hundred years, Cartesian dogmatizing had been set against cautious Baconian advance, or plodding, as its critics preferred to see it, and Huxley's remarks about Comte seem to be part of the long-running Anglo-French wars. Again like most of his British contemporaries, Huxley himself was an admirer of Germany, where he saw research and intellectual life given due emphasis. He had found his great hero in Karl von Baer, the embryologist; and whereas Darwinism made rapid headway in Germany (see Kelly), it never did in France. When Huxley was born, Paris had been the world's scientific capital; by 1860 it was so no longer. We can see Huxley's distrust of centralized, dogmatic and now provincial French science in his review of two reviews of the *Origin of Species,* one German (by an opponent Huxley considered worthy of respect) but the other by M.-J.-P. Flourens, Permanent Secretary of the French Academy of Sciences and an eminent physiologist (see Crosland).

Flourens received the sort of treatment Huxley gave Owen or Wilberforce: "while displaying a painful weakness of logic and shallowness of information," Huxley complained, he "assumes a tone of authority, which always touches upon the ludicrous, and sometimes passes the limits of good breeding" (*CE* 2:98). England was fortunate in being denied the "blessings of an Academy" if it meant dogmatism; and Flourens, "missing the substance and grasping at a shadow[, is] . . . blind to the admirable exposition . . . which Mr Darwin has given." He had "utterly failed to comprehend the first principles of the doctrine which he assails so rudely," and used language so "preposterous" that it simply gave away his ignorance, especially of embryology (104–6). Lucky Britons had never had academies to tell people what to think, like the French under the ancien regime. The Royal Society's motto, "Nullius in verba" [Take nothing on authority], aptly expresses Huxley's deepest convictions: while happy to expound a worldview, he would have been pained at the idea that he was promulgating a paradigm, or dogmatic normal science, which would have been counter to his agnosticism.

Unsympathetic as he was to the vanity of dogmatizing, Huxley was impressed by Descartes. In a talk of 1870 he traced from him two strands (*CE* 1:166–98). One is exemplified in the Critical Idealism of Immanuel Kant, an "idealism which refuses to make any assertions, either positive or negative, as to what lies beyond consciousness" (2:178). The other strand leads towards materialism, and here Huxley has another go at the Roman Catholics: "The Cardinals are at the Ecumenical Council [Vatican 1], still at their old business of trying to stop the movement of the world" (2:180), just as in Galileo's and Descartes's time. Descartes is thus a great precursor, leading "by way of Berkeley and Hume, to Kant and Idealism" and also, "by way of De La Mettrie and Priestley, to modern physiology and Materialism" (2:190). Huxley was happy to go with materialists as far as Descartes's path might lead them. He concluded with a description of the "Extrachristian" world in which he lived "a good deal," trying to distinguish truth from falsehood as Descartes had done (2:195). Huxley hoped that, unlike the Inquisition of the seventeenth century, the Christianity of his time would not seem to future generations to have recognized leading thinkers of the day simply as "objects of vilification."

David Hume and Skepticism

The doubting Descartes was thus attractive to Huxley, for whom science had to be an agnostic enterprise, but his real philosophical hero was the skeptical empiricist Hume. He even wrote a book about him, which tells us as much about its author as its subject. His excuse for writing, he said, "must be an ineradicable tendency to make things clear" (*CE* 6:51–52). Psychology was Huxley's, and he believed Hume's, route into philosophy. It differed from physical sciences only in its subject matter, and not in its method; and the essential prerequisite to philosophy was the application of scientific method to less abstruse subjects: "The laboratory is the fore-court of the temple of philosophy" (6:61). Huxley's Hume had recognized that philosophy was based on psychology, meaning experiment in moral subjects, and that all science has to start with hypotheses, to be tested and criticized until only the "exact verbal expression remains" and the scaffolding has gone (6:65). "Mitigated skepticism" was his recipe for keeping down dogmatism and superstition alike (6:67–68). Hume's physiological speculations (about animal spirits) Huxley updated as, "The key to the comprehension of mental operations lies in the study of the molecular changes of the nervous apparatus by which they are originated. . . . Operations of the mind are functions of the brain" (6:94). Huxley's favorite science of physiology was thus the basis for psychology and philosophy, and while this might indeed be materialism, Huxley believed it was compatible with the purest idealism.

Huxley explored Hume's account of miracles, deciding that while experience leads us to a firm belief in order, and demands very strong evidence indeed for breaches of it, we could never really say that an event exceeded the power of natural causes (*CE* 6:153–55). This kind of robust thinking goes with the practice of science: blank misgivings about the rule of law would lead to dropping out, which Huxley would never want to do. The book includes challenges to the clergy. Indeed, by his later years, Huxley, the "unconquerable champion and literary swordsman" had become a Goliath rather than a David, though he was never a philistine (Freeman 170). He points out that belief in necessity and determinism is as much a feature of Jonathan Edwards and other orthodox Calvinist divines as of freethinkers like Hume. When Huxley published the book in 1878, Hume was in the news; he had not usually been seen as a canonical philosopher, but there had been an edition of his writings in 1874–75, and T. H. Green of Balliol College had engaged with it in his enterprise of bringing German philosophy to Britain. Huxley brought Hume into the scientific tradition, as a great empiricist rather than just as a paradoxical thinker.

Bacon and Balfour

Examining Huxley's philosophy of science in his practice, we begin with his attitude to Darwin's idea of development with modification, through natural selection. Mario di Gregorio explored the differences between Darwin and Huxley: whereas Darwin saw the unifying value of his theory, for

Huxley it remained a hypothesis because a change of one species into another, no longer interfertile with it, had not yet been demonstrated. It is striking how in his textbook *The Crayfish,* only the last 10 percent of the book is concerned with its evolutionary history. Its anatomy and physiology, and its use as a "type" of invertebrate in his course, are given prominence; the final section, on its evolution, is presented as though it were an optional extra, for those who wanted to wrestle with such questions. This went well with the Baconian tone of much older textbooks, such as William Nicholson's on chemistry, where he "formed the determination of confining the theory, for the most part, to the ends of chapters" (vii).

Huxley's view that science was organized common sense echoed Davy's earlier statement about applied science (Knight, *Davy* 44). Many physical scientists since Galileo had, however, admired those prepared to defy common sense when experiment and mathematical reasoning indicated that it was right to do so (Banks 77–86). They had indeed expected that science and common sense would be at odds (see Wolpert). In 1879 A. J. Balfour, later prime minister, president of the British Association for the Advancement of Science, and the man who declared Palestine a homeland for the Jewish people, published *A Defence of Philosophic Doubt.* Through his brother-in-law, Lord Rayleigh, who had followed James Clerk Maxwell as Cavendish Professor of Physics at Cambridge, Balfour was well informed about science. His thesis was that all knowledge rested upon belief (Knight, "Balfour"); and his later philosophical writings, as he became more famous, attracted more attention to this message. The crisis in physics at the turn of the century seemed to support his view that science could not be separated from metaphysical doctrines: beliefs about causation, space and time, matter and energy, changed as classical physics was abandoned. The new beliefs were once again in defiance of common sense.

Balfour's skepticism placed religion and science on the same footing. Following the via negativa of Henry Mansel (1859), who had proposed a kind of Christian agnosticism (see Lightman ch. 2), such doubters outraged Huxley for whom Christians' deviations from what he saw as the fully orthodox position seemed an intolerable evasion: "[doubt as] 'the first and most essential step towards being a sound believing Christian,' though adopted and largely acted upon by many a champion of orthodoxy in these days, is questionable in taste, if it is meant as a jest, and more than questionable in morality, if it is to be taken in earnest. To pretend that you believe any doctrine for no better reason than that you doubt everything else, would be dishonest, if it were not preposterous" (*CE* 6:173n). This is rather the tone of one who believes that his clothes have been stolen by the opposition while he had been happily swimming in the ebbing sea of faith.

At the very end of Huxley's life, in 1895, Balfour published his *Foundations of Belief,* with a strong and witty attack upon what he saw as the dogmatic "scientific naturalism" of Huxley and his associates, which left no room for beauty or for morality. Huxley felt that he must reply, writing and revising a review on his deathbed. He wrote a spirited letter (full of contempt for inappropriate authority) to one of his daughters about it: "I think the cavalry charge in this month's *Nineteenth [Century]* will amuse you. The heavy artillery and the bayonets will be brought into play next month. Dean Stanley told me he thought being made a bishop destroyed a man's moral courage. I am inclined to think that the practice of the methods of political leaders destroys their intellect for all serious purposes" (*Life* 2:421). He added in a note to the editor that he was "rather pleased with the thing myself, so it is probably not very good!" He was thus outwardly in excellent form to the end, drawing upon Hume not for a deep skepticism like Balfour's, but for a critical empiricism.

Huxley's Method in Practice

Unlike his French contemporary Claude Bernard, whose *Introduction to the Study of Experimental Medicine* (1865) is a classic work on method, Huxley never wrote anything like a book on philosophy of science. *The Crayfish* was a textbook, where controversy and wit would be inappropriate; and it is in his more popular writings that we find his contentious remarks about science, its power, and its method. *Man's Place in Nature,* where we see serious comparative anatomy attractively presented, is midway between a monograph and a popularization (di Gregorio pt. 2). The idea that we should, to remove all prejudice, imagine ourselves scientific Saturnians, is an amusing way of making the Baconian point that we all have preconceptions or "idols" from whose pervasive influence we must escape. Science in *Man's Place* is indeed presented as superior common sense, in accordance with Huxley's characteristic views. Its subject and its style made it of wide interest; our relationship with the apes, and our common ancestry with them, is an exciting topic about which there were and are different strongly held opinions.

Other publications were more austere. Huxley had written up his researches on marine invertebrates done during his voyage as a surgeon on board HMS *Rattlesnake* in Australian waters. After various delays, which were not Huxley's fault, they were published in a handsome format but addressed to experts. The plates, engraved rather than lithographed (which would have been cheaper), are delicately and attractively done from drawings by Huxley; all science, but especially natural history, depends on both visual and ordinary language. The work is primarily descriptive: "a scientific classification is, after all, nothing more than a convenient mode of expressing the facts and laws established by the morphologist" (*Oceanic Hydrozoa* 20). Huxley pointed out a "general law of structure" among the Hydrozoa and "extended the generalization" from some to the whole group—an idea others adopted and confirmed (1). In classifying, Huxley had relied heavily on embryology: "the Hydrozoon travels for a certain distance along the same great highway of development as the higher animal, before it turns off to follow the road which leads to its special destination" (2). This did not mean that at this stage Huxley believed in what we call evolution; the book is in fact an excellent example of inductive science, with cautious generalizing following careful observation and wide reading. Making his way in science, he always needed to be very sure of his ground in order to maintain his reputation.

Relationships among groups of animals remained Huxley's concern, especially as he shadowed Owen (Rupke 71). In 1867–68 he explored the relationships between the reptiles and the birds, looking notably at dinosaurs (see Desmond 124–31), because the Pterodactyls were not particularly close to birds. By now, he could begin a Royal Institution lecture with the declaration "Those who hold the doctrine of Evolution (and I am one of them)" (*SM* 3:303). He demonstrated how the newly discovered fossils *Archaeopteryx,* clearly a bird, and the small dinosaur *Compsognathus longipes,* helped link two groups that most people had considered to be very far apart; lizards and tortoises do not seem very like sparrows or eagles, and dinosaurs had been perceived as creeping like alligators rather than running on their hind legs like rheas.

Thus bold in lectures not only to working men but also to the elite audience at the Royal Institution, he was still the careful inductivist in more formal writings, though by now self-confident. His paper on the classification of birds begins: "The members of the class Aves so nearly approach the Reptilia in all the essential and fundamental points of their structure, that the phrase 'Birds are greatly modified Reptiles' would hardly be an exaggerated expression of the closeness of that resemblance" (*SM* 3:238). Starting with illustrations of their skulls, he draws a family tree for the birds, with the *Ratitae* or struthious birds (the ostrich kind) as a group near the bottom, and considers carefully the geographical distribution of the various families. But in the formal papers there is not the enthusiasm for Darwinian theory that informs the contemporary Royal Institution lecture on the same subject. In his scientific practice, Huxley remained the Baconian inductive reasoner, separating the generalizations from the hypotheses, which were necessarily at the back (or even at the front) of his mind. His kind of science was far from what was going on in physics. He had begun as an explorer, and as an intellectual explorer he continued, if we may adopt a comparison from the new physics gaining momentum as Huxley died.

Thus J. J. Thomson, Rayleigh's successor at Cambridge, distinguished two sorts of scientists, comparing William Crookes's work on cathode rays with his own: "In his investigations he was like an explorer in an unknown country, examining everything that seemed of interest, rather than a traveller wishing to reach some particular place, and regarding the intervening country as something to be rushed through as quickly as possible" (*Recollections* 379). Crookes, a contemporary of Huxley's, described and demonstrated new and strange phenomena; Thomson had a counterintuitive theory of unobservable corpuscles (later called electrons) to test. Crookes's science was not far from Huxley's organized common sense; Thomson's cast of mind was deductive like Darwin's. With the triumphs of physics and of biochemistry, philosophy of science in the twentieth century has generally focused upon the traveler rather than the explorer in these voyages through strange seas of thought, and plumped for a hypothetico-deductive method.

Actually engaged in research, Huxley in his writing and his practice exemplified a view of science and its method that was close to the general view of his day, as propounded by Herschel and Mill. His originality was shown in his science, and not in his philosophy of science. When expressing and defending his general view of the world, he was altogether bolder, and his writings are most alive when he propagates his scientific naturalism.

Thomson's successor, Ernest Rutherford, is alleged to have said that there were two kinds of science, physics and stamp collecting. Our conception of physics as the fundamental science goes back to Huxley's contemporary Hermann Helmholtz (see Cahan), whose trajectory from medicine led him through physiology to physics. Physics became the key to power and influence in the scientific establishment. Huxley, against Helmholtz's friend William Thomson, Lord Kelvin, had stuck out for the autonomy of biology and geology (W. Thomson 6–127); but physicists' defiance of common sense has in the end carried the day, in the tradition of Galileo and Descartes rather than Bacon.

We certainly need Huxleys today to promote a critical and informed understanding of science as an exciting and imaginative activity based on freedom of inquiry. It may be that science has now become the kind of dogmatic system Huxley would have hated and that the time has come again for his Humean (and humane) skeptical empiricism. Mitigated skepticism is probably the right spirit in which laypeople especially should approach science and its claims. Similarly, organized common sense (in the form of public health, for example, or intermediate technology) may be much more valuable for solving humanity's problems than more high-flying science. Above all, Huxley's stand against inappropriate authority is always worth imitating. His staunch moral commitment is also something we could learn from. For him, science was to improve our lives at all levels. Faraday was portrayed as a saint. Nobody thought Huxley a saint, but in becoming a sage he showed how to move from knowledge to wisdom. Not all sages have practiced what they preached; on the whole, Huxley did.

REFERENCES

Balfour, Arthur J. *A Defence of Philosophic Doubt.* London: Macmillan, 1879.

———. *The Foundations of Belief.* 2d ed. London: Longmans Green, 1895.

Banks, Rex E. R., et al., eds. *Sir Joseph Banks: A Global Perspective.* London: Royal Botanic Garden, Kew, 1994.

Barton, Ruth. "An Influential Set of Chaps: The X-Club and Royal Society Politics, 1864–1885." *British Journal for the History of Science* 23 (1990): 53–81.

Cahan, David, ed. *Hermann von Helmholtz and the Foundations of Nineteenth-Century Science.* Berkeley: U of California P, 1993.

Comte, Auguste. *The Catechism of Positive Religion.* 3d ed. Trans. R. Congreve. London: Kegan Paul, Trench, Trubner, 1891.

———. *The Positive Philosophy.* Trans. H. Martineau. London: John Chapman, 1853.

Crosland, Maurice P. *Science Under Control: The French Academy of Sciences, 1795–1914.* Cambridge: Cambridge UP, 1992.

Desmond, Adrian. *Archetypes and Ancestors: Palaeontology in Victorian London, 1850–1875.* Chicago: U of Chicago P, 1984.

Desmond, Adrian, and James Moore. *Darwin.* London: Michael Joseph, 1991.

di Gregorio, Mario A. *T. H. Huxley's Place in Natural Science.* New Haven: Yale UP, 1984.

Fisch, Menachim. *William Whewell, Philosopher of Science.* Oxford: Oxford UP, 1991.

Fisch, Menachim, and Simon Schaffer, eds. *William Whewell: A Composite Portrait.* Oxford: Oxford UP, 1991.

Freeman, Richard B. *Charles Darwin: A Companion.* Folkestone: Dawson, 1978.

Haight, Gordon S. *George Eliot: A Biography.* Oxford: Clarendon Press, 1968.

Herschel, John F. W. *A Preliminary Discourse on the Study of Natural Philosophy.* 1830. Intro. M. Partridge. N.Y.: Johnson, 1966.

Huxley, Leonard. *Life and Letters of Thomas H. Huxley.* 2 vols. N.Y.: Appleton, 1901.

Huxley, Thomas H. *Aphorisms and Reflections.* Ed. Henrietta A. Huxley. London: Macmillan, 1907.

———. *Collected Essays.* 9 vols. 1894. N.Y.: Greenwood Press, 1968.

———. *The Crayfish: An Introduction to the Study of Zoology.* London: Kegan Paul, 1880.

———. *Lay Sermons, Addresses, and Reviews.* 6th ed. London: Macmillan, 1877.

———. *The Oceanic Hydrozoa.* London: Ray Society, 1859.

———. *The Scientific Memoirs of Thomas Henry Huxley.* Ed. Michael Foster and E. Ray Lankester. 4 vols. plus supplement. London: Macmillan, 1898–1903.

Kelly, Alfred. *The Descent of Darwin: The Popularization of Darwinism in Germany, 1860–1914.* Chapel Hill: U of NCP, 1981.

Knight, David. "Arthur James Balfour (1848–1930), Scientism and Scepticism." *Durham University Journal,* 87 (1995): 23–30.

———. *A Companion to the Physical Sciences.* London: Routledge, 1989.

———. *Humphry Davy: Science and Power.* Oxford: Blackwell, 1992.

Lightman, Bernard. *The Origins of Agnosticism: Victorian Unbelief and the Limits of Knowledge.* Baltimore: Johns Hopkins UP, 1987.

Lodge, Oliver, ed. *Huxley Memorial Lectures.* Birmingham, Eng.: Cornish Brothers, 1914.

Mansel, Henry L. *The Limits of Religious Thought.* 4th ed. London: Murray, 1859.

Nicholson, William. *The First Principles of Chemistry.* 3d ed. London: Robinson, 1796.

Owen, Richard. *The Hunterian Lectures in Comparative Anatomy, 1837.* Ed. P. R. Stone. London: Natural History Museum, 1992.

Rupke, Nicolaas A. *Richard Owen: Victorian Naturalist.* New Haven: Yale UP, 1994.

Thomson, Joseph J. *Recollections and Reflections.* London: Nelson, 1936.

Thomson, William. *Popular Lectures and Addresses.* Vol. 2. London: Macmillan, 1894.

Watson, J. Richard, ed. *Everyman's Book of Victorian Verse.* London: Dent, 1982.

Wolpert, Lewis. *The Unnatural Nature of Science.* London: Faber & Faber, 1992.

Yeo, Richard. *Defining Science: William Whewell, Natural Knowledge, and Public Debate in Early Victorian Britain.* Cambridge: Cambridge UP, 1993.

“A Good Impudent Faith in My Own Star”

Thomas Henry Huxley’s Odyssey in the South Seas

JOEL S. SCHWARTZ

When Thomas Henry Huxley learned of his appointment as assistant surgeon on the HMS *Rattlesnake,* it was confirmation that his belief in himself was not misplaced; he felt he was destined to make a significant mark in the world. Brimming with such confidence in his future, he wrote to the person he confided in the most, his eldest sister, Lizzie:

> Having a good impudent faith in my own star (Wie das Gestirn, ohne Hast, ohne Rast), I knew this was only because I was to have something better, and so it turned out; . . . Sir J. Richardson . . . told me that he had received a letter from Captain Owen Stanley, who is to command an *exploring expedition* to New Guinea . . . requesting him to recommend an assistant surgeon for this expedition—would I like the appointment? As you may imagine I was delighted at the offer, and immediately accepted it. (*Life* 1:26–27)

Thus, following in the footsteps of Charles Darwin and Joseph Hooker, Huxley’s career in the natural sciences was sparked by a voyage of exploration on a ship of the Royal Navy.

Unlike Darwin and Hooker, Huxley did not come from a wealthy or prominent family. Nor did he have the training or breadth of interest of either man at a comparable stage of life. His grandson Julian Huxley contrasted his early intellectual spirit with Darwin's: "In the young Huxley there is little trace of the variety of scientific interests displayed on the *Beagle*" (*Diary* 71). Huxley had set sail with precious few resources other than his artistic talent and an interest in engineering. He had received an ordinary education and had no formal training in natural history. By contrast, Darwin, seasickness notwithstanding, soon demonstrated acute powers of observation; "almost everything [was] grist" for his "mill" (72).

Uninterested in collecting and classifying different organisms the way other natural historians of the time did, Huxley received his greatest satisfaction from studying anatomy and comparative morphology. He examined living things by carefully dissecting them in order to discover their structure and function, rather than concentrating on how they interacted with each other and their physical environment. In spite of his viewing natural history the way an engineer would, his early travels helped him develop an appreciation for the natural world, allowing him to see the natural world not as a machine but more like a poem. As a consequence, Huxley always believed science and literature to be "not two things, but two sides of one thing" (Stephen 907). Examples of his artwork show how readily he navigated between the two cultures; his meticulous scientific illustrations and his well-executed sketches and drawings of landscapes and people represent a significant part of the record of his early adventures. They demonstrate his anatomical and observational skills, his artist's eye for detail, and his engineering ability, offering a perspective of nature very distinct from the views of such naturalists as Darwin, Alfred Russel Wallace, and Hooker. Huxley's "commanding knowledge of the structure of animals" was utilized by Darwin subsequently "to illustrate and amplify his arguments by a thousand anatomical proofs," as Chalmers Mitchell recognized a few years after Huxley's death (29).[1]

The ship's official naturalist, John MacGillivray, spent his time cataloging the different organisms that crossed their path, without paying careful attention to their structure or function. This suited Huxley, for it allowed him to concentrate his efforts on conducting a thorough examination of the unusual organisms he collected, work that was well suited to his talents. Huxley's travels to exotic and different regions of the world played a critical role in shaping his development as a naturalist. Traveling also helped determine the future course of his career; most obviously it directed him away from the practice of medicine when he returned home to England after being away for four years.

Huxley's Opportunity

The details of Huxley's early life offer few clues about his future role in helping to shape nineteenth-century biology or in founding one of Europe's most important scientific and intellectual families. As the seventh child in a large family, he was allowed to fend for himself. As a result, he developed a fiercely independent streak, fueled by a fiery temperament he inherited from his father, George Huxley, the assistant master at a private school at Ealing. He also inherited his father's artistic and engineering skills. Nevertheless, he was still more like his mother, Rachel Withers Huxley, possessing her intellectual ability and quick wit.

In spite of his intellectual gifts, Huxley was handicapped by meager formal schooling. The school he attended for a short while (and where his father taught) went out of business. Out of necessity, he had to educate himself—initially by unsystematically reading the family library. He decided to become a doctor rather than follow his natural inclination and train to become an engineer. Two of his brothers-in-law were doctors, and his brother James intended to study medicine, so his family background had some influence in shaping his choice of career. The cost of becoming an engineer also entered into his decision; the study of engineering was too much of a financial strain for people of modest means like the Huxley family. At the age of twelve, he was apprenticed to his brother-in-law Dr. John Godwin Scott, and this experience taught him the rudiments of anatomy and medicine, how to dress wounds, and some pharmacology.[2]

Although he did not pass the entrance examination for the University of London, he did win the silver medal of the Pharmaceutical Society, in recognition of his skills in that area. He was admitted with his brother to the medical school at Charing Cross Hospital, with a scholarship sufficient to defray the costs of his training. Unlike Darwin at Edinburgh medical school, Huxley enjoyed his anatomical and physiological studies; he was influenced particularly by Dr. Thomas Wharton Jones, a lecturer at Charing Cross, who had made significant advances in the study of the ovum and blood corpuscles.

When he completed his medical studies in 1845, he was only twenty-one, too young to qualify as a surgeon. Even though he was qualified to practice general medicine, his limited means prevented him from thinking about starting his own practice. With the need to support himself a pressing concern, he decided to seek a position as a doctor in the Royal Navy. In order to secure such a position, he wrote directly to Sir William Burnett, the director-general for medical service of the navy, and asked for an appointment. Surprisingly, his boldness succeeded. He was accepted into the navy with the title of assistant surgeon and assigned to the naval hospital at Haslar, where he was entered on the books of Nelson's old ship, *Victory*. During the time he spent there, Huxley caught the eye of the hospital chief, Sir John Richardson, who enjoyed a fine reputation as a naturalist and Arctic explorer. In spite of his brashness and relative inexperience in natural history and naval matters, Huxley was able to favorably impress people many years his senior, and this ability worked to his great advantage here. Richardson was instrumental in changing the course of Huxley's life when he recommended him for the position of assistant surgeon and unofficial naturalist on the HMS *Rattlesnake,* which was to embark on a survey of the northern Australian coast and New Guinea (primarily the Torres straits and the Louisiade Archipelago). Its mission was to make a detailed survey of the Inshore Passage between the coast of Australia and the Great Barrier Reef. In addition, it was to survey the unchartered waters of the Coral Sea and conduct a careful exploration of the southern coast of New Guinea and the Louisiade Archipelago.

Prior to his departure, Huxley had already made a contribution to the natural sciences. After leaving medical school, he delivered a short paper before the Royal Society on the structure of the human hair—"On a hitherto undescribed structure in the human hair sheath"—and published it in the *London Medical Gazette* (July 1845; *SM* 1:1–3). Although an unremarkable work, it is noteworthy because of his youth and the acute powers of observation he brought to bear in discovering a previously overlooked layer of cells in the inner root-sheath of the hair. This layer was subsequently named "Huxley's layer," earning him recognition from such figures in natural science as Edward Forbes and Richard Owen. His relationship with Forbes was particularly fortunate; he was a pioneer in the art of dredging, and he showed Huxley how to collect marine life in preparation for his voyage.[3] This knowledge was invaluable to him during his adventures on the *Rattlesnake*.

The Voyage on the HMS Rattlesnake

In spite of the importance of Huxley's experience on the *Rattlesnake* in shaping his life and career, most of the work dealing with his early life and career has been based on limited material; the principal sources have been MacGillivray's *Narrative of the Voyage of the HMS* Rattlesnake and Huxley's *Life and Letters,* compiled by his son Leonard. In 1935 his grandson Julian uncovered and published *T. H. Huxley's Diary of the Voyage of HMS* Rattlesnake, but studies written earlier were unable to use this valuable resource. The material contained in the *Diary* provides much information about the experience that helped to establish his reputation as a naturalist and that had such a profound effect on his personal life.[4] Huxley's notebooks and the other data he gathered during the voyage have never been given the same attention as similar data collected by Darwin on the *Beagle*. Unlike other Victorian naturalists, particularly Darwin, Huxley never made full use of the rich material he accumulated during the four years he was away. He made no effort to publish his diary, due, in part, to the antipathy that he had developed toward the navy.

Huxley's situation was a good deal more difficult than Darwin's on the *Beagle*. He did not get on very well with Owen Stanley, the captain of the *Rattlesnake,* and the crew did not share his enthusiasm for the research he was conducting. They did not treat Huxley with the same bemused tolerance the crew and officers of the *Beagle* had shown Darwin. On several occasions, they chucked his half-completed dissections overboard, complaining about the mess he had created. Also, the *Rattlesnake* was an old ship, in the navy class called "donkey" or "jackass" frigate, which had spent most of the years since its launching in 1822 chasing pirates and privateers on the high seas. In his obituary of Huxley, Michael Foster summarized the hardships faced by the young navy doctor:

> Working amid a host of difficulties, in want of room, in want of light, seeking to unravel the intricacies of minute structure with a microscope lashed to secure steadiness, cramped within a tiny cabin, jostled by the tumult of a crowded ship's life, with the scantiest supply of books of reference, with no one at hand of whom he could take counsel on the problems opening up before him, he gathered for himself during these four years a large mass of accurate, important, and in most cases novel observations, and illustrated them with skilful pertinent drawings. (xlviii)

After leaving Plymouth, England, on December 12, 1846, the *Rattlesnake* sailed to Madeira, where the ship stopped briefly (December 18, 1846), and where Huxley had his first glimpse of a tropical setting. His diary entry for December 26 notes: "A fair wind is carrying us fast away from Madeira. . . . The feelings produced by the rival beauties of its scenery would require for their fit expression more eloquence than I possess. . . . Mountain scenery is new to me, and the semi-tropical vegetation, the bananas, the cactuses and the palm trees, call to mind all that I have ever read of foreign countries and give me a foretaste of the far south" (*Diary* 17–18).

Huxley also reported that while in Madeira he had "made two excursions into the country" (with MacGillivray) and noted the beautiful tropical plants in the courtyard of the U.S. Consulate (*Diary* 19). He observed "strong evidence of the destructive nature of the torrents here, in three half-ruined arches stretching into the bed of what must be at times a wide river. . . . These arches were evidently the remains of a strong narrow bridge" (21). While the ship was making its way to Rio de Janeiro, Huxley remarked: "In passing the Cap de Verd I looked particularly after the dust mentioned by Darwin, but although the atmosphere is never particularly hazy, no dust has been seen" (22). Huxley was beginning to "take soundings" at this time; with the help of a primitively constructed towing net, he dredged up life in the waters of the South Atlantic. He reported gathering Diphydes (siphonophores), Ascidians (tunicates), Entomostraca (crustacea "nearly allied to Daphnia"), and "great numbers of transparent lanceolate bullheaded animals about ¾ in. long [which] resemble Epizoa" (sponges). Huxley observed that the latter had "a pair of strong comblike jaws moving similarly to those in the gizzard of Rotifera," possessing also "a simple intestinal canal, with a pair of long tubular glands filled with cells," which he speculated might be ovaries. The tail had "radiating filaments," but no other appendage or eyes were present. These creatures, Huxley noted, were "exceedingly voracious and swim about frequently with their heads buried in some unfortunate ascidian." He was also fascinated by the phosphorescence of the sea at night, noticing that this phenomenon had diminished since they left Madeira; he speculated that "it arose from a small medusa" (22).

On January 13, 1847, Huxley crossed the equator for the first time and was given the customary initiation rites (*Diary* 26). His January 24 letter to his mother illustrates how such incidents shaped his feelings toward the navy and its archaic customs: "We crossed the line on the 13th of this month, and as one of the uninitiated I went through the usual tomfoolery practised on that occasion. . . . I had the good luck to be ducked and shaved early. . . . I enjoyed the fun well enough . . . but unquestionably it is on all grounds a most pernicious custom. It swelled our sick list to double the usual amount, and one poor fellow . . . died of the effects of pleurisy then contracted" (*Life* 1:34).

Several days later, after catching his first Physalia (Portuguese man-of-war) with the tow net he and MacGillivray had improvised, he referred in his diary to the archaic system of classification of these organisms: "Made a most satisfactory examination of the said Physalia, [which] has completely disgusted me with the papers on the same subject in the *Suites de Buffon* and puts in a much clearer light the true analogies of these animals. These Frenchmen, [Cuvier particularly] except Milne-Edwards, are horridly superficial" (*Diary* 26). The weaknesses in the earlier taxonomic schemes sparked his interest in revising how these organisms were classified.

On January 23, the ship docked in Rio de Janeiro. During the stay there, he and MacGillivray collected aquatic life, first with their hands and later with a net they improvised from a wire meat-cover. Using only an ordinary microscope, Huxley found the transparent jellyfish easy to examine; their transparency allowed him to make detailed observations of their anatomy, with little dissection. Having the advantage of viewing these marine organisms in a much fresher state than other zoologists—who because they were confined to land, relied exclusively on preserved animals—he made many drawings of these organisms. He also observed that the jellyfish were composed of a complicated system of canals and organs, which were formed from two "foundation membranes," and he included this important observation in his first paper.

While in Rio, Huxley also had the opportunity to explore the Brazilian interior. He was impressed, like Darwin, by the abundance and variety of life in the dense forests of South America. Although this experience did not have the same impact on him as it had on Darwin, nevertheless it gave him the chance to observe a natural setting far different from the well-structured world he had left behind. He noted: "There are two things that forcibly strike anyone going into the country here. 1st the enormous number, variety and beauty of the butterflies, 2nd the noise of the cicadas. These fellows are as big as a man's thumb and fly about the trees like locusts. They emit a very acute loud continuous note, and when numbers of them are together it is really deafening" (*Diary* 30).

After its stay in Rio, the ship left for the tip of South Africa, where it was to remain for a month. Huxley used his free time to write on the marine life he had observed so far. While the ship was anchored in Simons Bay (near Cape Town), Huxley wrote on Physalia and jellyfish, using this opportunity to discuss the problems created by the rather clumsy system of classification that had previously handicapped naturalists. He saw that the jellyfish and related forms had little in common with such creatures as starfish and sea urchins (Echinoderms). They had all been grouped together with other very different organisms (e.g., sponges), into a large disparate group, Radiata (after Cuvier). Another subgroup of Radiata, Acalephae (sea-nettles)—a collection which included such different forms as jellyfish, Siphonophores (including Portuguese man-of-war), other colonial organisms, and comb jellies—had to be reconstructed because the entire rationale for making such a taxonomic scheme was flawed. After Huxley demonstrated that jellyfish had a similar body plan to polyps, coral, and sea anemones, he placed them in a separate phylum (later named Coelenterata). True Acalephae, such as the comb jellies, belonged in an entirely different phylum, Ctenophora.

Captain Stanley offered to use his influence to help Huxley get his paper published. He sent the paper in his official dispatch, remarking: "I have enclosed a paper on 'The Anatomy and Affinities of the *Medusa,*' by Mr. Huxley, our Assistant Surgeon . . . which he is very anxious to have read out at one of the learned societies—the Royal in preference, if it could be managed. I have written to my father [then President of the Linnean Society] to do what he can toward getting the paper read, and if you would send it to him, he will save all further trouble" (qtd. in Lubbock, *Stanley* 206).

Stanley's father forwarded the paper to the Royal Society for presentation and publication.

Huxley's account of this episode recorded in his diary (March 16), does not differ substantially from the captain's version, but mentions to whom he preferred his article to be sent: "I finished up my paper on Physalia at Simons Bay . . . and sent it home. I had at first thought of transmitting it to [Edward] Forbes. . . . But the Captain suggesting that it should go to his father as President of the Linnean . . . I stipulating only that it should go to Forbes in the second place. I think it is not improbable that the Bishop [the Captain's father] will get it printed in the *Linn. Trans.*, by no means on account of any inherent merit, but because it is the first-fruits of his son's cruise" (*Diary* 36). Resenting "the slightest hint of patronage," Huxley did not embrace Stanley's assistance warmly. A later entry in the diary, referring to this incident, illustrates that although he knew how to make himself more accepted by his shipmates, he was quite unwilling to change his behavior in order to be liked. Huxley recalled:

> I was finishing the drawings for my paper yesterday in the chart room when Capt. Stanley came in, and after looking at my work for some time asked me what I was going to do with them. . . . He suggested that I should send the paper to his father. . . . I must say this for the skipper—oddity as he is, he has never failed to offer me and give me the utmost assistance in his power, in all my undertakings, and that in the readiest manner. Indeed, I often fancy that if I took the trouble to court him a little we should be great friends—as it is I always get out of his way and shall do so to the end of the story. That same stiffneckedness (for which I heartily thank God) stands in my way with others, my "superior" officers in this ship, who if I consulted their tastes a little more and my own a little less would I am sure think me what is ordinarily called a "capital fellow" i.e. a great fool. (*Diary* 108–9)

Huxley was to exhibit this same "stiffneckedness" often in his career.

The Royal Society received "On the Anatomy and the Affinities of the Family of Medusae" on March 24, 1849, and it was read before the society on June 21. It helped establish Huxley's reputation as a promising young zoologist. Unaware that his paper had been published while he was away at sea (in *Philosophical Transactions of the Royal Society*, in 1849; *SM* 1:9–32), Huxley spent a good part of the voyage concerned about his career prospects. He did not know until he returned home that the favorable reaction to this work eventually would lead to his membership in the Royal Society (at the age of twenty-six) and to his being awarded the society's gold medal the following year. He actually expressed disappointment that his paper on Physalia had not been as comprehensive as he had originally intended. He confided in his diary that he had "a grand project floating through my head, working up a regular monograph of the Mollusca, anatomy, physiology and histology, based on examination of at least one species of every genus," but fearing that "my eyes are bigger than my belly," decided that this "grand project" would have to be deferred (*Diary* 35). His notebook of drawings and notes on marine animals also contain different taxonomic schemes, reflecting his interest in writing a more comprehensive work (HP 50:1–7).

When the *Rattlesnake* reached Mauritius (on May 4, his twenty-second birthday), Huxley was struck by the natural wonders he saw. He described the island in detail, recalling vividly the observations of earlier naturalists who had stopped there: "The aspect presented by the island as you approach its shores, is very beautiful, but will not give that idea of luxuriant fertility I had formed from the descriptions of Darwin and others . . . . But as we gradually neared the land the rich plain of the Pamplemousses became a prominent feature of the landscape, and its bright green richly wooded undulations, fully redeemed the character of the country" (*Diary* 37).

Huxley went ashore and explored Mauritius, noting his first glimpse of a fringing coral reef. He did a great deal of walking on the island, sketching the scenery and the people he came in contact with. One highlight for him was his visit to the tombs of Paul and Virginia, characters in Bernardin de Saint Pierre's *Chaumière Indienne,* a story he had long admired. He took delight in confirming that these two lovers had actually existed. Obviously, his undisciplined but voracious reading as a boy—including Darwin's adventures aboard the *Beagle* as well as the work of other naturalists and adventurers who preceded him—was an inspiration and guide as he explored this new world.

The ship left Mauritius at the end of May and headed across the Indian Ocean to Australia. During this time, Huxley fleshed out his ideas for his next papers concerning Diphydae and their relations with Physophoridae. He found many examples of different forms belonging to these families of invertebrate siphonophores in his trawlings conducted in the waters of the

South Atlantic and Indian Oceans. He discovered that these long trailing forms, with their swimming bells and tubular feeding-organs, were actually colonies of hydralike organisms, and were composed of the same two fundamental membranes he observed in the medusae. Upon closer examination, he noticed that the mouth and tentacle of these organisms had the same body plan as the jellyfish, and he properly placed these colonial polyps in the same taxonomic group as the free-living, solitary jellyfish. This finding further confirmed the point he established in his first paper, that these forms belonged in the phylum Coelenterata.

Before arriving in Australia, Huxley added to his published work in comparative invertebrate morphology with his "Über die Sexualorgane der Diphydae und Physophoridae" (eventually published in *Müllers Archiv für Anatomid, Physiologie und Wissenschaftliche Medicin* in 1851; *SM* 1:122–25), and "Notes on Medusae and Polypes" (which appeared in 1850 in *Annals and Magazine of Natural History; SM* 1:33–35). He was uncertain whether his earlier work had been published, and he wanted to make sure that his view—about the similarity of the fundamental plan of the medusa, the polyp, and the siphonophore to the very early stages of the chick embryo—was properly advanced. Von Baer's theory of the germ plasm was an important influence in helping him develop this idea.[5]

In early July 1847 the ship landed in Van Diemen's Land (Tasmania), where it stayed briefly before proceeding on to Sydney. Huxley—displaying a naïveté that may have been the result of his narrow education—described Van Diemen's Land as "one of the best places we have sojourned in," oblivious to the horrid conditions that existed on the island's penal colony (*Diary* 80). In Hobart, its capital, Huxley witnessed an ether operation for the first time. Although his work in natural history had given him much satisfaction, at this point he remained committed to continuing his career in medicine, and was much impressed by this medical advance.

When the ship and crew arrived in Sydney on July 16, they enjoyed the usual whirl of parties and receptions. At one such affair, held in the home of a prominent Sydney merchant, William Fanning, Huxley met Henrietta Heathorn, the sister-in-law of the host, and they promptly fell in love.[6] Huxley was not able to marry Henrietta for more than seven years after their initial meeting; their youth and Huxley's lack of prospects at the time mitigated against an early marriage. In spite of the strains of such a long engagement, their marriage was successful, and she was an important influence in his life.

During his first stay in Sydney, Huxley met William MacLeay—"one of the few scientific naturalists resident in Sydney"—and this renowned systematist provided him with valuable information and assistance in the preparation of his papers for publication (MacBride 25–26). He found himself in agreement with MacLeay over many points on the taxonomy of the transparent forms he had collected and studied. MacLeay's library was an enormous help after suffering the severe limitations of the sparsely endowed library of the *Rattlesnake.* In his paper "Notes on Medusae and Polypes," written in the form of a letter to Professor Forbes from Cape York in October 1849, Huxley noted how important MacLeay's assistance had been: "I have a great advantage in the society and kind advice (to say nothing of the library) of Mr. William MacLeay in Sydney. . . . I was astonished to find how closely some of my own conclusions had approached his, obtained many years ago in a perfectly different way. I believe that there is a great law hidden in the 'Circular system' if one could but get at it, perhaps in Quinarianism too; but I, a mere chorister in the temple, had better cease discussing matters obscure to the high priests of science themselves" (*SM* 1:34).[7]

Cruises of Exploration

Huxley remained in Sydney for three months before embarking on the first of the four surveying trips that were the primary goal of the expedition. In October 1847 the *Rattlesnake* sailed to Moreton Bay, and from there Huxley, the captain, and several members of the crew traveled up the Brisbane River in a smaller vessel (the *Asp*) to Brisbane. This relatively short trip did not provide Huxley with any fresh observations comparable to Darwin's eventful journeys in the interior of South America or his discoveries in the Galapagos and in Australia. However, the trip up the Brisbane River was noteworthy in that it gave him a chance to explore the interior of this sparsely populated region. In crossing this rugged terrain, he had to make a difficult climb through a pass covered with thick brush. He noted: "I thought my heart would have burst." At the same time, the contrast in scenery was quite worth the effort, and the sight of "beautiful, dense green foliage and whimsical festooned creepers clinging to the huge trees and thrown from branch to branch like a fantastic drapery," and stagshorn ferns growing on the trunks of bare gum trees standing out "like a corinthian capital," made a strong impression on him. Upon reaching the top of the range, he observed "quite another scene—a wide spreading view over the distant mountains" (*Diary* 91). The variety he witnessed had some impact on his development as a naturalist, although less than it had with Darwin and Wallace.

After spending a dreary Christmas off Curtis Island, Huxley returned to Sydney on January 14, 1848. His stay there was quite short; after several weeks the *Rattlesnake* set sail again, in a southerly direction, for the Bass Straits and Melbourne, ultimately arriving in Port Dalrymple in Tasmania. Huxley noted that he had "procured some exceedingly interesting animals," including "a new Diphyses and two species of the genus Phacellophora," which he examined in detail (*Diary* 100). He explored the area around Melbourne, a recently established community situated on the Yarra Yarra River. In March 1848 the ship returned to Sydney to prepare for the third cruise, a nine-month-long survey of the Inner Passage, the area located between the Great Barrier Reef and the northern Australian coast (from Rockingham Bay northward to Cape York).

Huxley, preoccupied about his future and worried that he would wind up a failure, did not make the most of the opportunity to examine the Great Barrier Reef in detail and study the abundant aquatic life that flourished there. His unpleasant physical surroundings also worked against his making significant discoveries here as his diary entry for May 28 indicates: "Rain, Rain! The ship is intensely miserable. Hot, wet, and stinking. One can do nothing but sleep. This wet weather takes away all my energies" (*Diary* 128). His March 14, 1849, letter to his sister—written while he was waiting in Sydney for the fourth cruise to begin—describes this voyage in some detail:

> We started from here last May to survey what is called the inner passage to India. . . . Now to get to India from Sydney, ships must go either inside or outside the Great Barrier. The inside passage has been called the Inner Route in consequence of its desirability for steamers, and our business has been to mark out this Inner Route safely and clearly among the labyrinth-like islands and reefs within the Barrier. And a parlous dull business it was for those who,

> like myself, had no necessary and constant occupation. Fancy for five mortal months shifting from patch to patch of white sand in latitude from 17 to 10 south, living on salt pork and beef, and seeing no mortal face but our own sweet countenances. (*Life* 1:47–48)

Huxley's unhappiness was in sharp contrast to his ebullient mood of several years before, prior to his departure from England. His May 28 diary entry also reveals the depth of his concern about his future prospects:

> Have I the capabilities for a scientific life or only the desire and wish for it springing from a flattered vanity and self-deceiving blindness? Have my dreams been follies or prophecies? . . . There is something noble, something holy, about a poor and humble life if it be the consequence of following what one feels and knows to be one's duty and if a man do possess a faculty for a given pursuit. If he have a talent intrusted to him, to my mind it is distinctly his duty to use that to the best advantage, sacrificing all things to it. But if this capacity be only fancied, if his silver talent be nothing but lead after all—no Bedlam fool can be more worthy of contempt. The man who has mistaken his vocation is lost and useless. He who has found it is, or ought to be, the happiest of happy. (*Diary* 128–29)

At the time Huxley made these comments, he was angry that the captain would not allow him to accompany the Australian explorer and government surveyor for New South Wales, Edmund B. Kennedy—who had conducted a number of previously successful missions—in exploring the Cape York peninsula (between Rockingham Bay and Cape York). Huxley was not granted permission to join this expedition because, as the captain explained, his duties on the ship would make such a venture impossible. However, Stanley did allow Huxley to join Kennedy on a short preparatory trip, and Huxley enjoyed this brief experience with the skilled explorer. It whetted his appetite for further adventure. He noted that Kennedy was "without pretension" and "well up to his work . . . admirably fitted to manage the party well," adding, "I believe he [Kennedy] will succeed if his undertaking be possible" (*Diary* 136).

When the *Rattlesnake* eventually reached Cape York, there was no word about Kennedy's expedition. In December 1848 they were to learn that Kennedy along with all but two in his group had been killed. Kennedy's native guide, Jackey Jackey, also survived and escaped to Cape York, where he brought news of the disaster to the settlement there. Huxley, although chastened by the knowledge that he could have lost his life had he been granted permission to go along, did not change his opinion of the captain, when, during the fourth cruise, Stanley showed even more caution about taking trips into the interior of the lands they were surveying.[8] Huxley's March 14 letter to Lizzie related the grim details about the fate of Kennedy and most of his party:

> We convoyed a land expedition as far as the Rockingham Bay in 17 south under a Mr. Kennedy, which was to work its way up to Cape York in 11 south and there meet us. A fine noble fellow poor Kennedy was too. I was a good deal with him at Rockingham Bay, and indeed accompanied him in the exploring trips which he made for some four or five days in order to see how the land lay about him. In fact we got on so well together that he wanted me much to accompany him and join the ship again at Cape York, and if the Service would have permitted of my absence I should certainly have done so. But it was well I did not. Out of thirteen men composing the party but three remain alive. The rest have perished by starvation or the spears of the natives. (*Life* 1:48)

Huxley made fewer entries in his diary during the third cruise, which occupied most of 1848. His journal contains little about the abundant marine life in the waters in the vicinity of the Great Barrier Reef and the adjoining areas. Rather, there is considerable detail about the numerous islands that fringed the northeastern coast of Australia as the ship headed northward toward Cape York and the Torres Straits (e.g., Albany Island, Barnard's Islands, Frankland's Islands, Fitzroy Island, Cape Grafton, Low Island, Lizard Island, Howick's Islands, and Calvin Cross Island, where he observed that the most conspicuous objects were the white Pisonia trees). When it was possible, he explored these islands, as well as the small rivers and estuaries along the mainland, with some of the crew and officers like MacGillivray. He had his first contact with the native people there, sketching some of them. This experience made him more curious about the indigenous people of the region, initiating a lifelong interest in anthropology and non-Western ethnology. Although there were few chances for him to pursue this interest on the third cruise, during the fourth voyage there was more opportunity to do so; "on this voyage it was the natives who kindled Huxley's interest" (*Diary* 155). Years later, he was able to apply this information in the lectures he delivered and in his essays. The voyage had not only honed his skills as a naturalist but also provided him with an educational experience he could not have received from his voracious reading alone. As his grandson suggested: "Only those who have been privileged to see primitive peoples leading their own strange lives in their natural haunts can realize what a revelation of novelty, what a stimulus to interest and reflection, such a contact can provide. . . . Without it, his ethnological work would inevitably have suffered in some measure from the aridity and unreality of purely armchair studies" (160–61).

When the ship arrived in Cape York in November 1848, Huxley suffered one additional unpleasantness before the ship could return to Sydney. They docked at the British military outpost of Port Essington, where the climate was hot and steamy and the conditions were altogether quite miserable. Huxley's attitude toward the navy certainly was not improved by his stay there. He wrote: "I can't say more for Port Essington than that it is *worse than a ship,* and it is no small comfort to know that this is possible" (*Diary* 149). When he returned to Sydney on February 1, 1849, he wrote his mother demonstrating the same rhetorical talents that marked his entire career: Port Essington "is about the most useless, miserable, ill-managed hole in Her Majesty's dominions. . . . It is the most insufferably hot and enervating place imaginable. . . . The commandant is a litigious old fool, always at war with his officers, and endeavoring to make the place as much as a hell morally as it is physically" (*Life* 1:47).

Between the third and fourth cruises, from February to May 1849, Huxley enjoyed a pleasant interlude with Henrietta in Sydney. In May the ship began the fourth voyage by sailing to Moreton Bay, near Brisbane, and then headed in a northeasterly direction across the Coral Sea to Cape Deliverance, the easternmost tip of the Louisiade Archipelago of New Guinea. The passage from Moreton Bay to the Louisiade Archipelago was marked by some of the roughest conditions the ship had encountered since leaving England two and half years before. Huxley's entries for May 27–29 reveal these difficulties: "The worst gale of wind we have had yet. . . . Nothing could exceed the extreme discomfort of the ship—plunging and rolling in the heavy seas like a log. To add to our disgusts she began to leak, owing to the bad caulking . . . consequently the after gun-room, gun-room and cabins thereto adjoining were all flooded, mine about the worst as I was to leeward.

Everybody felt qualmish and headachey and there was no sleep for any but the lucky possessors of hammocks" (*Diary* 172–73).

Fortunately, conditions cleared and all on board then enjoyed the "exquisite" weather. Huxley's mood became rhapsodic; his complaints about the poor job that was done caulking the ship were replaced by his comments about the beautiful, smooth sea. After a spell of hot, muggy weather, they arrived at the islands in the vicinity of Cape Deliverance. On June 13 they reached Rossell Island. Huxley observed that its richly luxuriant forest made it the most beautiful in the group. With the aid of the *Rattlesnake*'s tender ship, the *Bramble,* they attempted to find the best spot to anchor. Huxley was eager to go ashore and explore the territory that lay beyond what he observed from the ship, but the captain was uneasy about undertaking any such ventures, perhaps with Kennedy's experience in mind. As a result, Stanley sent few reconnoitering parties into the interior of the islands. Huxley had observed that when the natives outnumbered the combined crews of the *Rattlesnake* and the *Bramble* their behavior became more aggressive, somewhat justifying Stanley's excessive caution. Nevertheless, the few contacts they had with the natives only piqued Huxley's curiosity and made him eager to learn more about them. He made many drawings of the native people he met, their dwellings, their boats, and the scenery on these islands. Many of these illustrations were included in his diary and in MacGillivray's *Narrative,* and are an important record of the expedition.

Huxley's entry for August 31, 1849, records his feelings about how the voyage was being conducted and his frustrations with the captain:

> We spent a fortnight at Brumer Island, and have been allowed to touch the shore twice. . . . The natives have shown us the friendliest possible disposition. . . . The island has a very similar appearance to the main and a careful investigation of its natural products might have given us some idea of what is to be expected on the mainland, where examination must always be so much more difficult and dangerous. But we knew as much of its botany, similarly zoology, when we anchored, as we do now. . . . If this is surveying, if this is the process of English Discovery, God defend me from any such elaborate waste of time and opportunity. (*Diary* 231–32)

Little attention has been paid to Stanley's attitude about all this because he died during the voyage, and his complete journal of the voyage has never been published. Stanley's journal notes are located in the National Library of Australia in Canberra. Adelaide Lubbock has published some passages in order to shed light on the captain's cautious behavior and his attitude toward some of his officers, particularly his young assistant surgeon. In one passage from his account of the journey, he expressed his worry about the dangers when he speculated that a white man's skull would be highly prized by the natives. Then, in an attempt to make light of these concerns, he suggested that some of the naturalists on board would enjoy the prospect of having the captain's head prepared with "Doctor Huxley superintending the boiling down and subsequent steps required to prepare a skull" (Lubbock, "Owen Stanley" 53).[9]

As they headed west along the southern coast of New Guinea, contacts with the natives increased. They were able to conduct some trading with friendlier tribes who did not bear arms (e.g., in August 1849). After leaving Brumer Island—at the western end of the Louisiade Archipelago—in late August, they explored the mainland coast of New Guinea. On September 27 they sailed for Cape York and spent the remainder of the year in the Cape York and Torres Straits region. In February they returned to Sydney, and Huxley's entries in the diary became less numerous. Even the death of the captain, which occurred while the *Rattlesnake* was anchored in Sydney Harbor, was not noted by Huxley. Stanley had suffered an epileptic paralytic fit and died in Huxley's arms on the deck of the ship. Lieutenant Yule was named captain, and the *Rattlesnake* and its crew were ordered to sail home to England.

In spite of his harsh feelings concerning the captain and the way the voyage was conducted, Huxley, in his 1854 review of MacGillivray's *Narrative* of the voyage—containing many of his own watercolors and sketches—described Stanley in a kindly and respectful manner, far differently than he had in his private journal:

> Of all those who were actively engaged upon the survey, the young commander alone was destined by inevitable fate to be robbed of his just reward. Care and anxiety, from the mobility of his temperament, sat not so lightly upon him as they might have done, and this, joined to the physical debility produced by the enervating climate of New Guinea, fairly wore him out, making him prematurely old. . . . But he died in harness, the end attained. . . . He has raised an enduring monument in his works, and his epitaph shall be the grateful thanks of many a mariner threading his way among the mazes of the Coral Sea. ("Science at Sea" 103)

Was Huxley hypocritical in writing this tribute to Stanley, a man he previously had ridiculed privately in his diary and in his letters? The passage of time allowed him to reassess some of his harsh feelings, and perhaps he had matured. Also, he realized the importance of what Stanley had accomplished as far as the British navy and commercial interests were concerned. A clear channel had been found, consisting of at least thirty miles in width along the southern shores of the Louisiade Archipelago and New Guinea, in an east-west direction from Cape Deliverance to the northeastern entrance of the Torres Straits, and this accomplishment had an important impact on navigation for many years.[10] Moreover, when he wrote the review, he was in the process of leaving the navy, because of his increasing frustrations with the service since beginning his tour of duty on the *Rattlesnake.*

Huxley left for home on May 2, 1850, quite ambivalent about his departure and his future. He was anxious to return home and take leave of the ship, but he was reluctant to leave his fiancée. He was not sure when he would be able to send for her so they could be married. The concluding part of his diary was written for her in the form of a long letter, and it is quite different in style and substance from the previous portions.

Huxley's Return

When Huxley arrived in England in the autumn of 1850, he was quite different from the callow youth who had left four years before. The opportunity to travel to lands far different from well-cultivated England and to make observations in wild and underdeveloped regions had a significant impact on his intellectual development. He had observed nature's ferocity, as well as its variety and abundance—not the static and structured world most Victorian biologists were accustomed to. His ability as a morphologist and his natural skills as an engineer enabled him to describe not only the anatomy of the different forms he dissected, but his work with Molluscs and other marine invertebrates helped him to revise substantially Cuvier's design. The many fine illustrations of the pelagic creatures he observed were included in his published work. Huxley's unusual opportunity to examine closely the viscera of the transparent floating marine life, as well as his keen artist's eye, gave him a singular advantage over other zoologists who were confined to land.

Huxley spent the first few years after his return attempting to get the Royal Navy to bear the cost of publishing his research (including his drawings and observations), and also in securing a steady teaching and research position that would enable him to marry. Richardson was particularly supportive in these efforts. He wrote Huxley on June 24, 1851, in order to strengthen his resolve in his relations with the navy: "The ministry neither in a body nor collectively care about encouraging natural sciences, but they would like I daresay to appear to do so. . . . I think you are supported by the best people. It will be a shame if your observations are . . . unpublished here until their important points became known in other countries and are published by other naturalists" (HP 25:70).

Huxley, however, was in no mood to try to develop a better relationship with the navy. The four years he had spent on the *Rattlesnake* convinced him that such attempts would be futile. In spite of Richardson's advice, he resigned his commission in 1854. This hampered his efforts to publish work dealing directly with specific aspects of the voyage, as Darwin had done with his accounts of the geology, natural history, and journal of the *Beagle*'s expedition. His "Zoological Notes and Observations made on board the H.M.S. *Rattlesnake* during the Years 1846–50" is a notable exception. MacGillivray's *Narrative,* although limited, became the prime source of information concerning the voyage, with much of Huxley's material going unpublished. The favorable attention created by the work he did publish eventually assisted him in establishing a career in the natural sciences. It made it possible for him to form useful friendships with other naturalists, such as Darwin, who played a positive role when he searched for steady employment. With the help of Richardson, Forbes, and others—including Charles Darwin, who wrote testimonials for him—he was eventually able to gain full-time employment and settle into a successful career.[11]

The Scientific Benefits of Huxley's Experience

Huxley's observation that jellyfish are a complicated system of canals, organized from two basic or "foundation membranes," ectoderm and endoderm, helped establish his early reputation in the field of invertebrate morphology. In addition, he was able to extend his investigation significantly by demonstrating that the two cell layers in these medusae were composed of thread cells with external generative elements. In doing so, he determined that they were closely related to the hydroid polyp forms, and this allowed him to group these organisms in the same phylum. He also noted the similarity of the two foundation membranes of this phylum, the Coelenterates, to the serous and mucous layers of the vertebrate embryo. Inspired by von Baer's germ theory, he was able to go further than von Baer by suggesting that the adult form of the medusae resembled the larva of higher animals.[12]

Huxley's important discoveries in studying an order of tunicate worms, the Ascidians—a group of primitive nonvertebrate chordates—helped verify the finding of botanist and poet Adelbert von Chamisso that two well-known Ascidians, *Salpa* and *Pyrosoma,* have an alternation of generations. In addition, he showed that the endostyle was a structure common to all members of this group. He discovered that another primitive group of nonvertebrate chordates, the appendicularians, belonged with the tunicates and not, as Chamisso had thought, with the comb jellies (Ctenophores). Others at the time had suggested that they were Molluscs. Huxley's work with the tunicates must have provided him much satisfaction. Chamisso's explorations, conducted along the Pacific shores of South America in the early 1800s, as well as his literary contributions and his interest in botany and ethnology—he had brought back a splendid array of botanical specimens—were a source of inspiration to Huxley and other young men who shared similar experiences of exploration and discovery.

Huxley also made significant advances in the study of Cephalopods, a class of Molluscs including octopus and squid. He showed that the arms of Cephalopods were homologous to the foot of another class of Molluscs, the gastropods (snails and whelks). His keen eye and general ability as a morphologist, in addition to his understanding of what an archetype was (i.e., the fundamental structural plan of a group), allowed him to advance this idea. Earlier taxonomists relied on an idea of the archetype which was based on superficial resemblances of structure or lifestyle among organisms. Huxley's idea of the fundamental structural plan of a group rested on careful anatomical study. His mollusc archetype, for example, was bilateral symmetry, crawling locomotion accomplished by a flat muscular foot, possession of a dorsal shell and mantle cavity, and concentration of nerve cells into three main centers or ganglia. Julian Huxley noted that "Huxley's archetypal mollusc of 1853 resembles the ancestral type of mollusc as deduced by zoologists eighty years later" (*Diary* 65).

Huxley's use of the archetype to explain affinities among different groups, as well as his acceptance of recapitulation in development, should not be misconstrued as early support of evolution, even though Darwin subsequently used the concept of archetype to support the idea of common descent.[13] Huxley's friendship and the respect he had for Darwin made him privy to the discussion among the leading exponents of evolution. Also, his understanding of recapitulation theories may have allowed him to be more receptive to the theories that challenged the prevailing idea of stasis in the form and function of species, because recapitulation could be considered as evidence that species had changed or had been significantly modified.

Huxley's development as an evolutionary biologist took place very gradually, not because he was a committed creationist—he strenuously resisted belief in most doctrines—but because his limited, or narrow, training mitigated against his constructing broad theoretical statements out of miscellaneous data as Darwin or Wallace had done. He left this sort of speculation to other naturalists, while he devoted himself to a careful examination of the structures of the different forms of life he collected, dissected, and drew. However, the observations he made during the voyage on the variety and ferocity in nature influenced him in reaching the conclusion that species were not immutable. He had rejected the idea that organisms were placed into separate groups or types at creation, and that each of these discrete groups was distinctly different from the others. His experience on the *Rattlesnake* also showed him that living things were interrelated. But he was not inclined to advance any theory of evolution on his own, nor was he initially very much interested in the subject.[14] Even in his June 25, 1859, letter to Charles Lyell, written less than six months before the publication of the *Origin,* he expressed the standard, or non-evolutionary, view of change: "The fixity and definite limitation of species, genera, and larger groups appear to me to be perfectly consistent with the theory of transmutation. In other words, I think *transmutation* may take place without transition" (*Life* 1:185).

Huxley's Support for Darwin

As Huxley's reputation was enhanced through his work and public lectures, his contacts with such members of the British scientific establishment as Darwin and Lyell increased, and he became more confident and self-assured. His friendship with Darwin made him more receptive to the views of those naturalists advancing evolution. He and Darwin exchanged many letters during the period between his return to England and the publication of the *Origin.* Darwin, for his part, respected Huxley's opinion and sought out his views whenever possible. Huxley's anatomical research often provided corroboration for Darwin's findings. He helped furnish material on the cirripedes for Darwin's monographs on barnacles. Another example of the assistance he gave was his 1853 paper "On the Morphology of the Cephalous Mollusca" (*SM* 1:152–93). In his March 8, 1855, letter to Huxley, Darwin thanked him for his "preparations" and for showing him how the ovaria of the metamorphosing larva adapted to the cement organs (HP 5:25). Darwin respected Huxley's abilities. On September 5, 1855, he reported to his cousin, H. A. Wedgwood, Huxley "is a very clever man" (qtd. in Poulton 52).[15]

Shortly before completing the *Origin,* in his letter of June 2, 1859, Darwin expressed his concerns about this work to Huxley: "It is a mere rag of an hypothesis with as many flaws & holes as sound parts," but he felt it was worth it nevertheless (HP 5:65). Some months later (October 15, 1859), Darwin wrote to inform him that the *Origin* was finished: "I shall be *intensely* curious to hear what effect the book produces on you. I know that there will be much in it which you will object to, and I do not doubt many errors. I am very far from expecting to convert you to many of my heresies; but if, on the whole, you and two or three others think I am on the right road, I shall not care what the mob of naturalists think" (qtd. in Darwin, *Life and Letters* 2:172–73).

When the *Origin* was published, Huxley was impressed by the massive evidence Darwin had marshaled in defense of his theory and found himself in agreement with many of his arguments. He wrote his congratulations to Darwin on November 23, 1859, informing him that he looked forward to the expected battle:

> I finished your book yesterday. . . . Since I read Von Bär's essays, nine years ago, no work on Natural History Science I have met with has made so great an impression upon me. . . . As for your doctrine, I am prepared to go to the stake, if requisite, in support of Chapter IX and most parts of Chapters X, XI, XII, and Chapter XIII contains much that is most admirable, but on one

> or two points I enter a *caveat* until I can see further into all sides of the question. . . . As to the first four chapters, I agree thoroughly with all the principles laid down in them. I think you have demonstrated a true cause for the production of species, and have thrown the *onus probandi* that species did not arise in the way you suppose, on your adversaries. . . . I must read the book two or three times more before I presume to begin picking holes. . . . You have earned the lasting gratitude of all thoughtful men. . . . I am sharpening up my claws and beak in readiness. (*Life* 1:188–89)[16]

This letter signaled Huxley's readiness for the task that gave him his greatest recognition with the lay public. Shortly afterward, even before his famous confrontation with the Bishop of Oxford, Samuel Wilberforce, at the Oxford meetings in 1860, he substantially assisted Darwin's cause. He anonymously reviewed the *Origin* December 25, 1859, for the *Times* (London), and his praise for the work had a significant impact amid the negative criticism Darwin was receiving at the time.

Although Huxley is widely known as Darwin's main advocate and spokesman, as his "bulldog," his scientific reputation does not rest on the contributions he made in evolution; he actually contributed few original ideas to this field. This did not hamper his efforts in acting as an effective proponent of Darwin's ideas. With his vigorous, no-holds-barred support for Darwin, Huxley was able to receive the acclaim that his own careful observations could not bring him. In defending evolution, he called on his considerable rhetorical skills, his ability as a polemicist, and most of all, the knowledge he gained on his voyage of discovery, thereby solidifying his place in the history of science.

NOTES

I thank the American Philosophical Society Library in Philadelphia for their hospitality and Elizabeth Carroll-Horrocks, assistant librarian and manuscripts librarian, and Martin L. Levitt, assistant manuscripts librarian there, for their assistance.

1. Julian Huxley discusses the different approaches of Darwin and Huxley in the study of natural science in his notes accompanying the text of his grandfather's diary (*Diary* 71–74). Erling Eng indicates that even though Huxley shifted toward a more mechanistic philosophy in his later years, he never gave up his view of

nature as a poem (293). David A. Roos also questions whether, contrary to popular belief, Huxley ever gave up this idea (402).

2. The original surname of Huxley's brother-in-law, Dr. John Scott, was Salt. Ronald W. Clark speculates that he ran into some trouble with the law, possibly from debts, more likely from a charge of grave robbing, and that he and his wife, Huxley's sister Lizzie, had to leave the country in a hurry, eventually settling in Tennessee. They also thought it prudent to change their name (Clark 15). This affected Huxley deeply because when he returned to England his relationship with Lizzie could be sustained only by their letters to one another. Since mail delivery across the ocean took so much time, he missed the close relationship he had previously had with her.

3. Forbes went out of his way to assist Huxley. His interest in Huxley's work was instrumental in helping to shape the latter's early development as a naturalist. Shortly before the *Rattlesnake* departed on November 11, 1846, he wrote Huxley, "I should like to have had a talk with you in our museum when many things might have been explained which can not well be told in a letter. . . . If you are in town next week perhaps you would favor me with a call" (HP 16:151).

4. Julian Huxley recalls that his father, Leonard Huxley, "never had the heart to go through the mountain of material thoroughly" (*Diary* 1). He had already gone through much material after Huxley's death in 1895 and thought this was sufficient. Earlier, when Thomas Huxley reviewed MacGillivray's *Narrative* for the 1854 *Westminster Review* ("Science at Sea"), he used his notes in the diary extensively. Leonard also used these notes when he compiled *Life and Letters.* Julian speculates that his father knew that the diary existed but was unable to find it. A. J. Marshall's *Darwin and Huxley in Australia* frequently refers to Huxley's *Diary* but does not discuss the natural history of the voyage.

5. In his annotations in the *Diary,* Julian Huxley states that his grandfather anticipated Haeckel's doctrine of recapitulation in drawing such comparisons: "He further suggested that their [tunicates'] peculiar tail might be regarded as a primitive larval feature retained throughout adult life, in this anticipating not only much of Haeckel's theory of recapitulation, but some of the modifications, which, under the stress of modern criticism, it has recently undergone" (57–58).

6. Huxley was attracted by Henrietta's interest in his work and other intellectual pursuits. They also shared a fondness for the German language and its literature, which she had developed as a student in Germany. However, she had no background in science. She recalled that before they married, "I had the least idea of the true meaning of Science—Neither by education—nor from those about me" (HP 62:13).

7. MacLeay's quinary taxonomic system with its inosculating circles had an in-

fluence on Darwin and other naturalists at the time. Huxley's notebook, containing his drawings and notes on marine animals, has several pages of circles with the names of different phyla inside or on the circumferences of the circles. It appears that Huxley was trying different combinations of classification schemes by drawing these circles and writing alternate lists of phyla (HP 50:3–5). He refined his revision of Radiata further by including one such circle in "An Account of Researches into the Anatomy of the Hydrostatic Acalephae." He announced "it will be necessary to break up the class Radiata of Cuvier into four groups" (*SM* 1:101; see fig. 4.1).

8. His fiancée, Henrietta Heathorn, was concerned about the dangers of such an expedition, and she wrote him, August 6–23, 1848, expressing her relief that he was not allowed to go (HP, T. H. Huxley Correspondence with Henrietta Heathorn, 1848, no. 30).

9. In Lubbock's "Owen Stanley in the Pacific," the relations between Huxley and the captain are discussed in detail. Referring to the captain's remarks, Lubbock comments: "This dig at Huxley, although facetiously worded, was doubtless aimed in mild retaliation against the Assistant-Surgeon's evident contempt for his shipmates in general, and for his Captain in particular. Stanley does not allude again personally to Huxley in his Journal. Not so Huxley. His narrative contains much bitter and sarcastic criticism of Stanley, and whether deserved or not, his vituperative comments leave an impression of petty rancor unbecoming a man of his calling, although perhaps excusable on the grounds of his youth" (53).

10. Less than a year after Huxley returned home, Eliza Dolby Stanley, Captain Stanley's sister-in-law, wrote Huxley on behalf of the captain's widow to "beg your acceptance of a few books that belonged to my dear brother Owen which formed part of his library on the *Rattlesnake.*" There is no record of Huxley's response, but the letter reveals that apparently Stanley's family did not bear him any ill feeling (HP 2:253)

11. Darwin, in his testimonial in support of Huxley's candidacy for a professorship of natural history at the University of Toronto, wrote, on October 9, 1851: "I have much pleasure in expressing my opinion from the high character of your published contribution to science, & from the course of your studies during your long voyage, that you are excellently qualified for a Professorship in Natural History. You have my best wishes for success in your present application." (HP 31:68). Other prominent figures who wrote letters for Huxley were Edwin Lankester and Richard Owen. In spite of this support, he did not receive the Toronto position.

12. Julian Huxley noted that Huxley's observation about the similarity between the outer and inner membranes of a jellyfish or a polyp to the serous and mucous layers of the vertebrate embryo show that he "anticipated the Haeckelian doctrine of recapitulation in its most ambitious flights" (*Diary* 69).

13. Darwin told Huxley in his April 23, 1852, letter: "I am very obliged for your paper on the Mollusca; I have read it all with much interest. . . . The discovery of the type or 'idea' (in your sense, for I detest the word used by Owen, Agassiz & Co.) of each great class, I cannot doubt, is one of the very highest ends of Natural History" (qtd. in Darwin, *More Letters* 1:73). In his paper, Huxley defined his use of the word archetype as follows: "All that I mean is the conception of a form embodying the most general propositions that can be affirmed respecting the Cephalous Mollusca, standing in the same relation to them as the diagram to a geometrical theorem, and like it, at once, imaginary and true" (*SM* 1:176–77).

14. Mario di Gregorio indicates that "Huxley was persuaded of the value of evolution only as a working hypothesis in respect to the phenomena of organic nature; he supported Darwin's theory publicly as a kind of 'program' worthy of serious consideration but not yet ready for broad application because of natural selection, its basic tenet" (397–98). Huxley's commitment to the idea of evolution was stronger than a mere device to help Darwin in the ensuing debate. Although he had misgivings about the Darwinian mechanism of evolution, natural selection, by 1860 he believed in the general concept of organic evolution; otherwise, he would not have been able to defend Darwin so eloquently. Michael Bartholomew traces the development of Huxley's views in the years leading up to the publication of the *Origin* and the roles of von Baer and Lyell upon his own thinking. He suggests, "Lamarck, Owen and Agassiz all represented, in Huxley's opinion, bad science, while Baer, Lyell and, later, Darwin, represented good science" (526).

15. Darwin wrote Wallace on April 6, 1859, "Huxley is changed and believes in mutation of species; whether a *convert* to us, I do not quite know" (qtd. in Wallace 113) On November 13 the same year, he wrote Wallace, "Hooker is a complete convert. If I can convert Huxley I shall be content" (qtd. in Wallace 116).

16. Darwin was very interested in finding out Huxley's reaction to his work. In a letter dated November 24, 1859—obviously written after Huxley sent his critique to Darwin—Darwin wrote Huxley: "Remember how deeply I wish to know your general impression of the truth of the theory of Natural Selection. . . . I [should] be infinitely grateful for them" (qtd. in Osborn 476).

REFERENCES

Bartholomew, Michael. "Huxley's Defence of Darwin." *Annals of Science* 32 (1975): 525–35.

Clark, Ronald W. *The Huxleys.* N.Y.: McGraw-Hill, 1968.

Darwin, Francis, ed. *The Life and Letters of Charles Darwin.* 3 vols. London: John Murray, 1887.

Darwin, Francis, and A. E. Seward, eds. *More Letters of Charles Darwin.* 2 vols. London: John Murray, 1903.

di Gregorio, Mario A. "The Dinosaur Connection: A Reinterpretation of T. H. Huxley's Evolutionary View." *Journal of the History of Biology* 15 (1982): 397–418.

Eng, Erling. "Thomas Henry Huxley's Understanding of 'Evolution.'" *History of Science* 14 (1978): 291–303.

Foster, Michael. "Obituary Notice of Huxley." *Proceedings of the Royal Society of London* 59 (1895–96): xlvi–lxvi.

Huxley, Leonard. *Life and Letters of Thomas H. Huxley.* 2 vols. N.Y.: Appleton, 1901.

Huxley, Thomas Henry. "Science at Sea." *Westminster Review* 61 (1854): 98–119.

———. *The Scientific Memoirs of Thomas Henry Huxley.* Ed. Michael Foster and E. Ray Lankester. 4 vols. plus supplement. London: Macmillan, 1898–1903.

———. *T. H. Huxley's Diary of the Voyage of the HMS* Rattlesnake. Ed. Julian Huxley. London: Chatto & Windus, 1935.

Huxley Papers. Imperial College of Science, Technology, and Medicine Archives, London.

Lubbock, Adelaide. *Owen Stanley, R.N., 1811–1850, Captain of the* Rattlesnake. Melbourne: Heinemann, 1968.

———. "Owen Stanley in the Pacific." *Journal of Pacific History* 3 (1968): 47–63.

MacBride, Ernest William. *Huxley.* London: Duckworth, 1934. Vol. 34 of *Great Lives.*

Marshall, A. J. *Darwin and Huxley in Australia.* Sydney: Hodder & Stoughton, 1970.

Mitchell, P. Chalmers. *Thomas Henry Huxley, A Sketch of His Life and Work.* N.Y.: G. P. Putnam & Sons, 1900.

Osborn, Henry Fairfield. "A Priceless Darwin Letter." *Science* 64 (1926): 476–77.

Poulton, Edward Bagnall. *Charles Darwin and the Theory of Natural Selection.* London: Macmillan, 1896.

Roos, David A. "Neglected Bibliographical Aspects of the Work of Thomas Henry Huxley." *Journal, Society for the Bibliography of Natural History* 8.4 (1978): 401–20.

Stephen, Leslie. "Thomas Henry Huxley." *Nineteenth Century* 48 (1900): 905–18.

Wallace, Alfred Russel. *Letters and Reminiscences.* N.Y.: Harper & Brothers, 1916.

Convincing Men They Are Monkeys

SHERRIE L. LYONS

> By next Friday evening they will all be convinced that they are monkeys.—Huxley to his wife, March 22, 1861

THOMAS HENRY HUXLEY HAD INTERESTS THAT SPANNED virtually all areas of human knowledge. During his lifetime, he played a crucial role in educational reform, particularly in advocating the teaching of laboratory sciences in the schools. He wrote and lectured on ethics, as well as on theology and metaphysics, and he was involved in various political controversies. But fundamentally Huxley was a scientist. The pursuit of scientific knowledge, he thought, gave meaning to life. He believed that it was only through the use of empirical techniques that we could secure for ourselves a little more understanding of the previously unknown. As Huxley explained in his popular lectures:

> All human inquiry must stop somewhere; all our knowledge and all our investigation cannot take us beyond the limits set by the finite and restricted character of our faculties, or destroy the endless unknown, which accompanies, like its shadow, the endless procession of phenomena. So far as I can venture to offer an opinion on such a matter, the purpose of our being in existence, the highest object that human beings can set before themselves, is not the pursuit of any such chimera as the annihilation of the unknown; but it is simply the unwearied endeavour to remove its boundaries a little further from our little sphere of action. (*CE* 2:449)

In his unrelenting investigation of the natural world, Huxley made many important, original contributions to a variety of fields: paleontology, physiology, anatomy, and particularly developmental morphology. But Huxley is best remembered for his often highly polemical defense of Darwin. Calling himself Darwin's bulldog, he wrote Darwin that he was prepared to go to the stake if necessary in defense of Darwin's bold new theory.

Yet, in recent years Huxley's status as a Darwinian has come under attack. While the early historiography portrayed Huxley as the brilliant, courageous defender of Darwin, recent accounts have been much more critical, even hostile toward the man. In heavily contextualized sociopolitical economic analyses, Huxley becomes an aggressive, ambitious, manipulative opportunist (see Desmond; di Gregorio; Richards). Because Huxley did not make use of natural selection theory in his own research and was less than enthusiastic in his support of that doctrine, he has been labeled a pseudo-Darwinian, having anti- or pre-Darwinian attitudes (see Bartholomew; Bowler; Ghiselin). Certainly this recent scholarship has contributed to a fuller understanding of this many-faceted and controversial Victorian. But in toppling the romanticized image of Huxley as the heroic, idealistic defender of "the Truth," we are in danger of forgetting that Huxley was *the* premier advocate of science in the nineteenth century. That advocacy was the result of his being first and foremost a scientist, even if it is also true that he used science to push his other agendas. Huxley's rhetorical strategy to "convince men they were monkeys" drew on his independent anatomical work, which he began before the publication of the *Origin.* Within the context of Darwinian evolution, his scientific interests then become a vehicle to gain power in intellectual, institutional, and political arenas as well.

While many aspects of this story are well known, I think it is worthwhile to examine exactly how he chose to defend Darwin for a variety of reasons—not the least of which is that Huxley correctly perceived the most controversial aspect of Darwinism: the question of human ancestry. Thus, this is where he initially devoted most of his energy. While scientists argued whether natural selection could account for species change, whether change was saltational or gradual, and a host of other aspects of Darwin's theory, none of these engendered the fundamental quarrel with the *Origin.* Opponents of Darwin's theory attacked his book for being materialistic, atheistic, and worse. Darwin was telling Victorians that rather than being made in God's image, humans were little more than intelligent apes.

But Huxley was not merely interested in convincing men that they were monkeys. Examining Huxley's promulgation of evolution, it becomes apparent that he had a more ambitious agenda than just defending Darwin.

He used evolution theory to attack theology and to further his broad campaign to replace the power and moral authority of the church with that of the temple of science. If in the process he also managed to make fools of his enemies, this was the icing on the proverbial cake. In the controversy surrounding the ape-human question, science, politics, and personality conflicts were inextricably linked. Furthermore, the fracas spilled beyond the hallowed halls of science, providing an entertaining and illuminating episode in the acceptance of evolutionary theory.

Apes and Humans Pre-Origin

While Darwin clearly thought humans were a part of nature's evolutionary scheme, the *Origin* did not directly discuss the evolution of man. His cryptic comment promised that "light will be thrown on the origin of man and his history" (458). Huxley immediately saw the light. *Man's Place in Nature,* published in 1863, eight years before Darwin's *Descent of Man,* shows that Huxley fully grasped the implications of evolution for human origins and that he did not balk at those implications. *Man's Place in Nature* argued eloquently and powerfully that humans were not exempt from the workings of evolution. Yet many of the ideas that were so forcefully presented in *Man's Place* had their gestation in an earlier period that predated the publication of the *Origin.*

People had long been fascinated with the manlike apes. The relationship of humans to apes and of the apes to each other had become an issue as soon as such animals were discovered. Explorers greatly enjoyed the antics of what later were identified as chimpanzees and marveled at their ability to imitate the gestures of humans. But the early descriptions were a mixture of fact and fantasy, and it was difficult to separate the two. While cadavers of chimpanzees and orangutans had been dissected in the seventeenth century, the two species were often confused with each other as well as with the gibbon. The gorilla was not identified until the nineteenth century and "orang-utang" was often used as a generic term for all nonhuman primates well into the nineteenth century. Furthermore, the exact relationship of the "orang-utang" to humans was problematic. Edward Tyson in an excellent memoir entitled *Orang-Outang: Or the Anatomy of a Pygmie Compared with*

That of a Monkey, an Ape, and a Man, published by the Royal Society in 1699, detailed the remarkable similarities between the pygmy (clearly a chimpanzee according to Tyson's drawings and description) and humans. But he also recognized the differences: "[though it] does so much resemble *a Man* in many of its parts, more than any of the ape kind, or any other *animal* in the world that I know of; yet by no means do I look upon it as the product of a *mixt* generation—*'tis a Brute-Animal sui generis and a particular species of Ape*" (qtd. in *CE* 7:14–15). As Tyson pointed out in the dedication, the purpose of the monograph was to demonstrate the scale of nature in which the pygmy was the connecting link between the animal and the rational (see Greene 178–79). In his *System of Nature,* Linnaeus classified humans as quadrupeds, putting them in the same order as apes and sloths. The true orangutan (which he called *Simia satyrus*) was a member of the genus *Simia.* But in the genus *Homo,* Linnaeus included not only our own species but several others that in fact did not exit. He had never seen these creatures firsthand and his classification represented an amalgamation of traits based on old woodcuts and on a dissertation of his pupil Hoppius (*CE* 7:18). Georges Louis Leclerc, Comte de Buffon, in the fourteenth volume of *Histoire naturelle* (1766), also recognized the obvious physical closeness between humans and apes. But he was not very interested in classification, claiming that there was nothing "natural" about the divisions taxonomists set up. Nature's productions were always "gradual and marked by minute shades" (qtd. in Greene 182). Furthermore, while physically the apes might be intermediate between humans and animals, Buffon maintained that the ape "is in fact, nothing but a real brute, endowed with the external mark of humanity, but deprived of thought, and of every faculty which properly constitutes the human species" (qtd. in Greene 184).

In contrast to these previous views, James Burnett, later Lord Monboddo, much to the chagrin of his eighteenth-century Scottish and English Enlightenment colleagues, claimed that some men were born with tails and furthermore that the orangutan was a man. In his major treatise, *Of The Origin and Progress of Language,* published in 1773, Monboddo maintained that the orangutan deserved to be included in the family of man for a variety of reasons. Orangutans look like us and they also bear the "marks of humanity that I think are incontestable." He alleged such "facts" as "they live in society, they use sticks for weapons, they build huts out of trees, they carry off negroe girls for both pleasure and work, and joined in companies attack elephants" (qtd. in Cloyd 44). Although speechless, they had organs for speech. Monboddo's claim met with much criticism. He had not read Tyson's work and thus was quite incorrect when he maintained that the orangutan had the exact same form as humans and walked upright. While he corrected this oversight in the second edition, he nevertheless continued to argue that the orangutan was "one of us" to support his theory about the origin and development of language.

The point of this brief and incomplete overview is to demonstrate that there was considerable interest in the "man-like apes" well before the publication of the *Origin.* Their humanlike qualities captured the imagination of naturalists and of the general public alike. Do apes think? Can they develop language? Are they moral beings? With the exception of Lamarck, no one was suggesting the transformation of one species into another. But how to define boundaries between species was actively debated. And certainly the boundary between the manlike apes and humans garnered considerable attention. With Darwin's theory the discussion intensified, leading to debates not only on the origin of *Homo sapiens,* but also rekindling the eighteenth-century controversy over the origin of the different human races: Were human races separate species or just varieties of a single species? With the possibility that humans might actually be related to other primates the monogenist-polygenist debates became reformulated. Were present day human races derived from distinct species of apelike ancestors or just varieties that developed later from one common ancestor? The vigor of the debate notwithstanding, the vast majority of Victorians were deeply opposed to anything that broke down the barrier between humans and the rest of the animal world. Even Huxley claimed that his own mind was not definitely made up in 1857, when Richard Owen gave a paper to the Linnaean Society entitled "On the Characters, Principles of Division, and Primary Groups of the Class Mammalia" (*CE* 7:viii).

Huxley's Antagonist: Richard Owen

While the notorious disagreements between Huxley and Owen were real enough, the relationship between the two men had not always been so hostile. Huxley first came into contact with Owen as a result of difficulties he was having getting the navy to support the publication of his research from the voyage of the HMS *Rattlesnake.* Huxley needed time and money to finish his work. Owen wrote a glowing testimonial to the admiralty requesting that Huxley have at least a twelve-month appointment. While Huxley had other important men backing him, there is no doubt that Owen's name carried great weight in the admiralty's decision to give Huxley an appointment. In the next two years, Owen would write several letters of recommendation for Huxley, and the relationship between the two men was "amazingly civil." Yet glimmerings of mistrust and future rivalry existed as well. Huxley described Owen as "a queer fish" and "so frightfully polite," confiding to his sister that he never felt comfortable around the elder man. To William MacLeay he expressed his uneasiness about Owen more explicitly, writing that although Owen had helped him, he was a man "with whom I feel it necessary to be always on my guard" (*Life* 1:101). As several authors have noted (Desmond; Lyons, "Evolution"; Rupke), Huxley was not alone in his distrust of Owen and was astonished at the degree of animosity toward the man, claiming that he was both "feared and hated" (1:102).

Intrigues and rivalries existed throughout the British scientific community, in part because very few salaried positions existed. While Huxley claimed to be disgusted with all the scheming and strategies involved in making a successful career, he admitted to taking "a certain pleasure in overcoming these obstacles and fighting these folks with their own weapons." He was being initiated into the politics of science and was more than willing to play the game to ensure his own success. As for Owen, he claimed, "I am quite ready to fight half a dozen dragons. And although he has a bitter pen, I flatter myself that on occasions I can match him in that department also" (*Life* 1:106).

The relationship between the two men became further strained over the next few years. Huxley still needed Owen for testimonials; at the same time, on his own subjects Huxley already considered himself Owen's master. Huxley had been elected to the Royal Society at age twenty-five in 1851 and awarded its medal the following year. He had obtained a lectureship at the London School of Mines and was also appointed to be in charge of a coast survey for the Geological Survey. Shortly afterward he was offered a lectureship on comparative anatomy at St. Thomas's Hospital. In addition, several monographs on his research from the *Rattlesnake* had been published. Huxley was becoming more and more successful, and it appears that Owen was quite threatened by the younger man. The final strain in their association occurred in 1857. Owen was giving a paleontology course at the lecture theater of the School of Mines and was listed in the medical directory as professor of paleontology at the School of Mines. Huxley wrote John Churchill of the directory, pointing out that "Mr. Owen holds no appointment whatsoever at the Government School of Mines." Since Huxley was the professor of general natural history, which included comparative anatomy and paleontology, he wondered why Owen was listed with such a title (HP 12:194). Churchill's explanation was totally unsatisfactory, indicating that "the designations attached to Mr. Owen's name in the Medical Directory are inserted correctly from the return as furnished by that gentleman" (12: 196). Huxley was furious, perceiving Owen's actions as a direct threat to his own position at the School of Mines, and broke off his relationship with him (*Life* 1:153).

The obvious personal antagonism between the two men should not obscure the real scientific differences between them, which also existed from the beginning of their relationship. Owen had written a scathing article criticizing Lyell's 1851 anniversary address to the Geological Society while Huxley had been a supporter of Lyell's antiprogressionist view of the fossil record (Lyons, "Origins"). More important, Huxley disagreed with the philosophical underpinnings of Owen's science. Owen was firmly in the tradition of the idealistic German morphologists. Although Huxley had also been trained in this tradition, he quickly distanced himself from it. He had absolutely no use for the wider and sometimes mystical speculations of the transcendental anatomists (Lyons, "Evolution"). In "On the Cephalous Mollusca," Huxley pointed out that his use of the word archetype was quite distinct from that of the transcendentalists and it was apparent that his comments were directed against Owen's use of the word (*SM* 1:152–93).

It is in light of this fundamental disagreement that Huxley's 1858 Croonian Lecture "On the Theory of the Vertebrate Skull" should be examined. Describing Oken as a "fanciful philosopher," Huxley demolished his view—promulgated by Owen—that the skull was a modified vertebrae. Based on the work of Rathke, Reicher, and Hallmann, Huxley presented embryological evidence that demonstrated that the development of the skull and the vertebrae were quite different. "The spinal column and the skull start from the same primitive condition . . . whence they immediately begin to diverge. . . . It is no more true that the adult skull is a modified vertebral column, than it would be, to affirm that the vertebral column is a modified skull" (*SM* 1:585). Huxley undoubtedly took great pleasure in discrediting the views of a man he intensely disliked. But this should not overshadow the scientific disputes between the two men or Huxley's deep interest in the problem of development. Owen's science was outdated; his theological beliefs caused him to espouse views that were becoming increasingly untenable in the light of modern research. His religious beliefs, combined with his idealist morphology, led him to propose that humans should be placed in a separate division "Archencephala"—a separate genus that was superior to the others in the mammalian class. This resulted in his most famous conflict with Huxley—over the presence of a hippocampus in the brains of nonhuman primates.

The Hippocampus Debacle

While Huxley claimed his mind was not made up as to the classification of humans when he heard Owen's 1857 lecture, based on the work of earlier anatomists, he thought Owen was wrong and set about to reinvestigate the matter for himself (*CE* 7:viii). He found that the earlier investigators were correct and that humans had no structures not also found in the higher apes. Moreover, many human features were shared by the lower apes as well. Although Huxley did not make any public comment on his views at this time, he incorporated his findings in his teachings and continued to investigate the matter over the next two years.

Thus, Huxley was already investigating the taxonomy of apes and humans before the publication of the *Origin.* Michael Bartholomew has claimed that Huxley's work underwent no fundamental shift when the *Origin* appeared (525–35). But this was in part because the *Origin* strongly supported Huxley's findings regarding the structural relationships between humans and apes by providing a causal explanation as to why these similarities existed. Not only was a shift in thinking unnecessary to incorporate the ideas embodied in the *Origin,* but it inspired him to continue his investigation. In the preface to the 1894 edition of *Man's Place in Nature,* Huxley wrote, "And inasmuch as Development and Vertebrate Anatomy were not among Mr. Darwin's many specialties, it appeared to me that I should not be intruding on the ground he had made his own, if I discussed this part of the general question. In fact, I thought that I might probably serve the cause of evolution by doing so" (*CE* 7: ix). One could suspect that Huxley was engaging in a bit of revisionist history, but good reasons exist to accept Huxley's assertion at face value.

While recent scholarship has correctly criticized the overly simplistic analysis of the conflict between science and religion in such books as John Draper's *History of the Conflict between Religion and Science* (1875) and Andrew Dixen White's *A History of the Warfare of Science with Theology* (1896), one should not underestimate the importance of this conflict (see Lindberg and Numbers). Draper was at the Oxford meeting of the famous clash between Bishop Wilberforce and Huxley, and it was not a coincidence that he wrote his book in 1875 when many Victorians were still in an uproar over the implications of Darwinism for Christianity. The role of God, the question of "man's place in nature" was, in fact, precisely why the *Origin* was so controversial. Many of the scientific objections brought forth by people such as George Mivart, Louis Agassiz, and the Duke of Argyll boiled down, in the last analysis, to the question: "Where was God in the *Origin?*" Huxley was attracted to the *Origin* precisely because it removed investigations about the history of life from the realm of theology. This was a far more powerful reason for Huxley to become embroiled in the controversy over human origins than merely to engage in a personal vendetta against Owen. In this regard, I disagree with Mario di Gregorio who claims that the main impetus for Huxley's writing *Man's Place in Nature* "was to pursue the denigration of Owen" (129–53, particularly 152). Indeed, it was Owen's mixing of science with theology that led him into a completely untenable position regarding the structure of the human brain. Huxley could show the dangers of mixing theology with science; he could defend evolution; and he could discredit Owen all by investigating the ape-human question—an opportunity too good to pass up.

Huxley presented his findings on the relationship of humans to other primates in "On the Zoological Relations of Man with the Lower Animals." This paper embodied the strategy that Huxley used again and again in his support of evolutionary theory. He maintained that the question of classification was a purely scientific one. He did not want questions of the origins of man's morals or ethics mixed in with questions of anatomy. Huxley believed that an unequivocal demonstration of the close relationship between apes and humans would be the most powerful support for Darwin's theory because Darwin provided a clear logical explanation for the existence of those relationships. Huxley acknowledged that the question of human classification was controversial. Linnaeus had classified humans and apes in the same order of primates, with the genera *Homo, Sima, Lemur,* and *Vespertilio*—all having equal ranking. Cuvier believed that, because humans had two hands instead of four, as well as other distinguishing characteristics, this justified classifying them as a distinct order, which he named *Bi-mana.* Owen wanted *Homo* to be classified as a distinct subclass, and Terres thought humans were so distinctive that they should have their own kingdom. Some people even believed that humans should not be thought of zoologically at all. The classification of *Homo sapiens* had been complicated, Huxley complained, because "passion and prejudice have conferred upon the battle far more importance than, as it seems to me, can rationally attach to its issue" (*SM* 2:471–72). For Huxley, the question was strictly scientific, one that could be resolved by the facts of comparative anatomy and physiology, "independently of all theoretical views" (2:475). Darwin's theory, special creation, or any other theory need not even be mentioned in the investigation of the "facts" of anatomy and physiology. After such facts were established, then one could comment on how well a particular theory fit them. Huxley, moreover, believed the facts were well known. The researches of a variety of men, including Owen, all pointed to the same conclusion: that apes and humans were extremely closely related. But Huxley realized that before he could discuss anatomical and physiological evidence, he had to convince his audience that such a view in no way detracted from the dignity of humankind. It is this aspect of Huxley's writings that is so powerful, setting his work apart from other discussions on the ape-human question.

Huxley claimed that it did not matter whether man's origin was distinct from all other animals or whether he was the result of modification from another mammal: "His duties and his aspirations must, I apprehend, remain the same. The proof of his claim to independent parentage will not change the brutishness of man's lower nature nor . . . will the demonstration of pithecoid pedigree one whit diminish man's divine right of kinship over nature" (*SM* 2:472). Human dignity, according to Huxley, was not inherited, but rather "to be won by each of us, so far as he consciously seeks good and avoids evil, and puts the faculties with which he is endowed to its fittest use" (472). Thus, for Huxley, all aspects of human nature, both "brutishness" and "princely dignity," would have to be accounted for independently of the question of human origins. A vast chasm separated civilized humans from the "brute." Even if humans came *from* the brutes, they were not *of* them (*CE* 7:153). Huxley's strategy, then, was to convince his readers that the highly charged issues concerning humanity's morals and ethics, questions of good and evil, were not relevant to the question of human origins. This would allow the problem of classification to be investigated objectively and dispassionately.

This is not to say that Huxley was uninterested in such questions. He became quite embroiled in the question of evolutionary ethics. In his 1893 lecture on evolution and ethics, he discussed why he did not think the doctrine of evolution could give us an ethics to live by.

> The propounders of what are called the "ethics of evolution," when the "evolution of ethics" would usually better express the object of their speculations, adduce a number of more or less interesting facts and more or less sound arguments in favour of the origin of the moral sentiments, in the same way as other natural phenomena, by a process of evolution. . . . But as the immoral sentiments have no less been evolved, there is so far, as much natural sanction for the one as the other. The thief and the murderer follow nature just as much as the philanthropist. Cosmic evolution may teach us how the good and the evil tendencies of man may have come about; but, in itself, it is incompetent to furnish any better reason why what we call good is preferable to what we call evil than we had before. (*CE* 9:79–80; for a fuller discussion see Paradis and Williams.)

Thus, Huxley did not directly address the problem of whether humans were descended from an apelike ancestor. Instead, he asked how closely related were apes and humans, approaching the question just as a taxonomist would investigate how closely related were the cat and the dog. It is clear, however, from the rhetorical structure of the lectures that the close taxonomic relationship between humans and other primates was intended to be interpreted as support for human evolution from lower animals.

Like Darwin, Huxley appealed to the Victorians' love of their pets to demonstrate the unity of humans with the animal world. "The dog, the cat . . . return love for our love, and hatred for our hatred. They are capable of shame and sorrow; and though they may have no logic nor conscious ratiocination, no one who has watched their ways can doubt that they possess that power of rational cerebration which evolves reasonable acts from the premises furnished by the senses" (*SM* 2:473).

Thus, Huxley claimed a psychical as well as physical unity existed between man and beast. And whom did Huxley cite in support of such a view? None other than his arch-antagonist Richard Owen. Owen, in one of his more racist statements, claimed that any psychical differences between a chimpanzee, a Boschisman, and an Aztec with arrested brain-growth were only of degree. Furthermore, he noted that "every tooth, every bone was strictly homologous," and this "makes the determination of the difference between *Homo* and *Pithecus* the anatomist's difficulty" (Owen, "On the Characters, Principles of Division and Primary Groups of the Class Mammalia," n. 20, qtd. in *SM* 2:473). Owen removed this note in a later reprint of the paper, undoubtedly realizing it provided support for a position in direct opposition to the views he held. Owen believed that man's "psychological powers, in association with his extraordinarily developed brain, entitles the group which he represents to equivalent rank with the other primary divisions of the class *Mammalia,* founded on cerebral characters" which he named *Archencephala* (qtd. in *SM* 2:474). Huxley proceeded to demolish Owen's view that the human brain exhibited peculiar and unique characters.

Citing studies in embryology and comparative anatomy, Huxley claimed that all the evidence showed that the differences between humans and the higher apes were no greater than those between the higher and lower apes. Owen had gotten himself into a totally untenable position by claiming that *Homo sapiens* had a unique structure, the hippocampus minor, which justified ranking it in its own subclass. Unlike Huxley, he needed to ground his belief in the uniqueness of the human psyche in human anatomy. Huxley showed no mercy in his attack on Owen. He quoted Owen extensively, only to discredit him. Owen had asserted in his essay "On Characters":

> In man, the brain presents an ascensive step in development. . . . Not only do the cerebral hemispheres . . . overlap the olfactory lobes and cerebellum, but they extend in advance of the one, and further back than the other. . . . Their posterior development is so marked, that anatomists have assigned to that part the character of a third lobe; *it is peculiar to the genus Homo and equally peculiar is the posterior horn of the lateral ventricle, and the "hippocampus minor," which characterise the hind lobe of each hemisphere.* (qtd. in *SM* 2:476)

Huxley then proceeded to demonstrate in great detail that none of Owen's statements were true. The third lobe and the posterior cornu of the lateral ventricle were "neither peculiar to, nor characteristic of man." Both traits were found in all of the higher Quadrumana. The hippocampus was also not unique to humans, having been found in certain higher Quadrumana (*SM* 2:476). Huxley cited the work of Schroeder van der Kolk, Vrolik, and others in support of his views. He quoted Allen Thomson's letter of May 24, 1861, which summarized the result of his dissection of the brain of a chimpanzee: "There is, very clearly, a posterior lobe, separated from the middle one by as deep a groove between the convolutions on the inner side of the hemispheres as in man, and equally well marked off on the outer side. I should be inclined to say, that the posterior lobe is little inferior to that of man" (qtd. in 2:481).

Huxley concluded that "every original authority testifies that the presence of a third lobe in the cerebral hemisphere is not 'peculiar to the genus homo,' but that the same structure is discoverable in all the true Simiae . . . and is even observable in some lower Mammalia" (*SM* 2:481). But Owen would not give up. On March 19, 1861, he gave a lecture at the Royal Institution in which he still maintained that only humans had a hippocampus minor and that the difference between the human brain and the brain of a gorilla or chimpanzee was far greater than between the brains of other primates. The lecture was reported in the *Athenaeum,* and Huxley responded with a letter to Hooker pointing out the many errors in Owen's views, including commenting on the incompleteness and inaccuracy of the diagram of a gorilla brain that appeared in the article. A series of exchanges between Owen and Huxley printed in the *Athenaeum* occurred in which Owen seemed determined to make a fool of himself. Huxley described the absurdity of the dispute to Hooker: "A controversy between Owen and myself, which I can only call absurd (as there is no doubt whatever about the facts), has been going on in the *Athenaeum,* and I wound it up in disgust last week. . . . I do not believe that in the whole history of science there is a case of any man of reputation getting himself into such a contemptible position. He will be the laughing-stock of all the continental anatomists" (*Life* 1:206). The matter was finally put to rest in 1862, when Sir W. Flower, in a public dissection of ape brains demonstrated the existence of those cerebral characteristics that Owen claimed were unique to man.

Nicolaas Rupke has argued that Huxley's victory was not as complete as Huxley and the pro-Darwin forces would have us believe. He points out that Owen had never denied that the gorilla had a hippocampus minor, but that it was not as well developed as in humans and so did not deserve the name hippocampus minor. Yet Rupke also admits that by claiming the "posterior lobe," "posterior cornu," and "hippocampus minor" were features that were characteristic of the human brain, Owen implied that these features were not present in simian brains. Furthermore, Owen had not made clear that "indications" or "traces" of a posterior cornu and hippocampus minor were present in apes or monkeys in his original 1857 lecture (Rupke 293). Owen later retreated to a more moderate position, but Huxley's response was to this 1857 lecture. Rupke also confirms that new anatomic information, particularly Flower's study, showed that the cerebral characteristics did not vary in the distinct manner that Owen had argued and thus were not so useful for taxonomic purposes. Thus, while Rupke's account of this controversy is more sympathetic to Owen than my own, it basically confirms my own analysis. (See Lyons, "Evolution," and Rupke for a comparison.)

Man's Place in Nature

In spite of the heated debate between the two men, it is important to emphasize that Huxley was not just carrying on a dispute with Owen. The circumstances that led to the writing of *Man's Place in Nature* illustrate that scientific disputes, personality conflicts, and differences in worldview were inextricably intertwined in the question of human ancestry. As a result of the publication of the *Origin* and his clash with Wilberforce, in the spring of 1861 Huxley decided to devote his weekly lectures to working men to the relation of men to the rest of the animal kingdom. The lectures were quite popular. He wrote his wife, "Lyell came and was rather astonished at the magnitude and attentiveness of the audience," and he quipped that "by next Friday evening they will all be convinced that they are monkeys" (*Life* 1:205).

In the midst of all the controversy over the hippocampus, much to Huxley's surprise, he was invited to give two lectures on the relation of man to the lower animals to the Philosophical Institute of Edinburgh. He explicitly claimed that the comparative anatomy of primates provided powerful evidence in favor of Darwin's theory. If Darwin's hypothesis explained the common ancestry of the latter, then it followed that it also explained the origin of the former. The lectures were a success, but Huxley felt they had not reported quite accurately his comments about Darwin's theory. He complained to Darwin: "Nor have they reported here my distinct statement that I believe man and the apes to have come from one stock" (*Life* 1:209). To Hooker he complained of the same oversight: "I told them in so many words that I entertained no doubt of the origin of man from the same stock as the apes. . . . The report does not put nearly strong enough what I said in favour of Darwin's views. I affirmed it to be the only scientific hypothesis of the origin of species in existence" (210).

Nevertheless, Huxley was pleased with the reception he received in Edinburgh. These lectures allowed him to promulgate Darwinism, and in doing so, he was once again able to reiterate the importance of keeping scientific questions distinct from theological ones. That he was even invited to "saintly" Edinburgh and able to speak to an audience that applauded his belief that humans had an apelike ancestor, he found a "grand indication of the general disintegration of old prejudices which is going on" (*Life* 1:210). Huxley hoped to develop the lectures further, give them again in London, and then publish them. They did eventually form part 2 of *Man's Place in Nature.*

Huxley published *Man's Place in Nature* in 1863, in spite of warnings that it would be the undoing of his career. The book consisted of three parts. The heart of *Man's Place* both literally and figuratively was part two. Huxley had refined his earlier lectures, and the result was an educational tour de force documenting "the extent of the bonds which connect man with the brute world" (*CE* 7:81). He then used this evidence to argue for the superiority of the Darwinian hypothesis to all others regarding the origin of humans. Part 1, "On the Natural History of the Man-Like Apes," provided an overview of how the great apes had been regarded through history. It illustrated that the problem of the relationship of apes to humans had fascinated people for centuries. Do apes think? Are they capable of telling right from wrong? Do they have societies? Facts and mythology had intermingled, resulting in much confusion over the classification of these animals. Although not directly stated, Huxley was implying that an objective investigation of these animals, carefully examining their physical makeup, would clear away much of the confusion, thus laying the groundwork for part 2 of his book.

Part 3 discussed the recent discoveries of the fossil remains of man, including the Engis skull and Neanderthal man. Huxley concluded from these ancient skulls that they were not significantly different from modern humans and thus "do not take us appreciably nearer to that lower pithecoid form, by the modification of which he has, probably, become what he is" (*CE* 7: 208). But Huxley anticipated that the finds of paleontology would push the origin of humans back to a far earlier epoch than anyone had previously imagined. If humankind was that ancient, clearly this was evidence against the creation hypothesis. Once again, the Darwinian hypothesis was the only hypothesis that could make sense of these ancient human fossils.

In part 2 Huxley's presentation of the facts of development and comparative anatomy was superb. It was hard to imagine how any one could deny his or her ape ancestry after reading his discussion. But Huxley knew beliefs run deep. Whether scientists or members of the working class, whether they were religiously devout or not, most Victorians would still be appalled at the inevitable conclusions to be drawn from Huxley's analysis. They would argue that "the belief in the unity of origin of man and brutes involves the brutalization and degradation of the former" (*CE* 7:153). But Huxley questioned if that were really so. In a passionate entreaty, he claimed that human dignity did not depend on our physical characteristics or our origins.

> It is not I who seek to base Man's dignity upon his great toe, or insinuate that we are lost if an Ape has a hippocampus minor. On the contrary, I have done my best to sweep away this vanity. . . . Is it, indeed, true, that the Poet,

> or the Philosopher, or the Artist whose genius is the glory of his age, is degraded from his high estate by the undoubted historical probability, not to say certainty, that he is the direct descendant of some naked and bestial savage. . . . Is mother-love vile because a hen shows it, or fidelity base because dogs possess it?" (*CE* 7:152–54)

This was obviously a dig at Owen who claimed not only that the hippocampus minor, but the hallux, or big toe, were unique to *Homo sapiens.* But this was not just another example of Huxley's indulging in Owen bashing. His point was substantive: just because we share admirable traits with the lower animals does not make these traits less admirable. Furthermore, for Huxley, humankind's lowly ancestry was "the best evidence of the splendour of his capacities." Only man, after all, "possesses the marvellous endowment of intelligible and rational speech, whereby . . . he has slowly accumulated and organised the experience which is almost wholly lost with the cessation of every individual life in other animals; so that, now, he stands raised upon it as on a mountain top, far above the level of his humble fellows, and transfigured from his grosser nature by reflecting, here and there, a ray from the infinite source of truth" (*CE* 7:154–56). Huxley may have thought we descended from the brutes, but it is clear he also believed there was a vast gap between us and the rest of the animal world.

The Impact of Man's Place in Nature

Man's Place in Nature was an immediate success. Published in January 1863, by the middle of February it had sold two thousand copies. By July it was republished in the United States and was being translated into French and German (*Life* 1:217). It continued to be reprinted for the next forty years. The book made an important impression upon leading men of science. Darwin indicated to Lyell how "splendid some pages are in Huxley" and Lyell warmly agreed. If only, he speculated, Huxley had the advantage of their leisure to combine "the rigour and logic of the lectures," his Geological Society address and half a dozen other recent works into "one great book, what a position he would occupy!" (qtd. in F. Darwin 199 and Lyell 366). Clearly, Huxley's book was not going unnoticed.

An alternative and no less important means of assessing the impact of

Man's Place in Nature is to examine the popular literature surrounding the debate over human ancestry. The discussions over evolution and human origins had spread far beyond the staid halls of the Royal Society and the Geological Society. The hippocampus debate represented more than some esoteric scientific disagreement over an obscure part of the brain. Rather, "man's place in nature" was at stake. What made humans unique? What was the role of science in answering such a question? How could evolutionary ideas be made compatible with theology? It is certainly true that the personality clash between Owen and Huxley played a significant role in the popular accounts. London society revelled in the squabbles between the two men and satirical accounts soon appeared over the great hippocampus debate. But the fact that their disputes were even the subject of popular satire indicates how deeply the ape-human question had permeated society.

Punch printed a poem supposedly authored by a gorilla from the zoological gardens entitled "Monkeyana," with a picture of an apelike creature carrying a sign asking "Am I a Man and a Brother?" The poem was about Darwin's theory, but half of it spoofed the Owen-Huxley clash:

Then HUXLEY and OWEN,
With rivalry glowing,
With pen and ink rush to scratch;
'Tis Brain *versus* Brain
Till one of them's slain;
By Jove! it will be a good match!

Says OWEN you can see
The brain of Chimpanzee
Is always exceedingly small,
With hindermost "horn"
Of extremity shorn;
And no "Hippocampus" at all.

The Professor then tells 'em,
That man's "cerebellum,"
From a vertical point you can't see;
That each "convolution"
Contains a solution,
Of "Archencephalic" degree.

Then apes have no nose,
And thumbs for great toes,
And a pelvis both narrow and slight;
They can't stand upright,
Unless to show fight,
With "Du Chaillu," that chivalrous knight!

Next HUXLEY replies,
That OWEN he lies,
And garbles his Latin quotation;
That his facts are not new,
His mistakes not a few,
Detrimental to his reputation,

"To twice slay the slain,"
By dint of the Brain,
(Thus HUXLEY concludes his review)
Is but labour in vain,
Unproductive of gain,
And so I shall bid you "Adieu"!
(206)

This poem reveals that the debates over human ancestry also had implications for the relationship between the races. The United States was immersed in a civil war over the issue of slavery. Are not Negroes our brothers? Certainly if a gorilla is claiming to be related to humans, what does this say about all of humankind?

In 1863 an anonymously printed pamphlet, "A Sad Case" (it was in fact published by George Pycroft), appeared, which described the imaginary courtroom proceedings of Owen and Huxley, who had been charged with disturbing the peace. The two men had been about to come to blows when the police arrived. Huxley had called Owen a "lying Orthognathous Brachcephalic Bimaneous Pithecus," while Owen had accused Huxley of being "nothing else but a thorough Archencephalic Primate" (Pycroft 3). As the trial proceeded, it was apparent that the dispute was not just about whether apes had a hippocampus minor. Owen was outraged that he was being told in public that he was physically, morally, and intellectually only a little better than a gorilla (6). But the problem went deeper than that. Owen claimed that he had regarded Huxley as a "quiet well-meaning man," but since he had become successful he had become "highly dangerous" (5). Furthermore, Huxley and his new pals, Darwin, Rolleston, and others, were always ridiculing him. But Huxley countered that everything was fine, "as long as Dick Owen was top sawyer, and could keep over my head, and throw his dust down in my eyes. There was only two or three in our trade, and it was not very profitable; but that was no reason why I should be called a liar by an improved gorilla, like that fellow" (6). A Mr. Bull testified that there was bad feeling between the whole lot of these scientists. The Mayor suggested that perhaps the clergy might exert some influence over them, to which Mr. Bull replied that "no class of men paid so little attention to the opinions of the clergy as that to which these unhappy men belonged" (6). The story ended with the Mayor admonishing both men and telling them that they should be friends and help each other in their work. But as soon as they were out the door, an altercation ensued. That night Huxley was again attacking Owen in his lecture on man's place in nature. "But as the assemblage consisted of working men, and as they were very orderly they were not interfered with by the police" (8).

"A Sad Case" wonderfully illustrates the various issues surrounding the controversy over human origins. While the focal point is the clash between Huxley and Owen, their disagreements reflect the fighting for power and prestige that is occurring within the scientific community as a whole. Are Darwinian ideas going to prevail? Who will obtain the few paying jobs that are available? Huxley promulgates his ideas to the common man by means of his working men's lectures, which become an astute satire suggesting the decline of religion—especially in its conflict with science.

The most famous account of the feud between Huxley and Owen is undoubtedly by Charles Kingsley, who immortalized the dispute in his satirical fairy tale *The Waterbabies*, sparing neither man in his commentary. In the first part of the book Huxley and Owen were mentioned as experts who could help evaluate whether waterbabies really existed. Surely, if waterbabies existed, one would have been caught, put in spirits and cut in half. One half would have been given to Professor Huxley and the other to Professor Owen to see what each had to say about it (50). Later in the book, the two men appeared as caricatures of themselves. The empiricist Huxley was portrayed as the great naturalist Professor Pttmilnsprts of Necrobioneopaleonthydrochtho anthropopithekology. He held the view that "no man was forced to believe anything to be true, but what he could see, hear, taste, or handle." Furthermore, "he had even got up once at the British Association and declared that apes had hippopotamus majors in their brains just as men have" (107–9). But clearly, Kingsley was on Huxley's side. Ridiculing Owen's view, the narrator pointed out that to claim that apes had hippopotamus majors

> was a shocking thing to say; for if it were so what would become of the faith, hope, and charity of immortal millions? You may think that there are other more important differences between you and an ape, such as being able to speak and make machines, and know right from wrong, and say your prayers, and other little matters of that kind: but that is a child's fancy my dear. Nothing is to be depended on but the great hippopotamus test. If you have a hippopotamus major in your brain, you are no ape, though you had four hands, no feet, and were more apish than the apes of all aperies. But if a hippopotamus major is ever discovered in one single ape's brain, nothing will save your great-great-great-great-great-great-great-great-great-great-great-greater-greatest grandmother from having been an ape too. (109–10)

This passage not only lampooned the debate over the hippocampus, but also made light of the Wilberforce-Huxley clash, where Wilberforce supposedly asked Huxley whether he was related on his grandfather's or grandmother's side to an ape. (Oddly enough, no totally reliable version of what transpired at Oxford exists. Leonard Huxley reprinted most of the different accounts; see *Life* 193–204).

Both "A Sad Case" and *The Waterbabies* were insightful satires about the debates over evolutionary theory occurring at the time. Huxley was at the center of these intrigues and certainly relished the chance to topple the "British Cuvier" from his pedestal. Yet, in all of these accounts the technical aspects of the debate were portrayed as well. It is virtually impossible to disentangle the scientific disagreements from the larger social issues in which they were embedded. Still, following the chronology of Huxley's work between 1858 and 1863, it seems clear that the primary motive for his research on the relationship between apes and humans was the scientific issues involved. Huxley wanted the ape-human question to be regarded as a scientific one and to remove it from the realm of philosophy and theology. His involvement with this issue had begun before the publication of the *Origin,* but its publication further sparked Huxley's interest, resulting in a series of publications and public lectures. The investigation of these scientific questions brought him into conflict with Owen, not the other way round.

I HAVE ARGUED that Huxley wrote *Man's Place in Nature* specifically to support Darwin's theory. It is true that it also served as a popular vehicle to attack Owen's faulty anatomy and the conclusions he drew from that anatomy. But the grounds for that dispute were laid much earlier, and the issue had been pretty much resolved by 1861. Thus, I think we need to draw a distinction between Huxley's reasons for believing that humans had a pithecoid ancestry and how he chose to convince others that this was true. Huxley was a highly skilled polemicist and he realized that just the facts of anatomy would not be enough to convince men they were monkeys. Nevertheless, the reason Huxley believed that humans were not an exception to Darwin's theory of the origin of species can be found in the facts of primate anatomy and physiology. Certainly Huxley was naive in his belief that a clear distinction could be made between "facts" and the interpretation of them. The structures of the brain could not be as easily identified as either Huxley or Owen would have had us believe. Even Huxley's good friend Kingsley ridiculed Huxley's extreme empiricism in his portrayal of Professor Pttmilnsprts in *The Waterbabies.* In addition, Huxley controlled the boundaries of the discourse by framing the issues in particular ways, thus admitting some "facts" and not allowing others. However, this does not mean we have to look to social, economic, or political factors to explain why Huxley held the views he did.

While we do not need to turn to a social constructionist account to understand Huxley's view of "man's place in nature," to appreciate the significance of this episode we do have to place Huxley's argument in the context of the times. The hippocampus debate is an excellent example of how scientific discourse often operates at several different levels. "A Sad Case" is, among other things, a commentary on the political struggle for power going on within the scientific community. Moreover, the ape-human controversy provided a wonderful opportunity for Huxley to show the dangers of mixing theology with science. This served his larger political agenda of making science rather than the church the source of moral authority and power in society. His rhetorical strategy drew on his own anatomical work to advance a disciplinary argument that at the same time resulted in personal notoriety and promoted his institutional standing. But this should not overshadow Huxley's appreciation of Darwin's theory as a scientific achievement. He did not think transmutation had been proven, but as he wrote Lyell: "I view it as a powerful instrument of research. Follow it out, and it will lead us somewhere; while the other notion is like all the modifications of 'final causation,' a barren virgin" (*Life* 1:187). It had tremendous appeal for Huxley because it was a purely naturalistic explanation of the history of life, free from any argument of design or supernatural causation. Huxley knew that the most crucial aspect of having Darwin's theory accepted was "convincing men that they were monkeys" and he brought all his considerable talents to that task.

REFERENCES

Bartholomew, Michael. "Huxley's Defence of Darwin." *Annals of Science* 32 (1975): 525–35.

Bowler, Peter. *The Non-Darwinian Revolution: Reinterpreting a Historical Myth.* Baltimore: Johns Hopkins UP, 1988.

Cloyd, E. L. *James Burnett, Lord Monboddo.* Oxford: Clarendon Press, 1972.

Darwin, Charles. *The Origin of Species.* 1859. N.Y.: Avenel, 1979.

Darwin, Francis, ed. *Life and Letters of Charles Darwin.* Vol. 1. N.Y.: D. Appleton, n.d.

Desmond, Adrian. *Archetypes and Ancestors: Palaeontology in Victorian London, 1850–1875.* Chicago: U of Chicago P, 1984.

di Gregorio, Mario. *T. H. Huxley's Place in Natural Science.* New Haven: Yale UP, 1984.

Ghiselin, Michael. "The Individual in the Darwinian Revolution." *New Literary History* 3 (1971): 113–34.

Greene, John. *Death of Adam.* 1959. Mentor Press, 1961.

Huxley, Leonard. *Life and Letters of Thomas H. Huxley.* 2 vols. N.Y.: Appleton, 1901.

Huxley, Thomas Henry. *Collected Essays.* 9 vols. 1894. N.Y.: Greenwood Press, 1968.

———. *The Scientific Memoirs of Thomas Henry Huxley.* Ed. Michael Foster and E. Ray Lankester. 4 vols. plus supplement. London: Macmillan, 1898–1903.

Huxley Papers. Imperial College of Science, Technology, and Medicine Archives, London.

Kingsley, Charles. *The Waterbabies.* 1872. London: J. M. Dent, 1985.

Lindberg, D., and R. Numbers, eds. *God and Nature.* Berkeley: U of California P, 1986.

Lyell, Mrs., ed. *Life and Letters of Charles Lyell.* Vol. 2. London: John Murray, 1881.

Lyons, Sherrie. "The Origins of T. H. Huxley's Saltationism: History in Darwin's Shadow." *Journal of the History of Biology* 28 (1995): 463–94.

———. "Thomas Huxley: Fossils, Persistence, and the Argument from Design." *Journal of the History of Biology* 26.3 (fall 1993): 545–69.

Paradis, James, and G. C. Williams. *T. H. Huxley's "Evolution and Ethics," with New Essays on Its Victorian and Sociobiological Context.* Ed. Paradis and Williams. Princeton: Princeton UP, 1989.

Punch, May, 18, 1861.

[Pycroft, George]. "A Sad Case, Mansion House." Owen Manuscripts, British Museum, Natural History, April 23, 1863.

Richards, Evelleen. "Huxley and Woman's Place in Science: The 'Woman Question' and the Control of Victorian Anthropology." *History, Humanity, and Evolution: Essays for John C. Greene.* Ed. James R. Moore. Cambridge: Cambridge UP, 1990. 253–84.

Rupke, Nicolaas. *Richard Owen: Victorian Naturalist.* New Haven: Yale UP, 1994.

Thomas Henry Huxley and the Reconstruction of Life's Ancestry

PETER J. BOWLER

IF WE ACCEPT THAT HUXLEY'S CONVERSION TO DARWINISM WAS a major step in his intellectual life, we are nevertheless forced to admit that the event plays a more problematic role in his scientific career. For all his efforts as "Darwin's bulldog," Huxley found it difficult to take on board some of the components of Darwin's theory that modern biologists find most innovative. As a biologist he was primarily a morphologist and systematist; he studied the structure of living and fossil species with a view to working out their relationships. Coleman has shown that in the case of Carl Gegenbaur, it was fairly easy for a morphologist to accept the basic concept of evolution. The old idea of idealized relationships between species had to be replaced by the theory that homologous structures are signs of common descent: two species have an underlying similarity because they inherit much of their structure from a common ancestor. Late nineteenth-century morphologists and paleontologists were inspired to undertake a comprehensive effort to reconstruct the course of evolution, explaining the relationships they observed by postulating evolutionary trees defining the key branching points in the history of life on earth. Huxley played a leading role in this enterprise, but we are left with an important question: To what extent did his efforts derive their inspiration from the Darwinian theory with which he is so closely associated in the popular mythology of science?

Fortunately, we have detailed accounts of Huxley's work in this area from Adrian Desmond and Mario di Gregorio. Many other areas of evolutionary morphology and paleontology remain unmapped by historians (see Bowler,

Life's Splendid Drama, for an attempt to plug this gap in the literature). It is thus less easy to assess Huxley's contributions in their contemporary context—yet if we are to understand the role played by Darwinism in creating the theoretical framework of his evolutionism, it is precisely this kind of overview we need. The work that has already been done suggests the kind of problem we must grapple with. Both Desmond and di Gregorio recognize that Huxley's initial conversion to Darwinism in 1859 had no effect on his detailed scientific work. Both accept that he only began to search for evolutionary relationships after he read Ernst Haeckel's *Generelle Morphologie* of 1866. But Haeckel was no straightforward Darwinian; his evolutionism contained a strong element of German Naturphilosophie, and he openly supplemented the selection theory with Lamarckism (the inheritance of acquired characteristics). Many commentators have pointed out that Huxley himself retained doubts about the efficacy of natural selection. More important, he openly proclaimed his support for non-Darwinian mechanisms of evolution. In his 1871 essay "Mr Darwin's Critics" he wrote of a "law of variation" that might somehow direct evolution (*CE* 2:181–82). Later, he envisioned directed evolution producing predetermined trends, with selection merely weeding out those trends that were nonadaptive (*CE* 2:223). These ideas were typical of those favored by many morphologists and paleontologists in the late nineteenth century (Bowler, *Eclipse*). But they would have been unacceptable to Darwin himself, and for this reason I have called Huxley a "pseudo-Darwinian" (Bowler, *Non-Darwinian Revolution* 73–74).

These suggestions are important in the context of Huxley's wider motivations. Desmond has portrayed the debates between Huxley and scientific rivals such as Richard Owen and St. George Mivart in terms of the social tensions of the time. Huxley's Darwinism was meant to uphold a liberal philosophy of economic and social progress, while Owen and Mivart's idealistic view of organic relationships was more in tune with a conservative political agenda. Desmond suggests that Huxley's actual scientific work was on occasion deflected by his opposition to Owen's idealism: certain kinds of evolutionary relationships were unacceptable to Huxley precisely because they were too obviously anti-Darwinian and thus gave support for the claim that evolution was driven by mysterious built-in trends. But the anti-Darwinian ideas mentioned above seem to count against Desmond's interpretation; they suggest that Huxley too thought of evolution as directed along predetermined channels. David Hull has seized upon this point to argue that there was no fundamental theoretical difference between pseudo-Darwinists such as Huxley and outright anti-Darwinists such as Mivart. They disagreed merely over their willingness to accept Darwin as the figurehead for the evolutionary movement, and if the personal circumstances had worked out differently, Mivart could have remained in the Darwinian camp.

My own feeling is that Hull's reduction of the whole debate to the personalities of the scientific community goes too far in its attempt to sideline the theoretical debates of the time. Desmond's sociological interpretations are often difficult to pin down (however suggestive they might be), but they do correspond to genuine differences in the interpretation of how evolution might work. Huxley may not have taken on board all of Darwin's thinking, and he may have actively favored some ideas that Darwin rejected. But he was not out to construct a model of the history of life on earth that was totally incompatible with Darwin's theory. Mivart's *On the Genesis of Species* was an attempt to articulate exactly such an anti-Darwinian model, and we need to understand the differences between the two approaches. The testing ground for this analysis must be Huxley's detailed work on what Haeckel called "phylogeny"—the reconstruction of life's ancestry. The account that follows contends that Huxley's contributions, while they do include some reference to non-Darwinian mechanisms, nevertheless show that he accepted more of the Darwinian worldview than Mivart was able to tolerate.

There are several key issues that must be understood as we look into the details of Huxley's work. First, to what extent was evolution directed along predetermined lines by internal biological forces that are unrelated to the demands of the environment? Mivart appealed over and over again to the existence of parallel trends in the history of life as evidence that there were built-in directing forces. For him, any evidence of the same structure being evolved independently in two parallel lines was incompatible with the element of "random" variation in Darwinism. It was simply too much of a coincidence for the same structure to have been independently invented by two different groups of animals. He was particularly impressed with evolutionary developments that could not possibly have occurred if natural selection were operating, especially the production of characters with no adaptive value. Huxley occasionally allowed for parallel or directed evolution—but always under circumstances where there was an adaptive benefit resulting from the change. Darwinians could accept an element of what was called "convergent evolution" because it was not impossible that a really advantageous change could have been hit upon independently by two different groups. What they could not accept was evolution being driven in totally nonadaptive directions by internal forces.

The element of adaptation is critical here. True, Huxley was by no means as concerned about this issue as Darwin himself. Huxley often paid scant attention to the details of how an organism would respond to changes in its environment, because he was more concerned with the relationships between species than with the actual process of change. But in general he was reluctant to postulate totally nonadaptive evolution, and his support for parallelism never took him into the realm of active opposition to the Darwinian worldview. Symptomatic of his support is his interest in an area of research that is characteristic of Darwin's own approach to evolution: geographical distribution. Morphologists in general were unimpressed by Darwin's efforts to use biogeographical evidence as a means of reconstructing episodes in the history of life. Again, Huxley did not get involved to the same extent as Darwin himself, but he *did* work on biogeography and showed some interest in the possibility that the current distribution of species in a group is the legacy of a historical process influenced by environmental factors. In the end, Huxley shared at least some of Darwin's conviction that the current state of life on earth is the product of an unpredictable historical process, rather than the unfolding of a rigidly predetermined pattern.

Horses and Crocodiles: Patterns in the Past

One of the reasons Huxley was at first reluctant to get involved with the reconstruction of phylogenies from fossil evidence was his prior commitment to the idea of "persistence of type." In his preevolutionary days he had opposed the idea of a simple progressive trend in the history of life by pointing out that many forms of life have persisted with only minor changes since the very early stages of the fossil record. His 1862 address to the Geological Society entitled "Geological Contemporaneity and Persistent Forms of Life" stressed the same point despite his conversion to Darwinism (*CE* 2:272–304). Huxley still insisted there is little evidence of progress in the fossil record—that is why Darwin's is the only acceptable theory of evolution, because it does not demand constant progress. But the same principle makes it unlikely that we shall ever find the fossil evidence of evolutionary transformations. At this point Huxley still believed that most groups of animals would have originated at a very early point in the history of life, long before there was an adequate fossil record by which to trace their origins. He also insisted that there was no evidence of a gradual specialization within any group in the course of the known fossil record (2:303).

Most authorities would have said that Huxley was going too far in making these claims. Richard Owen had been tracing sequences of gradual specialization in the fossil record since the 1850s (Ospovat; Bowler, *Fossils and Progress* ch. 5). By the time Huxley gave his 1870 address "Palaeontology and the Doctrine of Evolution," he was prepared to admit that "there is much ground for softening the somewhat Brutus-like severity with which, in 1862, I dealt with a doctrine, for the truth of which I should have been glad enough to be able to find a good foundation" (*CE* 8:347–48). This was particularly the case, he argued, for the later evolution of mammalian groups. He made the important point that it was necessary to distinguish between "intercalary types," which merely illustrated intermediate stages between earlier and later forms, and genuine evolutionary transitions. This was necessary because "it is always probable that one may not hit upon the exact line of filiation, and, in dealing with fossils, may mistake uncles and nephews for fathers and sons" (8:349–50).

Huxley appealed to the work of Albert Gaudry and others who had been turning up fossils that seemed to fall naturally into an evolutionary sequence leading toward the modern horse (Rudwick 245–49). This was a classic example of evolutionary specialization: an early, generalized form of mammal seemed to have evolved by a series of recognizable intermediate stages toward the modern horse, with its specialized teeth and hooves adapting it for life on the open plains. The sequence led back to the three-toed *Anchiotherium* and *Plagiohippus* of the Eocene. Not surprisingly, Huxley and others had sought the ancestry of the horse in European fossils—the horse had been extinct in modern North America and had been introduced by the Europeans. But when Huxley went to America in 1876 he visited O. C.

Marsh and was shown a far more convincing sequence of fossil horses discovered in the American Midwest. In his "Lectures on Evolution," Huxley announced his conversion to Marsh's theory of the American origin of the horse and reproduced Marsh's diagram showing a series of six stages leading from the four-toed *Orohippus* of the Eocene to the modern horse (*CE* 4:130). This was, he proclaimed, "demonstrative evidence of evolution" (4:132). He was also able to announce, in a footnote to the printed text, Marsh's discovery of the even more primitive *Eohippus,* or "dawn horse," which carried the sequence back almost to the hypothetical five-toed ancestor that would link the horse family to the general mammalian type.

Marsh and Huxley presented the fossil horses in a tabular form that invites the reader to visualize a linear process of evolution running toward the modern horse as a goal. Modern evolutionists would certainly be wary of such a means of presentation, because we now know that the family tree of the horse has many side-branches leading off to extinction. It is certainly true that some late nineteenth-century paleontologists saw in the simplified linear sequence depicted by Marsh a linear process of evolution that was incompatible with the undirected nature of natural selection. This was the case for Marsh's great rival Edward Drinker Cope and other members of the neo-Lamarckian school of thought (Bowler, *Eclipse* ch 6). But the sequence is not one of "orthogenesis" as di Gregorio claims (99), because this term was reserved for evolutionary trends that were supposed to be driven by nonadaptive processes. The neo-Lamarckians' claim that evolution was directed by the habits taken up by the animals in a new environment was not fundamentally incompatible with Darwinism because the resulting trend was still adaptive. Darwin himself expected natural selection to produce specialization in most evolutionary lineages, and Huxley thought he was merely providing factual evidence of such a trend. Given his warning about mistaking uncles for fathers, we can be sure he would not have wished to commit himself to an absolute linearity of development that would be incompatible with the overall Darwinian view of evolution.

Huxley had originally opposed the view that any evolutionary changes could be traced in the fossil reptiles, but in his 1875 paper "On Stagonolepis Robertsoni and the Evolution of the Crocodilia," he again reversed his original position (*SM* 4:66–83). He now arranged the crocodiles into three orders, the Parasuchia, Mesosuchia, and Eosuchia, arguing that the sequence represented a transition from a generalized, still lizardlike form to the specialized modern crocodiles. Since this was also the order in which they appeared in the fossil record, this was again a fossil sequence offering clear evidence of evolution. Desmond notes that Huxley's thesis mapped nicely onto his views on human progress: the crocodiles had undergone an evolutionary spurt and then stagnated, just as social progress has been interrupted by occasional dark ages (4:170–74). More significant for our present theme, Desmond also notes that Huxley made no effort to explain *why* the crocodiles evolved in the way they did. In this case, no obvious reason for the specialization was given. Indeed, Desmond argues, it was Huxley's great rival Richard Owen who tried to understand crocodile evolution in terms of the changing ecological balance through geological time. In this case, Owen's search for the "purpose" of evolution led him to a more modern interpretation of what was going on than was possible with Huxley's purely morphological approach, which merely demonstrated relationships.

Fish, Birds, and Mammals: Vertebrate Innovations

Some of Huxley's most important contributions came in the reconstruction not of patterns in the evolution of particular groups, but of the major transformations by which new vertebrate classes were established. His most successful innovation was the proposed link between dinosaurs and birds. This interpretation of avian origins is still taken seriously today, although some aspects of Huxley's thesis have not stood the test of time. Less successful in the long term, but highly influential at the time, was his interpretation of the origin of mammals, which bypassed the reptiles and derived the mammals directly from amphibians. As Desmond has noted, Huxley's refusal to acknowledge the significance of the mammal-like reptile fossils being found in South Africa skewed the evolutionists' approach to this question until well into the 1890s (193–201).

Both of these theories confronted the problem of parallelism, and the question of whether each class has a monophyletic origin. As Haeckel had argued, the Darwinian was automatically inclined to assume that major innovations in the history of life had occurred only once. It was the anti-Darwinians like Mivart who wanted to argue that mysterious forces might lead two or more lines of evolution independently to advance from one class up to the level of the next. But Huxley was not dogmatic on this point, partly because to exclude parallelism in the origin of a particular class often required the postulation of the independent invention of similar structures in two different classes. As long as the parallel developments had adaptive significance, they were plausible within a generally Darwinian view of the history of life.

In the case of the birds, Huxley excluded any phylogenetic link between birds and the flying reptiles, the pterosaurs. Mivart tried to argue that the striking parallels in structure between birds and pterosaurs suggested that some kind of non-Darwinian directing force had been at work. Huxley simply ignored this claim, on the grounds that any good Darwinian would expect two groups independently acquiring the habit of flying to develop the appropriate adaptations. Huxley's own interpretation is, however, striking for the lack of interest it shows in the precise reasons why a running dinosaur should have found it advantageous to fly. In the case of mammalian origins, Desmond argues that Huxley excluded the reptiles because to accept that both birds and mammals had independently evolved from reptiles would have given too much away to the anti-Darwinians. Yet Huxley himself opted for several parallel lines of development leading independently to the monotremes, the marsupials and the placentals. This delighted Mivart, who openly proclaimed that the mammals were polyphyletic and argued that this was incompatible with Darwinism. In this case, though, Huxley seems to have felt that the advantages of developing mammalian characters might have been influential in several lines of (originally amphibian) evolution.

Before exploring these points in detail, I briefly note Huxley's involvement in debates on the transition from fish to amphibians. In a preevolutionary memoir on the fossil fish of the Devonian epoch, Huxley created the ganoid suborder Crossoptrygidae to include both these fossils and the living lungfish (*SM* 2:417–60). The lungfish were soon separated off into a distinct order, the Dipnoi. Most early efforts to explain the origin of the amphibians derived them, not surprisingly, from the lungfish. Eventually, however, paleontologists began to recognize that the amphibians were more plausibly derived from the fossil crossopterygians.

In the meantime, Huxley became involved in the debate over the relationship between the paired fins of fishes and the limbs of tetrapods. Carl

Gegenbaur had eventually decided that the fin of the newly discovered Australian lungfish *Ceratodus* was the "archipterygium" or basic form from which the fins of other fishes and the amphibian limb had been derived. But few, apart from Gegenbaur's own disciples, were entirely happy with the theory. The archipterygium consisted of a central rod of bones with rays branching out symmetrically on either side. Yet the tetrapod lower limb consists of *two* bones, the radius and ulna in the anterior limb or arm, the tibia and fibula in the posterior or leg. These in turn must articulate in a particular way. Gegenbaur certainly tried to identify the bones of the leg and arm with elements of the symmetrical archipterygium. He believed that the homologues of the main axis in the archipterygium were (for the forelimb) the humerus, the radius, and the first digit (*Grundriss* 497). The pentadactyle limb was thus derived from only one side of the archipterygium.

In 1876 Huxley published a study of *Ceratodus* in which he evaluated Gegenbaur's theory (*SM* 4:84–124). He noted the problem that in fish and tetrapods the limbs rotate in different directions with respect to the trunk (4:109–10). While accepting that the archipterygium of *Ceratodus* was the fundamental form of the limb, he was forced to dissent from the rest of Gegenbaur's theory. As he understood the homologies of the limb bones in fish and tetrapods, the rotations required by the theory would create a torsion of the humerus that he found quite implausible (4:118). Gegenbaur thought that the tetrapod limb was produced by a continuation of the same process as that which generated the asymmetrical fins of other fish, but Huxley argued that by abandoning this assumption a simpler explanation became possible. The tetrapod limb, or cheiropterygium, and the fish fin were developed by different kinds of specialization starting from the archipterygium. He suggested that further investigation of Paleozoic amphibians was needed to test the different suggestions that had been put forward to explain the relationship.

Gegenbaur accepted Huxley's criticism, later editions of his works showing the main axis running through to the fifth digit (e.g., *Elements* 480). Little further progress was made, however, while the majority of biologists continued to believe that the lungfish were the starting point for amphibian origins. Only when it was recognized in the 1890s that the crossopterygians offered a more plausible ancestral form did new developments became possible, and this came too late for Huxley himself to participate.

The attempt to reconstruct an evolutionary link between dinosaurs and birds was Huxley's most significant contribution. In this case the evolutionists had important fossil evidence, especially *Archaeopteryx*—the reptile-bird intermediate from the Jurassic limestone of Solnhofen. The story of this discovery has been told over and over again in the context of the debates between Huxley and his archrival Richard Owen. But both Desmond (124–46) and di Gregorio (76–96) stress that the classic image of Huxley, the triumphant evolutionist, seizing upon the fossil's intermediate character to checkmate the antievolutionist Owen is a vast oversimplification. Huxley's views on bird origins were skewed by his own strange ideas about the timing of major evolutionary events, and *Archaeopteryx* was by no means crucial to his case. Owen, far from being an antievolutionist, proposed a rival theory in which birds were derived from pterosaurs.

The first specimen of *Archaeopteryx,* showing feathers but minus the head, was described by Owen as a primitive bird. Huxley was already a bitter foe of Owen, but he was slow to make evolutionary capital out of the fossil. His commitment to the idea of persistence of type led him to assume that the birds would have appeared much earlier than the Jurassic. This alone led him to doubt that *Archaeopteryx* could be the genuine transitional form leading to the birds, even if it shared the characters of the two classes. When he began to look for the ancestry of the birds he was encouraged not by *Archaeopteryx* but by the apparently birdlike character of some dinosaurs. In January 1868 he wrote to Haeckel: "In scientific work the main thing just now about which I am engaged is a revision of the Dinosauria, with an eye to the 'Descendenz Theorie.' The road from the Reptiles to Birds is by way of Dinosauria to the Ratitae [flightless birds]. The bird 'phylum' was struthious [ostrichlike], and wings grew out of rudimentary forelimbs. You see that among other things I have been reading Haeckel's *Morphologie*" (*Life* 1:325). In the same year Huxley published his own account of *Archaeopteryx* (*SM* 3:340–45), ridiculing Owen for describing the fossil upside down, but insisting that it was a bird—and would still be a bird even if a head with teeth were found (a prediction that would be fulfilled in 1877).

In the same year Huxley published his classic paper "On the Animals which are Most nearly Intermediate between Birds and Reptiles," based on a popular lecture at the Royal Institution (*SM* 3:303–13). Here he certainly used the fact that *Archaeopteryx* was more reptilian than any living bird to confound those who claimed that the fossil record was devoid of intermediates (3:305). But he passed straight on to a different topic: the birdlike feet and limbs of some dinosaurs. Huxley suggested that many large dinosaurs, including *Iguanodon,* had walked at least part of the time on their hind legs. More especially, he drew attention to a small dinosaur, *Compsognathus longipes,* whose hind limbs were so birdlike that it could be counted as a "missing link" between reptiles and birds (3:311). Huxley concluded with a warning that the known fossils were merely illustrative of what the much earlier transitional forms would have looked like, but insisted that "surely there is nothing very wild or illegitimate in the hypothesis that the *phylum* of the class *Aves* has its roots in the Dinosaurian reptiles; that these, passing through a series of such modifications as are exhibited in one of their phases by *Compsognathus,* have given rise to the *Ratitae;* while the *Carinatae* [flying birds] are still further modifications and differentiations of these last" (3:312).

Huxley explicitly ruled out a link between birds and pterosaurs. The flying reptiles were equivalent to bats in the mammalia; their wings were built on entirely different principles to those of birds and they had acquired the power of flight independently. In a paper on the classification of dinosaurs published in 1870, Huxley insisted that those characters in which the birds did resemble pterosaurs had relation "to physiological action and not to affinity" (*SM* 3:494). In other words they were convergences, adaptive modifications evolved separately in two groups with similar needs. The same year saw the appearance of another paper offering "Further Evidence of the Affinity between the Dinosaurian Reptiles and Birds" (3:465–86). Here Huxley brought forward evidence from John Phillips based on dinosaur fossils in the Oxford museum. In the discussion following the presentation of this paper at the Geological Society of London, Harry Govier Seeley challenged several of Huxley's points. Seeley was making a study of pterosaur fossils and was already moving toward a view that saw this group as distinct from the other reptiles. He also dismissed the similarities between the hind limbs of dinosaurs and birds as due to the similar functions they performed in running, not to genetic affinity—a point that would become central for later critics of Huxley's theory (qtd. in *SM* 3:486).

In the lectures he gave on evolution in America in 1876 Huxley introduced—in addition to *Archaeopteryx*—the toothed birds described from the Cretaceous rocks of the American Midwest by O. C. Marsh. Huxley insisted that these were "intercalary types" rather than truly transitional forms; they blurred the distinction between classes without necessarily marking the actual course of evolution. The true line of descent was from the birdlike dinosaurs, and Huxley suggested that the fossil tracks from the Triassic of Massachusetts, long attributed to birds, may have been made by dinosaurs (*CE* 4:109). The legacy of the concept of the persistence of type was still apparent, however; Huxley thought that the actual transition to birds had probably taken place in the Paleozoic.

The toothed birds, or "Odontornithes" as Marsh called them, had been discovered in Kansas in 1872. Like *Archaeopteryx,* they seemed to confirm that in the earlier part of their existence the birds had still retained reptilian characteristics such as teeth. Marsh partially endorsed Huxley's theory of bird origins in his 1880 monograph *Odontornithes.* The birds were an ancient class, he claimed, dating back at least to the Triassic, although the birds of that epoch would have been considerably more reptilian than any so far discovered (Huxley and Marsh 188). Marsh doubted that the reptilian ancestor would have been a dinosaur—it was more likely a generalized sauropsid from which both birds and dinosaurs had diverged. But certainly the early birds had been flightless, even when they had developed feathers and warm blood. The modern Ratitae were the surviving remnants of this early form (189). Marsh thought the flying birds had evolved when the smaller of these early types had taken to an arboreal life and had begun jumping from branch to branch.

Marsh's position anticipated the view that became popular in the early twentieth century (see, for instance, Heilmann). The birds had evolved not directly from the dinosaurs, but from a predinosaurian reptile. But Marsh retained one aspect of Huxley's theory that did not stand the test of time: the idea that the flightless birds were the ancestors of the whole class. Most authorities came to doubt that the similarities between the running dinosaurs and the flightless birds were a sign of evolutionary ancestry. They argued that the reptiles must have taken to the trees first and then evolved into flying birds, while the flightless birds are degenerate types that lost the power of flight at a later point in the class's evolution. The problem with Huxley's theory was that it was based solely on morphological resemblance. He made no attempt to explain why a running dinosaur should have found it advantageous to develop wings that would later be used for flying. Francis Nopsca later tried to plug this gap by arguing that the ancestral birds had oared the air with their wings to speed themselves along (234). But most biologists found this scenario implausible and argued that flight must have evolved in creatures that jumped from branch to branch in the trees.

On this view, the similarities between the ratitites and the dinosaurs are, as Seeley claimed, adaptive convergences produced by the fact that both have independently adapted to a running gait. This point was made against Huxley by Richard Owen (*Monograph* 87–88), although Owen also insisted that it was the bird-pterosaur resemblances that were the true sign of an evolutionary relationship. Mivart thought the Darwinians were caught in a trap in any case: either the bird-pterosaur or the ratitite-dinosaur similarities were independently evolved, both cases being incompatible with natural selection (83–86). Most later authorities would side with Huxley in arguing that such adaptive convergences are *not* incompatible with the selection theory. "Random" variation does not rule out the independent invention of similar characters where there is some good adaptive purpose involved.

Far more controversial, even by the end of Huxley's own life, was his claim that the mammals could not have evolved from any known living or fossil reptilian type. The most decisive statement of this view came in his 1879 paper "On the Character of the Pelvis in the Mammalia, and the Conclusions respecting the Origin of Mammals which may be based on them" (*SM* 4:346–56). Here Huxley took issue with Gegenbaur on the interpretation of the pelvis, insisting that the reptilian modifications were opposite to those found in mammals. The fossil Dicynodontia had a backward extension of the subsacral part of the ilium characteristic of the mammals, but had no obturator fontanelle. Thus, argued Huxley, "it seems to be useless to attempt to seek among any known Sauropsida for the kind of pelvis which analogy leads us to expect among those vertebrated animals which immediately preceded the lowest known mammals" (4:350). The "promammalia" would have had a pelvis intermediate between that of *Orthnithorhynchus* (the platypus) and a land tortoise, and this would have been the type from which all modifications of the Sauropsida and Mammalia had diverged. This type, he claimed, could be seen in the pelvis of the salamander. Hence, "These facts appear to me to point to the conclusion that the Mammalia have been connected with the Amphibia by some unknown 'promammalian' group, and not by any of the known forms of Sauropsida" (4:354). There was other evidence pointing in the same direction, especially the fact that the main conduit of arterial blood from the heart is the right aortic arch in the Sauropsida, but the left in the Mammalia. This could be explained if there were two lines of evolution diverging from the heart of the Amphibia.

Huxley concluded his article by suggesting that the successful establishment of the reptile-bird link should encourage the search for the unknown ancestors of the mammals in the fossil record. Yet, as Adrian Desmond has pointed out, the significance of the fossil mammal-like reptiles of South Africa was already being stressed by Richard Owen (193–201). Huxley pointedly ignored Owen's suggestions that these fossils provided evidence of the 'missing' line of evolution leading to the mammals. Desmond suggests that Huxley's refusal to follow up this lead reflects his antagonism to Owen's almost idealist philosophy of evolutionism, in which the history of life has been governed by divinely implanted trends. The problem was that the South African fossils were being described as *reptiles,* not as members of a hitherto unknown transitional type. This would mean that two lines of evolution leading from the amphibians had independently acquired the characteristics typical of reptiles, implying a degree of parallelism that was favorable to Owen's anti-Darwinian view of evolution. But Huxley himself insisted that there were parallel lines of evolution running through the main stages of mammalian evolution. One is left to wonder if his refusal to acknowledge the mammal-like reptiles was merely a product of his hostility to Owen: it would simply have been too galling for him to admit that his great rival had solved this major problem in phylogeny. Having committed himself to the claim that the ancestors of the mammals were not reptiles, he was unable to admit that any new fossil being described as a reptile could fit the bill.

The consequence of Huxley's refusal to consider a modification of his position was that the mammal-like reptiles did not receive the attention they deserved until much later in the century. Even when the case for a distinct line of reptile evolution leading toward the mammals had become generally accepted among paleontologists, some morphologists continued to support Huxley's original position. In the meantime, though, there were still active debates on the origin of the mammals, centered principally on the question of whether the monotremes and marsupials represented stages in the ascent toward the placentals. This was certainly Haeckel's position, which George Gaylord Simpson would later dismiss as "a recrudescence of the old naive conception of a *scala naturae,* masked by its application to great groups instead of small ones" (162). Huxley himself became associated with this position—rather unfairly, as Simpson notes, because his interpretation differed significantly from Haeckel's. Huxley certainly saw a sequence of developmental stages in the evolution of the mammals that he quite explicitly compared to the old chain of being. But for him, the living monotremes and marsupials were only illustrative of the stages; they were not directly related to the ancestors of the placentals, who had passed through the same stages quite independently. Huxley thus set up a system of parallel evolution in which several different lines of evolution had advanced to different levels of the same developmental hierarchy.

This interpretation was outlined in his 1880 paper "On the Application of the Laws of Evolution to the Arrangement of the Vertebrata and more particularly of the Mammalia" (*SM* 4:457–72). Huxley called the stage now reached by the monotremes the "Prototheria," while the marsupials were at the stage of the "Metatheria." The placentals were the "Eutheria." Huxley made it clear that the living monotremes and marsupials were specialized in ways that meant they could not represent the actual ancestors of the placentals. He suspected that the primitive marsupials had evolved the technique of preserving their young in pouches because they had been arboreal in habit, a point taken up by several later writers including Louis Dollo. He hoped for the discovery of fossil representatives of the metatherian stage from the Mesozoic that would not have borne their young in pouches. A diagram given to illustrate his idea shows a series of parallel lines of evolution passing through the three stages. Even the main placental groups each had a separate line that had passed through the prototherian and metatherian stages independently (4:469; see di Gregorio 102–3). The living monotremes and marsupials were additional lines that had never advanced all the way up the scale. Huxley again noted that the prototherian stage would have had an amphibian, not a reptilian, origin.

In Huxley's view, the monotremes were equivalent to, but not the superficially modified descendants of, an early stage in mammalian evolution. It was precisely this point that left his wider theory linking the mammals directly back to the amphibians open to challenge. For the zoologists of the 1880s, the strongest line of evidence suggesting that the reptiles were, after all, the parents of the mammals came not from the fossil record but from a more detailed study of monotreme reproduction. At the suggestion of the

Cambridge professor of morphology Francis Balfour, W. H. Caldwell went to Australia to study these elusive animals (see Gruber). As a result of his efforts, Caldwell was able to send his famous cable to the president of the British Association for the Advancement of Science to read at the 1884 meeting in Montreal: "Monotremes oviparous, ovum meroblastic" (see Caldwell 466). The significance of the latter point was that this was a reptilian kind of egg. If the monotremes were anything to do with the early phases of mammalian evolution, the discovery suggested that the class had indeed evolved from a reptilian ancestry. Partly as a result of this discovery, more attention now switched to Owen's mammal-like reptiles, with H. G. Seeley, Robert Broom, and others contributing to the erection of the modern theory in which the early reptiles split into two great stems, one leading to the mammal-like reptiles and the mammals, the other to the dinosaurs and birds.

Huxley's involvement in the debates over bird and mammal origins was thus productive of much controversy. But the positions he took up show a degree of flexibility that allowed him to postulate a certain amount of parallel evolution without ever going to the extremes that we find in Mivart's anti-Darwinian tirade. Huxley may have believed that evolution was to some extent directed by internal forces, but he also accepted that no significant development could occur that did not constitute an adaptation to a particular way of life, or an improvement in the organism's basic constitution. Huxley certainly thought the mammalian structure was superior to the reptilian and amphibian, hence it was not surprising that several related amphibian types might have independently developed at least part of the same improvement. This type of parallelism was not incompatible with the kind of worldview that Darwin had sketched: evolution was driven by adaptive benefits, not by built-in forces that produced rigid trends entirely independent of the external world.

Geographical Distribution: Evolution in Space and Time

The study of the geographical distribution of species was a characteristic aspect of Darwin's effort to see the present state of the world as the relic of historical circumstances governed by the interaction between living things and an ever-changing environment. Morphologists in general did not pay much attention to this line of evidence, since they were mainly concerned with purely structural relationships between species. Anti-Darwinians such as Mivart were especially indifferent, because their commitment to massive parallelism made it impossible for them to use geography as a key to the past. If two apparently "related" forms could easily be produced by independent lines of evolution, one could not use the distribution of a group as a clue to its past history. Where the Darwinian postulated migration to explain how members of a group came to be located in different parts of the world, the extreme anti-Darwinian could not make this inference. The fact that Huxley and Mivart stood on opposite sides of an important theoretical divide is shown by the fact that Huxley did take an interest in geographical distribution as a means of reconstructing a group's past history. He certainly did not share all of Darwin's views on the topic, but he appreciated the basic point that the present state of affairs must be explained in terms of changing geographical and environmental circumstances.

In his 1870 address to the Geological Society, Huxley addressed the problem posed by the unusual character of the inhabitants of Australia and New Zealand. In a paper the previous year on the fossil reptile *Hyperodapedon* (*SM* 3:374–90), he had already noted the close resemblance between this and the living New Zealand reptile *Sphenodon* (the tuatara). The New Zealand fauna must be a relic of that which had enjoyed a worldwide distribution in earlier times, protected from competition with later, more highly evolved types by geographical isolation (3:388–89).

The 1870 address extends the same line of argument to the history of the mammals. Huxley thought the mammals of Miocene Europe had invaded what is now Africa and India when a geographical barrier was broken. They had subsequently been wiped out in Eurasia by the Ice Age (*CE* 8:374). He explained the Australian marsupials as a relic of a once widely distributed population of Mesozoic mammals. He postulated an ancient Mesozoic continent, now sunk beneath the waters of the North Pacific and Indian oceans (8:387). This continent had once been linked to Australia, and its inhabitants lingered on there still, protected from the more highly developed placental mammals, which had expanded their territory after the land-bridge had sunk. In tune with his general idea of the antiquity of types, he thought that the placental mammals had evolved in the Mesozoic on another unknown and isolated continent. The sudden revolution in the fossil record that marked the end of the Mesozoic represented only the breaking of the barrier that had let the placentals flood out into the rest of the world.

Darwin, it must be said, did not approve of vast ancient continents being thrown up to explain problems of distribution (see Browne 196–202). He preferred to believe that the continents were fixed, and to explain migration by accidental transportation. In fact Huxley himself conceded that the present continents had always existed much as we know them today. But in any case, he and Darwin would have agreed that it was the geographical barriers existing in recent geological epochs that had protected the marsupials of Australia from extinction at the hands of the placentals. More highly evolved types did not necessarily replace primitive types; they could only do so in those areas to which they were physically able to migrate. Huxley was thus prepared to admit that the present state of the earth's fauna had been shaped by the accidental circumstances of geography.

Huxley had also addressed these problems in an 1868 paper "On the Classification and Distribution of the Alectoromorphae and Heteromorphae" (*SM* 3:346–73). Here he argued that the relationships between the various gallinaceous birds could be represented by a typical phylogenetic tree. But he then went on to explore the history of the group in terms of its geographical distribution. He evaluated the scheme of zoogeographical provinces suggested by Philip Sclater, suggesting that the northern regions should be grouped into a single province, Arctogaea. From this he distinguished Sclater's Neotropical region (Central and South America), which he preferred to call Austro-Columbia and Australasia. The pigeons and parrots were well represented in both Austro-Columbia and Australasia, but how had they come to gain this distribution? Huxley argued that the answer to the question lay in whether the Miocene rocks of Europe would yield fossils of these birds. If there were no such fossils, then the group must have originated in the southern provinces, and its representatives in the north would be recent immigrants. If the fossils *were* there, then the group had originated in the north and its currently flourishing state in the southern regions was due to the fact that immigrants there had found the climate highly suitable. He was inclined to suspect that the modern distribution of the parrots favored the latter hypothesis. Huxley was thus prepared to accept that the group had not evolved in the region to which it was best adapted, and that its modern distribution was a later phenomenon made possible by its expansion into new territory.

Something of the same concerns enter into the last chapter of Huxley's classic introductory biology text, *The Crayfish.* In his chapter on comparative morphology, he shows that the modern crayfish must be divided into two groups, a northern hemisphere family, the Potamobiidae, and a southern hemisphere group, the Parastacidae. Embryological evidence suggested that both groups had evolved from a common parent form, the Protostacidae, now extinct. When he turns to geographical distribution, Huxley notes that the pattern for the crayfish is quite different from that observed in birds and mammals, since the regions are determined by different factors. The European crayfish can be interpreted as the products of a series of westward migrations of an Aralo-Caspian stock following the Ice Age (321–22). The distinction between the northern and southern types can be explained on the assumption that the crayfish are the descendants of marine ancestors that migrated into rivers and lakes and adapted to freshwater. This is similar to the case of freshwater prawns, but the marine ancestor of the crayfish has vanished, leaving two groups of its descendants isolated at opposite ends of the globe. Huxley notes that the absence of crayfish from certain apparently suitable regions can be accounted for when we note that fluvatile crabs would be powerful rivals for this group (337). Wherever the crabs had flourished, the crayfish could not gain a foothold.

Here again we see Huxley being prepared to admit that the present must be seen as the product of a historical process governed by the interaction between the possibilities of evolution and the limitations of the environment. My conclusion is that this aspect of his work helps us to see the reasons why Huxley allied himself with Darwin against Owen and Mivart, even though he did not accept all of Darwin's thinking. Whether or not we accept the social agenda that Desmond sees behind the debate, there is a real distinction here between those biologists who were prepared to see the development of life on earth as a process governed to some extent by historical accidents and those who preferred to see it as the product of goal-directed trends.

Huxley was primarily a morphologist, and this meant that he often restricted his evolutionary work to the reconstruction of phylogenetic trees. He had little interest in the actual process of adaptation, and indeed was sometimes less aware than Owen of the ecological pressures that might prompt a certain evolutionary development. He was prepared to admit that evolution might be driven by internally programmed trends, something that was anathema to Darwin. Yet in the end he would not follow Mivart into the realm of pure anti-Darwinism, where the main driving force was some mysterious developmental force. For Huxley, only those trends that conferred adaptive benefit could succeed in the long run. Convergent evolution due to the independent invention of the same improvement by two distinct groups was possible, but nonadaptive parallelism was not. And even when improvement occurred, the subsequent history of the new groups was governed by historical factors such as changing environments, the opening and closing of migration routes, and the presence or absence of competitors. Huxley was certainly not a good Darwinian by modern standards, but he could not tolerate the completely predetermined historicism of Mivart's anti-Darwinian viewpoint. I submit that the designation of Huxley as a "pseudo-Darwinian" is not inappropriate.

REFERENCES

Bowler, Peter J. *The Eclipse of Darwinism: Anti-Darwinian Evolution Theories in the Decades around 1900.* Baltimore: Johns Hopkins UP, 1983.

———. *Fossils and Progress: Paleontology and the Idea of Progressive Evolution in the Nineteenth Century.* N.Y.: Science History, 1976.

———. *Life's Splendid Drama: Evolutionary Biology and the Reconstruction of the History of Life on Earth, 1870–1930.* Chicago: U of Chicago P, 1996.

———. *The Non-Darwinian Revolution: Reinterpreting a Historical Myth.* Baltimore: Johns Hopkins UP, 1988.

Browne, Janet. *The Secular Ark: Studies in the History of Biogeography.* New Haven: Yale UP, 1983.

Caldwell, W. H. "The Embryology of the Monotremata." *Philosophical Transactions of the Royal Society* 178.B (1887): 463–86.

Coleman, William. "Morphology between Type Concept and Descent Theory." *Journal of the History of Medicine* 31 (1976): 149–75.

Desmond, Adrian. *Archetypes and Ancestors: Palaeontology in Victorian London, 1850–1875.* Chicago: U of Chicago P, 1984.

di Gregorio, Mario. *T. H. Huxley's Place in Natural Science.* New Haven: Yale UP, 1984.

Dollo, Louis. "Les ancestres des Marsupiaux étaient-ils arboricoles?" *Traveaux de L'Institut Zoologique de Lille* 7 (1899): 199–203.

Gegenbaur, Carl. *Elements of Comparative Anatomy.* Trans. F. Jeffrey Bell. London: Macmillan, 1878.

———. *Grundriss der vergleichenden Anatomie.* 2d ed. Leipzig: Wilhelm Engelmann, 1874.

Gruber, Jacob W. "Does the Platypus Lay Eggs? The History of an Event in Science." *Archives of Natural History* 18 (1991): 51–123.

Haeckel, Ernst. *Generelle Morphologie der Organismen.* 2 vols. 1866. Berlin: Walter de Gruyter, 1988.

Heilmann, Gerhard. *The Origin of Birds.* London: H. F. & G. Witherby, 1926.

Hull, David L. "Darwinism as a Historical Entity: A Historiographical Proposal." *The Darwinian Heritage.* Ed. David Kohn. Princeton: Princeton UP, 1985: 773–812.

Huxley, Leonard. *Life and Letters of Thomas H. Huxley.* 2 vols. N.Y.: Appleton, 1901.

Huxley, Thomas Henry. *Collected Essays.* 9 vols. 1894. N.Y.: Greenwood Press, 1968.

———. *The Crayfish: An Introduction to the Study of Zoology.* London: Kegan Paul, 1880.

———. *The Scientific Memoirs of Thomas Henry Huxley.* Ed. M. Foster and E. Ray Lankester. 4 vols. plus supplement. London: Macmillan, 1898–1903.

Marsh, Othniel C. "Odontornithes: A Monograph on the Extinct Toothed Birds of North America." *Report of the Geological Exploration of the Fortieth Parallel* 7 (1880).

Mivart, St. George Jackson. *On the Genesis of Species.* London: Macmillan, 1871.

Nopsca, Francis. "Ideas on the Origin of Flight." *Proceedings of the Zoological Society of London* (1907): 223–36.

Ospovat, Dov. "The Influence of Karl Ernst von Baer's Embryology, 1828–1859: A Reappraisal in the Light of Richard Owen's and William B. Carpenter's Paleontological Application of Von Baer's Law." *Journal of the History of Biology* 9 (1976): 1–28.

Owen, Richard. "On the Archaeopteryx of von Meyer." *Philosophical Transactions of the Royal Society* 153 (1864): 33–47.

———. *A Monograph on the Fossil Reptilia of the Mesozoic Formations.* London: Palaeontographical Society, 1874–89.

Rudwick, Martin J. S. *The Meaning of Fossils: Episodes in the History of Paleontology.* 2d ed. N.Y.: Science History Publications, 1976.

Sclater, Philip L. "On the General Geographical Distribution of the Members of the Class Aves." *Journal of the Linnean Society, Zoology* 2 (1857): 130–45.

Simpson, George Gaylord. *A Catalogue of the Mesozoic Mammalia in the Geological Department of the British Museum.* London: British Museum, 1928.

Thomas Henry Huxley and the Status of Evolution as Science

MICHAEL RUSE

I BEGIN THIS DISCUSSION WITH A PARADOX. THOMAS HENRY Huxley is rightly remembered by his nickname "Darwin's bulldog." For over thirty years after *The Origin of Species* was published, Huxley never lost an opportunity to promote the cause of evolution or the greatness of Charles Darwin. Starting on Boxing Day 1859, with a glowing review in the *Times* (London) right up to his Romanes lecture on evolution and ethics in 1893, Huxley lived, breathed, thought, argued, promoted, glorified evolution. He debated bishops; he lectured to working men's clubs; he went touring in America; he wrote articles and books; he carried on an extensive correspondence, and throughout, his faith in evolution and his veneration of Darwin was absolute.

Yet when it came to one of the most important things in Huxley's life, his teaching and his interaction with his students, he excluded evolution entirely. A student once wrote: "One day when I was talking to him, our conversation turned upon evolution. 'There is one thing about you I cannot understand,' I said, 'and I should like a word in explanation. For several months now I have been attending your course, and I have never heard you mention evolution, while in your public lectures everywhere you openly proclaim yourself an evolutionist' " (*Life* 2:428).

This was no exaggeration. Huxley would give a one- or two-year course on biology, of about 165 lectures. Incredibly detailed, magnificently comprehensive, beautifully illustrated—Huxley was a fine blackboard artist—students of the life sciences would learn as had never before been possible in Britain. Students from Oxford and Cambridge would slip up to London to supplement their learning from the older institutions.

What they would not get would be evolution. I have located two complete sets of lecture notes (1869–71 and 1879–80, taken by P. H. Carpenter and H. F. Osborn, respectively), and there is barely a hint about evolution—half a lecture at most, and a mere sentence or two on natural selection ("There is no positive evidence that selective breeding alone will produce forms infertile with one another," November 16, 1869). Nor would there have been stimulus to mug up the subject on one's own, for the examinations that Huxley set simply had nothing on evolution in general or Darwinism in particular. A typical question, set by Huxley, from the first biology honors examination at London University in 1862 was "Describe the structure of the Eye in (1) a mammal, (2) a fish, (3) a cephalopod, (4) an insect." Darwin's bulldog was ever ready to bite; but, in front of the classroom, he had little inclination to bark.

Solving this paradox—for it is soluble—takes us right to the heart of Huxley's life work, and what he truly hoped to achieve. It also raises questions about whether what he hoped to do could ever be done and points down through history to the immense influence Huxley has had on the biological sciences in English-speaking countries, even to the present day. To begin, let me go right to the fundamental questions about Huxley's intentions.

Huxley and the Idea of a Professional Science

Huxley trained as a medical doctor, and then joined the navy where he spent several years traveling the globe as ship's surgeon on the HMS *Rattlesnake.* Both medicine and the navy were but stepping stones to the future, that of a full-time scientist, first as a comparative anatomist specializing in marine invertebrates and then with broader interests that encompassed vertebrate paleontology. Huxley was brilliant, both as a student and as a researcher, but his work was never deeply innovative or profound. Early in his career Huxley realized that German biology was significantly ahead of work in his own country, and he made major and successful efforts to inform and guide his own efforts accordingly. However, to speak in well-known philosophical categories of today, he was ever a man working within a "paradigm" rather than a revolutionary (see Kuhn).

Huxley was as capable of self-deception as the rest of us; but to his credit, I doubt that he would have disagreed with this assessment of his labors as a scientist. He was proud of his calling, and he was proud of his own work. Yet overall, he would have claimed no special merit, say of the degree that he and we would accord to the great German embryologist Karl Ernst von Baer. (My perhaps unkind suspicion is that Huxley was more concerned to see that his hated rival Richard Owen did not get more credit than was his due, than that he himself got as much credit as was his own due.)

Ultimately, the evaluation that he was not a scientist of the highest rank would probably not have upset Huxley; from a very early stage, he had set his sights on other game than straightforward scientific success. Like a huge number of those earnest midcentury, middle-class Victorians, Huxley's dream was to reform and improve British society—to take it from something still burdened by the preindustrial customs and rules of the eighteenth century and to set its face toward the integrated, bureaucratically run state of the twentieth century. For Huxley himself, his particular labors were to be in the realm of science—that is, science administration, science education, and science influence and application. It was here that Huxley saw his mission, and it was here that Huxley would have invited judgment, by his contemporaries and by us.

His behavior in the mid-1850s confirms my assessment of Huxley's true intentions. By then he was already earning respect and admiration for his achievements, not to mention really important offers of employment. For instance, the biology professorship in Edinburgh was his for the asking. But he preferred to stay closer to the action in London. And indeed, we find him setting out a program of self-improvement, almost as if he were a humble clerk or schoolteacher in a provincial town: "1856-7-8 must still be "Lehrjahre" to complete training in principles of Histology, Morphology, Physiology, Zoology, and Geology by Monographic Work in each Department. 1860 will see me well grounded and ready for any special pursuits in either of these branches—" (*Life* 1:162). But why? What need had Huxley of this broad training? Would he not have been better employed in forging ahead in those areas of biology, specifically the comparative anatomy of the invertebrates, of which he was already a master? There are, after all, a lot of animals without backbones. The answer comes in the next paragraph. Huxley's sights were on the reformation of biological science as a professional science, rather than on personal excellence in one particular branch: "In 1860 I may fairly look forward to fifteen or twenty years 'Meisterjahre,' and with the comprehensive views my training will have given me, I think it will be possible in that time to give a new and healthier direction to all Biological Science" (1:162). A point that should be added, acknowledging the deeply emotional and religious (if in a secular sense) nature of Huxley, is that this was written on the night of the birth of his first son, Noel, leading the father to add: "Waiting for my child. I seem to fancy it the pledge that all these things shall be" (1:162). When, some four years later, poor Noel died suddenly of scarlet fever, the pledge and commitment became sacred as never before: "My boy is gone, but in a higher and a better sense than was in my mind when I wrote four years ago what stands above—I feel that my fancy [to prepare himself for the tasks ahead] has been fulfilled" (1:163).

What did this all mean? What is to say that Huxley set about, with a right will, to the task of making a *professional* biological science in Britain? There are no fixed set of criteria by which we can offer an answer, but there are a number of features that distinguish professional science (see Ruse, *Darwinian Revolution*). What separates such a phenomenon from that which we might label *amateur* or *popular* science is that it requires proper training; it needs funding; it needs consumers—people prepared to pay for the products and, even more, to offer employment to the finished students; it sets standards with selective admission to specialized societies; it has its own line of journals that are probably understandable only to the initiated and that certainly restrict publication to those whose efforts fit certain, possibly unstated, criteria; and many more such marks along the same vein.

Huxley and his friends and collaborators—men like the botanist Joseph Hooker and the physicist John Tyndall—did all these things and more (see Caron). Huxley was ever a university professor, becoming dean of the science college in South Kensington. He was deeply involved in the science curriculum and was one of the moving forces in the development of biology as a university subject—offering courses, setting examinations, and ultimately writing what became the standard textbook. He made sure that his students, or men of whom he approved, were appointed to key posts in Britain and abroad. An unpublished letter from Trinity College, Cambridge, dated April 2, 1870, reflects his influence: "I read your letter to the Seniority yesterday. Your suggestion as to the Praelectorship of Physiology and the man to fill it [Michael Foster], was most favourably received. . . . If Dr. Carpenter, whom we have also consulted, gives the same advice, I have no doubt it will be followed" (HP 4:172). Likewise, at the newly founded and incredibly successful U.S. research university Johns Hopkins, Huxley saw to it that his student and textbook collaborator H. N. Martin got one of the key biology posts.

In the all-important question of consumers, Huxley was no less active. He was a key figure in persuading the medical profession that all of its students needed a proper grounding in physiology, a service that his group was happy to supply. And in his own particular area of anatomy, he made major efforts, for instance, serving on the London School Board, to see that it would be considered a fit and necessary subject for the developing mind. This required qualified school teachers, and especially through summer courses (staffed by himself and his own special students) Huxley answered the demand that he had created.

In science generally, Huxley was no less active. A member of the famous, or notorious, X-Club, Huxley and his close friends discussed and decided science policy in Britain, particularly in the prestigious Royal Society, of which Huxley became president. For example, in an unpublished letter of October 2, 1870 one member of the X-Club (THH) wrote to a second (Tyndall) about a third (John Lubbock). "No-one it seems to me would be a better, a fitter man for the treasurership [of the Royal Society]. And if you think so and will take care to bring his name forward at once no one can possibly oppose him" (HP 4:172).

The scientific societies—the Linnaean, the Geological, and above all the Royal—felt Huxley's influence also when he acted as an individual. Most particularly this came when, through the refereeing process, Huxley acted as a conduit for work of which he approved: "I have no hesitation in very strongly recommending the publication of the chief part of this memoir." "Well worthy of publication in the transactions of the Royal Society." "The memoir is a very valuable one" (Referee's reports, Royal Society RR.6.200).

In like vein, Huxley was an important contributor to the new journals that were founded; he was a major presence in the early years of *Nature.* And the Huxley touch was similarly evident elsewhere. He served for a number of years as a government-appointed fisheries commissioner, a significant area was to the economy of Britain. The involvement of a man like Huxley must have reflected credit back to his own discipline. He made regular people, especially including politicians, aware of the need for trained "experts."

How successful overall was Huxley in his aims? My assessment is that generally speaking he had much cause to be satisfied with his labors, although I make qualifications and distinctions. In Britain, biology has never achieved the general success of physics, and within biology certain subjects throve more than others. Thanks to its medical connection, physiology at once found a firm base and market. Anatomy was always a lesser sibling—teaching being less prestigious in itself than medicine and biology teachers having to compete with teachers of other science subjects. (I make this judgment on the basis of the number of honors students specializing in physiology and anatomy, respectively, at the older universities.)

But reservations notwithstanding, Huxley strove to make biological science professional, and to an impressive extent he was successful. The question I address concerns the place and status of evolution in Huxley's thought and in the professional biology he was building.

Huxley and the Status of Evolution

Let me start by making one thing absolutely clear. Thomas Henry Huxley was totally and completely committed to the truth of evolution, the development of living forms—including humans—from one or just a few simple organisms, which in turn he thought would have come ultimately from inorganic materials. "I can see no excuse," he declared, "for doubting that all [living things] are co-ordinated terms of Nature's great progression, from the formless to the formed . . . from blind force to conscious intellect and will" (*CE* 7:151). He thought this process was slow and (within bounds) gradual, and above all he thought it was natural, that is to say the causal end product of regular unguided laws. Nothing I say can or should challenge this fact.

However, this firmly stated, one must add some refinements to the picture. First, Huxley was not always an evolutionist. As he entered the scientific community, like Saul of Tarsus in an earlier era, no one was more shrill or violent in his opposition to evolutionism. Admitting that the critique Huxley wrote of Robert Chambers's evolutionary tract *Vestiges of the Natural History of Creation* was in effect being used primarily to harm Owen (some of whose speculations Chambers had used), the tone is quite exceptional in vehemence and nastiness: "The product of coarse feeling operating in a crude intellect" (*SM* supplement:5). He was, therefore, if not a latecomer to evolution, not an early one—as, for example, were Alfred Russel Wallace or Herbert Spencer. And unlike others who came to evolution after a time of reflection—Darwin for instance—Huxley's early career was marked by strong opposition.

Second, after he became an evolutionist—in the late 1850s, under the influence of Darwin—Huxley still thought, in significant places, that the evidence was uncertain and doubtful. He was convinced of evolution, for only then did he think one could explain such things as comparative anatomy and the isomorphisms ("homologies") between organisms; but, at the same time he thought that other areas of biology—notably paleontology—were lacking in conclusive evidence (see *CE* 3:272–304). Huxley was happy to use the newly discovered reptile-bird hybrid *Archaeopteryx* to support evolution (*SM* 3:303–13), and later he did become more confident of the strength of the fossil record, for vertebrates at least (3:510–50). In America in the late 1870s, he welcomed and made much of the evidence of horse phylogeny ("American Addresses"). Initially, however, he thought the rocks provided no definitive support.

Third, and perhaps most important, Huxley always drew back from full acceptance of Darwin's mechanism of natural selection. He accepted it in part, which is hardly surprising, since even non-evolutionists generally accepted selection in some fashion; but he never thought it proven, nor did he ever truly think that it could do all that Darwin claimed. At best it was going to be a secondary mechanism, and although Huxley was as agnostic about causes of change as he was about God, it seems clear that his real preference was for some sort of evolution by jumps, what have come to be known as "saltations." Intellectually, Darwin's bulldog was no ardent Darwinian—at least not when it came to a mechanism for evolution (see Ruse, *Darwinian Revolution*).

Huxley was a genuine evolutionist, albeit a late evolutionist and no enthusiast for what Darwin himself thought to be his greatest contribution to science. We come now to our big question. Or rather, we start to edge toward an understanding of the answer to our big question. We know how Huxley treated evolution. It simply did not figure in his professional science. It was not part of his life's work as an active scientist. Certainly, this was the case when it came to his students, and it was a pattern that held throughout his working life: in what he wrote, in what he accepted from others, in what he promoted in the societies. In these areas, evolution simply had no role. Huxley could have been a Six-Day Creationist for all that evolution counted in his professional activities.

Nor was this exclusion a matter of chance or oversight—how could it be? To his student who worried about the evolution-free content of the courses, Huxley was candid. "Here in my teaching lectures (he said to me) I have time to put the facts fully before a trained audience. In my public lectures I am obliged to pass rapidly over the facts, and I put forward my personal convictions. And it is for this that people come to hear me" (*Life* 2:428). And hear him they did, but always in the public or popular arena—at those working men's clubs, in the much-read journals and quarterlies of the day, and when the work was done at a conference and it was time for the president to wax philosophical.

Why did Huxley treat evolution as he did? At one level, what seems clear is that Huxley, as he said to his student, thought evolution a matter of opinion, and at some deep level lacking as science. Huxley took pride in his empiricism—he wrote a book on Hume—and it may be that evolution necessarily upsets the empiricist, for it takes one beyond the evidence and into the realm of speculation. Evolution belongs to the unseen world, like fairies and angels and phlogiston and those sorts of things. It is one thing to believe in them. It is another to claim them as science, especially when, as in the case of evolution, one thinks the evidence a little suspect.

But this cannot be the whole story. Huxley's enthusiasm for evolution went beyond "personal convictions." He thought that people who rejected evolution were wrong. He certainly did not think evolution on a par with angels or phlogiston. In any case, he himself certainly accepted unseen things in the name of science. For instance, he brought Darwin up short for speculating on the causes of heredity without taking account of then-modern thinking about cells (Olby). Moreover, although he may have started with doubts about the fossil record—doubts sufficiently strong to make him question the ultimate truth of evolution—these doubts were to be resolved: "when we turn to the higher *Vertebrata,* the results of recent investigations, however we may sift and criticise them, seem to me to leave a clear balance in favour of the doctrine of the evolution of living forms one from another" (*CE* 8:348).

Furthermore, there were respectable empiricist philosophies of science that Huxley could have embraced, if he had a mind to accept evolution as professional science. The fact that evolutionary theorizing deals with the past is no barrier. John F. W. Herschel (1830), a major philosophical forerunner of John Stuart Mill, had judged Lyell's geology to be the perfect exemplar of the way an empiricist science should be conducted (See Ruse, "Charles Lyell"). Nor should we think that Huxley qua philosopher was forever fixed in his thinking. As the evolution of his position on ethics well shows, he was certainly capable of searching around for substitute foundations and justifications as the need arose (see *CE* 9:46–116).

My feeling is that there were at least three other, more important reasons why Huxley positively and actively excluded evolution from the realm of professional science. First, not to be underestimated, is the fact that, when he became a biologist, he was no evolutionist. This point is highly significant because it was during this early stage that he formulated his philosophy of teaching—a philosophy that went unchanged throughout his life. Huxley's pedagogy was founded on the "type" system, where one introduces the student in succession to a series of representative organisms, each representing a branch of the animal kingdom: frog, crayfish, earthworm, and so forth. It is true that eventually Huxley's teaching assistants persuaded him, on grounds of intrinsic interest, to put humankind first rather than last—every college teacher knows about the importance of capturing the interest of your audience in the first weeks—but essentially his pedagogy never changed.

However, the essence of the type system, as conceived by Huxley, is deeply antievolutionary. He was well aware of this conflict: "It may indeed be a matter of very grave consideration whether true anamorphosis [evolution] ever occurs in the whole animal kingdom. If it does, then the doctrine that every natural group is organized after a definite archetype, a doctrine that seems to me as important for zoology as the theory of definite proportions for chemistry, must be given up" (*SM* 1:192). This was written in the preevolutionary days, and obviously, in some sense, as he became an evolutionist, Huxley had to come to terms with the contradiction. Since this process of accommodation could, apparently, in no way change his system of teaching, the pressure was to keep evolution out of the classroom. At some deep level, when Huxley was inducting would-be professionals into biology, the very process spoke against evolution. Hence, evolution had to remain at the popular level. As with Heisenberg's Uncertainty Principle, the awkward questions are avoided rather than solved.

Second, there is Huxley's indifference to natural selection. As an anatomist-paleontologist he simply did not need natural selection. He was not searching for adaptations and their purposes—at least not in some exclusive way. Rather, he was after similarities and disanalogies. Like his German mentors, he cherished form rather than function. Notoriously, he once denied that the colors of butterflies could have any adaptive value—the very things that Darwin and other selectionists, especially Wallace and Henry Walter Bates, were claiming as the ultimate proof of selection. "Is it to be supposed," he scoffed, that such beauty is "any good to the animals? That they perform any of the actions of their lives more easily and better for being bright and graceful, rather than if they were dull and plain?" (*SM* 1:311). You do not have to be a selectionist to be a professional, but if you are no selectionist—and not working on problems that cry out for selection as a solution—then you have omitted a major reason for using evolution in your active life as a professional scientist. Evolution in this respect becomes an option rather than a necessity.

Third, and perhaps most important of all, emotionally Huxley was not drawn to evolution as professional science. He wanted it in the public or popular realm. He himself worked to keep it on the podium rather than in the *Philosophical Transactions of the Royal Society*. For Huxley, evolution was always something much closer to secular religion than it was to professional science. It was something that grew out of his deep conviction (as a scientific "Calvinist") that the laws of nature rule supreme and that all that we can do is appreciate them and submit. For Huxley, evolution was the agnostic's answer to Christianity.

Consider, for a start, where it was that Huxley first learnt of evolution in a favorable light, in an atmosphere where he would have listened with respect if not with agreement. Before he became entangled with the Darwinians, he had already, in the mid-1850s, struck up a friendship that was never to be equalled. This was with the man who, far more than Darwin ever did, was to epitomize evolution for the Victorians. I refer of course to Herbert Spencer, with whom Huxley used to go for weekly walks, and whose brilliance Huxley openly praised again and again. "It seems as if all the thoughts in what you have written were my own and yet I am conscious of the enormous difference your presenting them makes in my intellectual state" (*Life* 1:229). And this was written on September 3, 1860, *after* publication of the *Origin.*

Although he certainly learnt of evolution from Spencer, what Huxley could not learn from Spencer was of evolution as professional science. From beginning to the end of his very long life, Spencer was spinning a world picture, a metaphysic, a secular religion, with the ideology of evolution at its heart. He could do this because he, like everyone else, saw evolution as having *meaning,* or MEANING, namely, the progressive upward rise from the simple to the complex, from the homogeneous to the heterogeneous, from the monad to the man (see Spencer; Richards). Spencer's vision was the epitome of the popular, the public, the "philosophical" (to use a term that professional scientists are wont to use pejoratively), and although in later life Huxley was to break from Spencer's particular vision of progress, he never ceased to regard evolution in the same light.

Nor did the coming alliance with Darwin change this. Indeed, it reinforced it. The link with Darwin and the Darwinians gave the emotionally insecure Huxley the intellectual and social home he desperately craved. He hated Owen, not because he hated Owen's science—indeed, much of Huxley's own science was much closer to that of Owen than it ever was to Darwin's—but because the man had been condescending and unkind and rejecting. (This was mainly, I suspect, because Owen in turn felt threatened by the younger Huxley.) Darwin, already a great man, welcomed and flattered Huxley. He made him feel wanted and appreciated. He made him feel like a somebody—although, in fact, privately, the Darwin family thought Huxley talked a lot of nonsense, although in an amusing way. This is made clear in a letter that Darwin's son, Leonard, wrote to Ronald Fisher on November 6, 1932 (qtd. in Bennett 158).

The welcomed Huxley responded. If there was work to be done, he was the Darwinians' man. I have spoken of Huxley as Saul in his attack on Chambers. This is no idle metaphor. Once in Darwin's camp, Huxley took on the world with the enthusiasm of the convert—so much so that often his supposed opponents protested that they were really no opponents at all. Not that this deflected Huxley one whit. His was a mission with the fervor of a religious crusade, and such crusades need an Antichrist, no matter whom. (Actually, Owen was ever the Antichrist. Speak of the other opponents as fallen angels.)

Evolution was the doctrine, and the unwashed masses were the heathen to be saved. It was not by chance that Huxley, unlike the more reticent Darwin, went straight for the jugular and spoke on evolution and humankind (see, for instance, *Man's Place*). This was where the real action lay. This was the resurrection story of evolution. It was irrelevant that Darwin protested, more than once, that Huxley's sermons had little to do with the content of the *Origin*. Like St. Paul before him, Huxley had a better idea than did the Messiah of what the people really needed.

These, then, are the reasons why Huxley kept evolution outside his professional science. It simply did not fit with what he wanted to do in upgrading British biology; it was not necessary for what he wanted to do with such biology; and, most importantly, Huxley had an altogether different path marked out for evolution. He wanted it in the public realm, where it could function as a secular religion.

Successful Strategy?

I could end my story now, for the paradox is resolved. Huxley's aim as an active scientist was to build a professional biology, and the inclusion of evolution as he conceived it was not only not necessary for this end, it would have been a positive obstruction. More than this, Huxley had another role for evolution, and this was as something inherently in the popular domain.

The impact of Huxley does not, of course, end with his death. Indeed, many (particularly non-evolutionary biologists) would think that one could readily underscore Huxley's story by bringing the tale rapidly down to the present and by showing how his influence has persisted even unto our time. Thanks to Huxley's students, people like Michael Foster in England and

H. N. Martin in America, professional biology throve in the Anglo-Saxon world, but evolution continued to be outside the pale. At the end of the nineteenth century, the trendy science was embryology; then, as we come into our century, we get the growth of genetics, culminating in the brilliant work of T. H. Morgan and his group. (See Maienschein for an excellent discussion of the kind of science that the top professionals did in the years after Huxley.)

In the 1930s and 1940s, thanks to population geneticists like Ronald Fisher and Sewall Wright, and to "synthetic theorists" like Julian Huxley and Theodosius Dobzhansky, it may have looked for a while as if evolution was going to cross the popular-professional divide. But then came molecular biology, and evolution slid right back down again. Today, it is but a sickly branch of the biological family, lagging behind when it comes to grants and students and a place in the academic sun. It is philosophers like myself who really get enthusiastic about the subject. As one eminent evolutionist and paleontologist, Anthony Hallam, recently said to me, rather sadly: "I'm still dismayed that only a small proportion of active, working biologists [in the U.K.] are really interested in evolutionary questions. The vast majority of biologists are physiologists or what I call 'cell counters.'" Talking of the work and status of Britain's leading evolutionist, John Maynard Smith, Hallam continued: "There's no question about the fact that my biological colleagues—in this university and I think they speak for many biologists—don't see the sort of thing that Maynard Smith does as too pertinent to the mainstream of biology and what really turns them on. As I say, physiologists and biochemists and the like—cell biologists and molecular biologists for the most part carry on—they assume evolution has happened—but their work isn't directed to the actual testing of evolution. So it is very difficult to get a debate going in this country" (spring 1991, interview with author).

Although there is truth in telling history in this way, it is a distortion. Rather than asking whether others specifically attempted to follow in the path of Huxley, it might be more useful to examine how successful Huxley's strategy was to separate evolution as popular science from biology as professional science. Facing this second question gives us a somewhat better sense of the richness of the story and a fuller appreciation of where we find ourselves today.

Could even Huxley himself really run with the hare of popular-science evolution and hunt with the hounds of professional-science biology? He could certainly do this at one level, but at another level there were tensions, to say the least. Most particularly, we must appreciate that if Huxley as the apostle of evolution was to have real authority, it could only derive from his being a well-respected professional scientist. Why would one take note of him otherwise? People, especially Huxley, laughed at Mr. Gladstone when he presumed to write on matters evolutionary, particularly since he did so from a biblical perspective. We must also appreciate that, since Huxley was speaking with the authority of a professional, it had to be conceded that at some level what he was talking about, namely evolution, was science, and not only science but professional science. Otherwise, anybody would have done as a spokesman—even a journalist or a politician.

The point is that the divide could not be absolute, and Huxley himself seemed to realize this; he was certainly prepared to talk on matters evolutionary before professionals, on such occasions as presidential dinners. It is true that these are the times for the cigars and brandy; but one is clearly edging a little closer to professionalism than, say, in an address before worthy artisans. But it was necessary to make this connection or else the status of evolution, even as Huxley used it in the popular domain, would be diminished. Evolution as secular religion had to be first and foremost the religion *of* professional scientists *for* professional scientists—that which gave a background meaning to their work. Only then could it be something passed on to the masses.

There is more, however, to concern us if once we start to think of some of the reasons why Huxley wanted evolution as popular science. He had developed his philosophy of education before he became an evolutionist, but this was not necessarily the case with his students. There was no reason—and certainly no reason at that all-important emotional level—why they should have thought that teaching biology excluded evolution. After all, one could have evolution between types. One might therefore have expected these students to try their hands at a professional-type evolutionism, especially since Huxley was forever urging them to look for guidance to Germany, a country in which there were certainly those (notably Ernst Haeckel and his friend Carl Gegenbaur) who were trying to put an evolutionarily informed comparative anatomy on a professional basis (see Nyhart).

Such a move toward an evolutionary professional biology is precisely what did come from Huxley's students. Most notably, Huxley's protege E. Ray Lankester ardently pushed a Germanic form of (natural-selection-indifferent) evolutionary biology, and since he was the lifelong editor of the *Quarterly Journal of Microscopical Science,* he was able to make sure that lots of his own contributions got published. Fascinatingly, so ardent were the young men around Huxley that, at the end of his own scientific career, even Huxley himself tried his hand at classifying the mammals on evolutionary grounds (see *SM* 4:457–72). It was not much; it was not repeated; and it certainly did not appear in a prestigious place; but it did happen.

This was perhaps a place where the earlier Huxley's instincts had been sound. By the century's end, it had become apparent to all—at least, to all who wanted to be successful professional biologists—that an evolutionary-based morphology was just not the stuff of worthwhile professional science. There were too many disanalogies, too many gaps. The ambitious therefore turned to other things like genetics, and the unambitious moved from the universities and frontline research to such places as museums, where one could fantasize about evolution to one's heart's content (see Rainger).

But the fact is that, although professional evolutionism may have failed at its first attempt, there was nothing in the logic of the situation that precluded another, perhaps successful, attempt. Huxley may have kept evolutionism (more or less) out of professional biology, but he did not do so on conceptual grounds that would forever preclude it. Which point brings us straight to the second way in which history answers our question. In considering the success of Huxley's strategy, we must remember a major scientific reason why he wanted no truck with professional evolutionism. He had neither time nor need for natural selection. For him, evolution really solved no interesting scientific problems.

But what of those scientists for whom evolution as represented by natural selection did solve interesting and important problems? History tells us two things. First, for various reasons—some of which I am sure were connected with the fact that no one could see a cash crop at the end—selection studies, professional or otherwise, took a long time to get started. The selectionists of Darwin's era, Bates and Wallace, for instance, were too concerned with making a living of any kind to do much unremunerated pure experimental science, let alone found a professional discipline.

Then, after the time of Darwin and Huxley, the selectionists had what one can only describe as bad luck. The most promising was probably Raphael Weldon, around the turn of the century. He did first-class work that was highly professional. But before he could start a school—probably something impossible anyway given his disputative nature—he died at the age of forty-six. His seminal achievements are now being rediscovered and praised by historians, but that is no substitute for contemporary influence (see Gayon).

Second, when people did start to find money through government grants and the like (innovatively, in the case of E. B. Ford, by persuading the money-heavy Nuffield foundation that evolutionary studies would throw light on human diseases), it is precisely in the area of selection studies of various kinds that would-be professionalizers of evolution struck hard and struck with most success. The population geneticists and the synthetic theorists were selectionists, as are the active professional evolutionists today. I do not mean that everyone was or is an ardent adaptationist, a so-called pan-selectionist, although some were. But it is true that adaptationism, pointing to the mechanism of natural selection, has much aided evolutionists in making a professional discipline of their subject.

It is true, speaking to the potted history offered at the beginning of this section, that compared to other areas of biology, especially molecular biology, professional evolutionism rates as pretty small beer. But naysayers notwithstanding, it has developed and does function. For all that evolution has the Huxley odor of popular, or nonprofessional, science hanging over it, the barrier has been cracked somewhat by those who have aims and needs different from those of Huxley himself.

Which brings me to my final point. I stress again a key to understanding Huxley: he was a deeply religious man. In another age, he would have been the Archbishop of Canterbury or the Pope. In the mid-nineteenth century this just could not be. Socially, the church stood for everything that Huxley opposed; intellectually, Christianity was crumbling, especially from the corrosive effect of the German-based Higher Criticism. But this did not exclude Huxley's need of a faith, a doctrine, a metaphysic. And this is precisely the role in which he cast evolution—that of secular religion—and why for him it could never truly be professional science. The science-religion dichotomy was as strong for him as it was for any defender of the U.S. Constitution.

But once again, history tells us that Huxley was only partially successful in what he achieved. Just as conceptually he did not preclude evolution from being professional science, so conceptually he did not forever establish evolution as popular science, meaning a kind of secular religion. Indeed, as you might expect, from Huxley's time on, we have had a constant thrust by would-be professionalizers of evolution to downplay and expel the secular-religious aspect of their science (see Ruse, *Evolutionary Naturalism*). Since the escatology of this secular religion is founded on progressionism—evolution has meaning because it leads to humankind—this thrust has centered on the expulsion of progress from evolution. This was the case at the turn of the century, when a new generation of evolutionists tried to build a new evolutionism based on natural selection. And it continued to be the case four decades later when the synthetic theorists made a discipline for their evolutionism and realized they had to throw off the old idealist ways of thought (see Gould; Nitecki).

Today, I think many would regard the battle as won. Certainly, a good number of today's evolutionists—especially the younger researchers—would resent strongly the suggestion that their science functions as a secular religion, for themselves or for anyone else. They would rightly note that their writings give no comfort whatsoever for the progressionist—or for the anti-progressionist for that matter. Such value-talk is simply excluded and irrelevant. Evolution may or may not be all that Huxley tried to prevent it from being, but it is certainly not all that Huxley wanted it to be.

Yet while agreeing in essence with this—while agreeing that the points made in this section do show generally that biological science has moved on since the days of Thomas Henry Huxley—let me swing the pendulum back a little in recognition of the continued influence of Huxley's strategy. The simple fact is that the thrust to expel the taint of progressionism from evolution—and with it the status of secular religion—has hardly been triumphant. It is true that some of the best younger evolutionists of today are not enthusiastic about the idea, but it is by no means extinct. Paradoxically, indeed, we often get a kind of internal conflict within people now whose very reason for taking up professional evolutionism as a lifetime's occupation is that they are looking for a meaning to life. Moreover, these are not just fringe figures. Some of today's leading evolutionists—the Harvard entomologist and sociobiologist Edward O. Wilson, for one—are as committed to their progressionism as was Herbert Spencer.

Naturally enough, realizing that talk of progressionism is threatening to professionalism, almost all evolutionists today (Wilson is an exception) tend to be shamefaced about their progressionism and strive to keep it out of their professional writings (see Dawkins). But professional or popular—and in the latter, even the most professional of evolutionists do still tend to give way to metaphysics—the spirit of Huxley is there. To end on an analogy with many dimensions: like St. Paul's influence on Christianity, Huxley's influence on evolutionism may not have been altogether good. But, like St. Paul, influence he certainly was and still is.

REFERENCES

Bennett, J. H., ed. *Natural Selection, Heredity, and Eugenics, including Selected Correspondence of R. A. Fischer with Leonard Darwin and Others.* Oxford: Clarendon Press, 1983.

Caron, Joseph. "'Biology' in the Life Sciences: A Historiographical Contribution." *History of Science* 26 (1988): 223–68.

Chambers, Robert. *Vestiges of the Natural History of Creation.* London: Churchill, 1844.

Dawkins, Richard. *The Blind Watchmaker.* N.Y.: Norton, 1986.

Dobzhansky, Theodosius. *Genetics and the Origin of Species.* N.Y.: Columbia UP, 1937.

Fisher, Ronald A. *The Genetical Theory of Natural Selection.* Oxford: Clarendon Press, 1930.

Gayon, Jean. *Darwin et l'apres Darwin: une histoire de l'hypothese de selection naturelle.* Paris: Kime, 1992.

Gould, Stephen Jay. *Wonderful Life: The Burgess Shale and the Nature of History.* N.Y.: Norton, 1989.

Herschel, John F. W. *Preliminary Discourse on the Study of Natural Philosophy.* London: Longman, Rees, 1830.

Huxley, Julian S. *Evolution: The Modern Synthesis.* London: Allen & Unwin, 1942.

Huxley, Leonard. *Life and Letters of Thomas H. Huxley.* 2 vols. N.Y.: Appleton, 1901.

Huxley, Thomas Henry. *American Addresses, with a Lecture on the Study of Biology.* 1877. N.Y.: Appleton, 1888.

———. *Collected Essays.* 9 vols. 1894. N.Y.: Greenwood Press, 1968.

———. *The Scientific Memoirs of Thomas Henry Huxley.* Ed. Michael Foster and E. Ray Lankester. 4 vols. plus supplement. London: Macmillan, 1898–1903.

Huxley, Thomas, and H. N. Martin. *A Course of Elementary Instruction in Practical Biology*. London: Macmillan, 1883.

Huxley Papers. Imperial College of Science, Technology, and Medicine Archives, London.

Kuhn, Thomas. *The Structure of Scientific Revolutions*. Chicago: U of Chicago P, 1962.

Lankester, E. Ray. "Limulus an Arachnid." *Quarterly Journal of Microscopical Science* n.s. 21 (1881): 609–49.

———. "Notes on the Embryology and Classification of the Animal Kingdom, Comprising a Revision of Speculations Relative to the Origin and Significance of the Germ Layers." *Quarterly Journal of Microscopical Science* 17 (1877): 395–454.

Maienschein, Jane. *Transforming Traditions in American Biology, 1880–1915*. Baltimore: Johns Hopkins UP, 1991.

Nitecki, Matthew H., ed. *Evolutionary Progress*. Chicago: U of Chicago P, 1988.

Nyhart, Lynn K. *Morphology and the German University, 1860–1900*. Diss., U of Pennsylvania, 1986.

Olby, R. C. "Charles Darwin's Manuscript of *Pangenesis*." *British Journal for the History of Science* 1.3 (1963): 251–63.

Rainger, Ronald. *An Agenda for Antiquity: Henry Fairfield Osborn and Vertebrate Paleontology at the American Museum of Natural History, 1890–1935*. Tuscaloosa: U of Alabama P, 1991.

Referees' Reports. Royal Society. Carleton Terrace, London.

Richards, Robert J. *Darwin and the Emergence of Evolutionary Theories of Mind and Behavior*. Chicago: U of Chicago P, 1987.

Ruse, Michael. "Charles Lyell and the Philosophers of Science." *British Journal for the History of Science* 9 (1976): 121–31.

———. *The Darwinian Revolution: Science Red in Tooth and Claw*. Chicago: U of Chicago P, 1979.

———. *Evolutionary Naturalism*. London: Routledge, 1994.

Spencer, Herbert. "The Development Hypothesis." *Essays: Scientific, Political, and Speculative*. London: Williams & Norgate, 1852. 377–83.

———. "Progress: Its Law and Cause." *Westminster Review* 67 (1857): 244–67.

Wilson, Edward O. *Sociobiology: The New Synthesis*. Cambridge: Harvard UP, 1975.

Wright, Sewall. "Evolution in Mendelian Populations." *Genetics* 16 (1931): 97–159.

Thomas Henry Huxley and German Science

MARIO A. DI GREGORIO

SINCE THE MID-1970S A NUMBER OF SCHOLARS HAVE ELABORATED Michael Bartholomew's view that there was no real turning point in Huxley's scientific production after the publication of Darwin's *Origin.* Many now think that the classic view he often presented and that was later spread by his son Leonard of Huxley as "Darwin's bulldog" should be revised. All this goes well with Peter Bowler's denial of a true Darwinian revolution in the nineteenth century, thus inducing historians to consider in detail nineteenth-century conceptions that appeared to be evolutionary but were not of the Darwinian kind.

If we carefully analyze Huxley's scientific production, we realize that he started applying evolutionary reasonings to his science only in 1868, nine years after the publication of the *Origin.* His *actual* scientific research seems to contradict his public statement that it was the *Origin* that had persuaded him to assign scientific value to the idea of evolution. Moreover, if we bear in mind, as Mayr proposes, that Darwin's exposition in fact consists of (at least) five distinct theories—evolution, gradualism, common descent, multiplication of species, and natural selection—we notice that natural selection is practically absent from Huxley's work throughout his career. Was Huxley a real Darwinian, or rather what Bowler calls a "pseudo-Darwinian?"

All things considered, we have reason to believe that in 1859, under Darwin's influence, Huxley became convinced of the value of evolution only as a working hypothesis: in public he supported the Darwinian *hypothesis* as a kind of program worth considering but not applicable on a large scale because of the uncertainty about natural selection, the core Darwinian concept. Huxley introduced evolutionary reasonings in the study of animal life only in the late 1860s, when he could comfortably relate evolution not to natural selection but to descent; he did so just when the descent theory was being accepted, applied, and elaborated in scientific circles, especially through the efforts of the Jena morphologists Ernst Haeckel and Carl Gegenbaur (see Krausse, *Ernst Haeckel;* Uschmann). Huxley's zoological research and development of evolutionism in his systematic work are both chronologically related to his German colleagues' researches. This suggests that Huxley ought to be studied in relation to the German-speaking scientific community.

Young Huxley had been deeply influenced by German science, especially by Karl Ernst von Baer's embryological typology, and only works in some way integrated into that tradition could be on his wavelength when he was confronted with evolutionism. Gegenbaur—and especially Haeckel—somehow claimed to be heirs of von Baer's embryological tradition in spite of fundamental differences with von Baer, as Gould has pointed out (184–85), and were staunch supporters of evolution and common descent. Haeckel's *Generelle Morphologie* came out in 1866; two years later evolution and common descent started appearing in Huxley's scientific memoirs. If Huxley ever converted to evolutionism he did so not in 1859 but in the late 1860s under Haeckel's influence rather than Darwin's.

Embryological Typology

To understand why it was Haeckel rather than Darwin who had such a decisive impact on him, one must consider Huxley's cultural and scientific roots. As a young man, Huxley was, like so many at the time, deeply attracted to German Romanticism, especially in the version represented in Britain by Thomas Carlyle (see Paradis). Through Carlyle's translation of *Wilhelm Meister,* Huxley approached Goethe's work and, using the original version of the novel, taught himself German. In the early stages of his career, Huxley was close to some aspects of German *Naturphilosophie,* though in its more acceptable version, and to an aesthetic conception of nature that was later to constitute a further point of agreement with Haeckel. (Both

Huxley and Haeckel combined artistic and scientific talent.) Furthermore, his friendship with Edward Forbes, who, as a Scot, was close to continental scientific themes, pushed him further toward a German-inclined scientific discourse. We now know there were different versions of *Naturphilosophie,* and Huxley was close to the interpretation that was of a Kantian kind (see Lenoir; Engelhardt). He had moved toward a use of embryological morphology in a manner that did not appear too metaphysical to the generation of younger researchers. Huxley's morphological research in the 1850s led him to analyze some of the fundamental theoretical problems concerning the order of nature, in search of a thread that would connect all the facts and observations he had collected during the *Rattlesnake* voyage (see his *Oceanic Hydrozoa; Diary*). The basic concept applied by Huxley is the "type," that is, a structural plan capable of unifying all aspects of the morphology of the time (Ràdl 2:22); however, there was more than one interpretation of that concept, Ospovat having identified two major tendencies in early nineteenth-century morphology in his *Development of Darwin's Theory.* On the one hand, there was Cuvier's anatomical tradition with its emphasis on teleological adaptations; on the other, there was the embryological interpretation of the type of von Baer, Pander, and Rathke (see Russell). Von Baer's embryology is *the* nineteenth-century reply to the question of generation, as Jacques Roger has shown with special relation to French science. Von Baer's research was based on rigorous method and experimentation (see Raikov), and the embryological typology that derived from it was greatly successful in the nineteenth century. Milne-Edwards, for example, believed embryology to be the key to systematics, and Louis Agassiz even propounded a connection between embryology and paleontology.

As was characteristic for continental things, von Baer's embryology reached Great Britain through the Scottish connection (see Ospovat, "Influence"), there first represented by Martin Barry and then quoted in W. B. Carpenter's *Principles of Physiology.* In 1853 Huxley translated the fundamental Fifth Scholion of von Baer's masterpiece into English. As Huxley explains in his brief preface to his translation:

> The present selections have therefore been made, and it is hoped that they embody all the important points of von Baer's doctrines. The translator will be more than gratified, if yet, during the lifetime of the venerable author, he should be the means of assisting to place in its proper position the reputation

> of one who had in the completest manner demonstrated the truth of the doctrine of Epigenesis three years before the delivery of Cuvier's *Leçons sur l'histoire des sciences naturelles* (in which he still advocates the Evolution theory, in its original meaning of development); and who had long recognized development as the sole basis of zoological classification while in France Cuvier and Geoffroy St. Hilaire were embittering one another's lives with endless mere anatomical discussions and replications, and while in Germany the cautious study of nature was given up for the spinning of *Naturphilosophies* and other hypothetical cobwebs. ("Fragments" 176)

Huxley was in search of a "great principle" able to account for the connections that relate living things and could find it only in von Baer's embryological typology. He followed von Baer in his refutation of J. F. Meckel's law of parallelism. With Cuvier he shared a more radically discontinuous view of the four fundamental types than that held by von Baer, but was unhappy with the concept of teleological adaptations. Teleology was also a foundation stone of British natural theology with which Huxley had nothing in common. Teleological considerations were present in von Baer's work as well, but Huxley believed that embryological typology could be utilized in a neutral form, as a technical device without its specific philosophical overtones. Thus he applied his favorite von Baerian typology in his important zoological researches in the early 1850s, especially to determine the homologies of jellyfish, which led him to an anticipation of the germ-layer theory, which later had important evolutionary consequences in the work of Alexander Kowalewsky, Haeckel, and E. R. Lankester (*SM* 1:80–95; Winsor 73–97). Again following von Baer's embryological method, Huxley established the archetype of the cephalous molluscs in "On the Morphology of the Cephalous Mollusca." To write about "archetypes" in the middle of the nineteenth century meant trespassing into the field of research of the formidable British morphologist Richard Owen, who was to become Huxley's main scientific enemy (see Rupke, *Owen;* Desmond, *Archetypes*). The idea of archetype that Owen developed in the theoretical works of his maturity, was essentially Platonic (see Owen, *On the Archetype* and *On the Nature of Limbs*) and one from which Huxley wanted to distance himself. Huxley's attempt to avoid an overtly idealistic philosophy without giving up the useful concept of type is clear: "In using this term, I make no reference to any real or imaginary 'ideas' upon which animal forms are modelled. All that I mean is the conception of a form embodying the most general propositions that can be affirmed respecting the Cephalous Mollusca, standing in the same relation to them as the diagram to a geometrical theorem, and like it, at once imaginary and true" (*SM* 1:176–77).

However, one can be an anti-Platonist and still a typologist, since there is more than one meaning to the term "type." There is a Platonic type, but also an Aristotelian one. Platonic types only formally underlie the plans of organization of real organisms and have no material reality; form is thus radically independent of history. Aristotelian types, on the other hand, are obtained from the observation of real organisms and do not subsist independently of material reality. Their eternity is an extrinsic feature, based on the perceived steady-state of existing species perceived as discrete, not a logically inherent one as with Plato. While Platonic types, purely geometric and abstract, are static and immutable, Aristotelian types are modifiable and usable in evolutionary perspectives, as the cases of Gegenbaur and Haeckel help us to understand (see di Gregorio, "A Wolf"). Moreover, Rupke claims that Owen's original concept of the archetype was Aristotelian, and only later was Platonically reinterpreted, which may help us also to understand why Owen subsequently claimed to be an evolutionist (Rupke, "Richard Owen's").

Aristotelianism, however, implies the deployment of teleological explanation, which Huxley (like Gegenbaur and Haeckel) considered "philosophically" inadequate and unfashionable. They therefore were in search of a comprehensive theory that would not be openly and excessively speculative. In the meantime, Huxley endeavored to present his concept of archetype in the most empirical and least metaphysical manner.

The Origin of Species

Before 1859 Huxley rejected any form of evolutionism, both in Lamarckian shape and as propounded by the anonymous author of the *Vestiges of Creation* (see his "Vestiges" in *SM* supplement). When Huxley read and reviewed the *Origin* he claimed to have recognized what he needed:

> The Darwinian hypothesis has the merit of being eminently simple and comprehensible in principle, and its essential positions may be stated in a very few words: all species have been produced by the development of varieties from common stocks; by the conversion of these, first into permanent races and then into new species, by the process of *natural selection,* which process is essentially identical with that artificial selection by which man has originated the domestic races . . . the *struggle for existence* taking the place of man, and exerting, in the case of natural selection, that selective action which he performs in artificial selection. (*CE* 2:71)

Huxley was prepared to support Darwin's view for two theoretical reasons. First, it was a vera causa of the production of species. The vera causa point was one of the central issues of nineteenth-century philosophy of science in Britain and a major reason for contention between the schools of John Stuart Mill and William Whewell. Here Huxley and Darwin are in agreement; both side with Mill. Second, in Huxley's opinion it was a valid hypothesis of the order of nature, which "assumes that the present state of things has had but a limited duration; but . . . supposes that this state of things has been evolved by a natural process from an antecedent state, and that from another, and so on; and, on this hypothesis, the attempt to assign any limit to the series of past changes is, usually, given up" (*CE* 4:50).

The term "hypothesis" is here Huxley's keyword, conveying his fundamental reservation about Darwin's viewpoint:

> There is no fault to be found with Mr. Darwin's method, then; but it is another question whether he has fulfilled all the conditions imposed by that method. Is it satisfactorily proved, in fact, that species may be originated by natural selection? that there is such a thing as natural selection? that none of the phenomena exhibited by species are inconsistent with the origin of species in this way? If these questions can be answered in the affirmative, Mr. Darwin's view steps out of the rank of hypotheses into those of proved theories; but, so long as the evidence at present adduced falls short of enforcing that affirmation, so long, to our minds, must the new doctrine be content to remain among the former—an extremely valuable, and in the highest degree probable, doctrine, indeed the only extant hypothesis which is worth anything in a scientific point of view; but still a hypothesis, and not yet the theory of species.
>
> After much consideration, and with assuredly no bias against Mr. Darwin's views, it is our clear conviction that, as the evidence stands, it is not absolutely

proven that a group of animals, having all the characters exhibited by species in Nature, has ever been originated by selection, whether artificial or natural. Groups having the morphological character of species—distinct and permanent races in fact—have been so produced over and over again; *but there is no positive evidence, at present, that any group of animals has, by variation and selective breeding, given rise to another group which was, even in the least degree, infertile with the first.* . . . [This would be resolved by experiments which,] conducted by a skillful physiologist, would very probably obtain the desired production of mutually more or less infertile breeds from a common stock, in a comparatively few years; but still, as the case stands at present, this "little rift within the lute" is not to be disguised or overlooked. (*CE* 2:73–75; my italics)

These passages reveal that Huxley subscribed to a criterion of definite experimental proof that would permit us to distinguish theories, those views capable of being proven, from mere hypotheses, which cannot or have not yet been proved. Since selection had not been proved experimentally, it was only a hypothesis and not the theory of species. Darwin disagreed on this point:

The principle of natural selection may be looked at as a mere hypothesis, but rendered in some degree probable by what we positively know of the variability of organic beings in a state of nature—by what we positively know of the struggle for existence, and the consequent almost inevitable preservation, of favourable variations—and from the analogical formation of domestic races. *Now this hypothesis may be tested—and this seems to me the only fair and legitimate manner of considering the whole question—by trying whether it explains several large and independent classes of facts;* such as the geological succession of organic beings, their distribution in past and present times, and their mutual affinities and homologies. *If the principle of natural selection does explain these and other large bodies of facts, it ought to be received.* (*Variation of Animals and Plants* 9; my italics)

It is a question of whether the final value of a theory can be treated as absolute, that is, if it can be isolated and evaluated individually in a context of truth or falsity provided by directly observed facts: if knowledge in a strictly Cartesian sense of indubitability can be obtained, piece by piece, epigenetically, and empirical methods are the means to reach it, then we conclude that a theory has—indeed, must have—an absolute value, which experimentation will discover. The role of experiments—hence a "modern" view of science—is guaranteed, thus avoiding the pitfall of wild speculations. This is Huxley's approach to scientific explanation. On the other hand, a different approach sees the extraction of pieces of knowledge as a secondary, even illusory, objective, whereas its main concern is to provide and account for the connection between phenomena until something better is available. The theory can be evaluated in relation to its precursors, rivals, and potential successors, in the context of a total conceptual scheme that provides the criteria of evaluation and itself undergoes modification. The existence of a mechanism and its manner of operation can be postulated, the phenomena classified, and the result compared with other theories. This seems to be Darwin's criterion, one that also allows the empirical aspect of science a decisive role but has a pragmatic concept of truth (see di Gregorio, "Order"). This is not surprising, if we follow Jonathan Hodge, in his *Origins and Species,* and see how Darwin's science was not a filiation of a Cartesian approach but had its roots—through Lyell's geology—in aspects of the philosophies of George Berkeley and David Hume. In later life Huxley carefully studied Berkeley and Hume, and published on them. What attracted him to Hume was the relationship between common sense and science, especially Hume's discussion of miracles, which could be used to separate theological from scientific arguments, a vital element in Huxley's "agnosticism." Hume, however, was not part of Huxley's original cultural background, whereas he was part of Darwin's. John Passmore is probably correct in arguing that it was the study of Darwin's work that led Huxley to consider Hume's philosophy (and its Berkeleian antecedents).

Another point on which Huxley disagreed with Darwin was gradualism. Huxley maintained his original discontinuous view of nature even after becoming an evolutionist. This discontinuous stance allowed him to play down natural selection, seen by Darwin as directly related to gradualism. Huxley's discontinuous typology is detectable in his paleontological work. His paleontological memoirs concerned the classification of fossil reptiles and Devonian fish (*SM* 2:286–312; 2:421–60), whose forms were classified according to purely anatomical criteria without any reference to evolution. His essay "On the Classification of Birds" is of great theoretical interest (3:239–97). Most of it is occupied with a long and detailed description of the morphological characters used for ornithic taxonomy, but some paragraphs, although apparently irrelevant to the rest of the work, establish the "typological proximity of birds and reptiles." This seeming digression provides the structural evidence on which the evolutionary analysis that follows will depend:

> The members of the class *Aves* so nearly approach the *Reptilia* in all the essential and fundamental points of their structure, that the phrase "Birds are greatly modified Reptiles" would hardly be an exaggerated expression of the closeness of that resemblance.
>
> In perfect strictness, no doubt, it is true that Birds are no more modified Reptiles than Reptiles are modified Birds, the reptilian and ornithic types being both, in reality, somewhat different superstructures raised upon one and the same ground-plan. (*SM* 3:238)

Enter Haeckel

It was only in 1867 that Huxley applied evolutionary notions to strictly scientific research. The essay entitled "On the Animals Which Are Most Nearly Intermediate Between Birds and Reptiles" opens with a striking sentence on evolution: "Those who hold the doctrine of Evolution (and I am one of them) conceive that there are grounds for believing that the world, with all that is in it and on it, did not come into existence in the condition in which we now see it, nor in anything approaching that condition" (*SM* 3:303). This said, Huxley does not concern himself with natural selection but rather with "missing links" in evolutionary history. His aim is to verify the plausibility of the idea of descent of forms, in order to validate the evolutionary interpretability of fossil data. Thus Huxley compares birds and reptiles, and to understand their evolutionary connection he turns to fossil forms. Dinosaurs are the most plausible candidates for the position of "missing link" between reptiles and birds. Huxley claims that

> there is nothing very wild or illegitimate in the hypothesis that the *phylum* of the class *Aves* has its root in the Dinosaurian reptiles. . . .
>
> The facts of Paleontology, so far as Birds and Reptiles are concerned, are not opposed to the doctrine of Evolution, but, on the contrary, are quite such as that doctrine would lead us to expect; for they enable us to form a conception of the manner in which Birds may have evolved from Reptiles, and

> thereby justify us in maintaining the superiority of the hypothesis, that birds have been so originated to all hypotheses which are devoid of an equivalent basis of fact. (*SM* 3:312–13)

While Huxley presented himself to the public as the most strenuous defender of Darwin's views, it is striking that in fact the core of Darwin's view is missing in his scientific works. What we find is the type concept, the search for missing links, the construction of phylogenies, the imprint of Huxley's inclination toward German science. No doubt Darwin rescued evolution from the bad repute into which in the eyes of the scientific establishment it had been thrown by such previous evolutionists as Lamarck and Chambers. However, there was not much application for natural selection, in Darwin's eyes the fundamental means of interaction among individuals in a given environment. This difficulty was especially visible in those who had occupied themselves with systematics, comparative anatomy, and more broadly, typological morphology. The German scientific community, and Huxley with it, reacted positively to the rehabilitation of the idea of evolution, where paleontology and morphological typology, especially in its embryological formulation, could be linked. This would allow the drawing of phyla, which were hypothetically possible but de facto necessary. The key was the introduction of descent within a typology no longer purely anatomical (which would have led to a system Gegenbaur would have called *starr,* like those of Cuvier and Owen), but rather embryological, thus capable of accounting for the dynamic aspects of natural science. The work that had connected descent and embryology, typology and evolutionism, comparative anatomy and paleontology, ontogeny and phylogeny, in the perspective of a new systematics, was Haeckel's *Generelle Morphologie* of 1866. Huxley's first clearly "Haeckelian" memoir is from 1868, although traces of this attitude are less explicitly present in 1866 and 1867. On January 21, 1868 Huxley wrote to Haeckel: "In scientific work the main thing about which I am engaged is a version of the Dinosauria—with an eye to the 'Descendenz Theorie.' The road from Reptiles to Birds is by way of Dinosauria to the Ratitae—the Bird 'phylum' was Struthious, and wings grew out of rudimentary forelimbs. You see that among other things I have been reading Ernst Haeckel's *Morphologie*" (*Life* 1:325).

How was it that Haeckel succeeded where Darwin had failed, namely, in causing Huxley to apply evolutionary reasoning to his actual scientific productions? Haeckel claimed to have introduced a von Baerian typology within his own evolutionary system, thus obscuring the profound divergence between his own views and those of the old embryologist, who complained vigorously, since he thought he could recognize in the new evolutionism a rebirth of Meckel's recapitulationism, which he had strenuously combated. Moreover, Haeckel, thanks to the *biogenetisches Grundgesetz,* established in a real and causal manner, and not metaphorically as with Agassiz, a link between embryological development and the other components of the morphological approach, that is, comparative anatomy (the field of Haeckel's friend and mentor Carl Gegenbaur) and paleontology—thus uniting Huxley's original and contemporary fields of research. Furthermore, Haeckel's development of efficient causes maintained typology but freed it from teleology. To do so he had introduced the Darwinian doctrine into his own system, since he saw selection as an explanatory concept of a mechanical-causal type and as such antiteleological.

The fundamental aspect of Huxley's research had consisted in applying the type concept to discover the orderly pattern of nature: if we want to understand nature we must compare the different types of organization with the different stages in the development of types. Now, if we replace type modifications with actual historico-genealogical relationships, we have preserved the concept of pattern of nature and we have created the *Descendenztheorie.* This was done by Gegenbaur and Haeckel, and the latter's *Generelle Morphologie* became both reference and incitement for more than one evolutionary naturalist, certainly for Huxley. Gegenbaur himself would not have been able to produce his reform of morphology without the support provided by Haeckel's work, in spite of its crudity and excessive dogmatism. Haeckel's evolutionism has characters distinct from Darwin's, with roots in the German tradition of Johannes Müller (see Krausse), and is founded on a thorough discussion of cell theory (see Weindling). These two kinds of evolutionism, the Darwinian and the Haeckelian, were not seen by the two anatomists as incompatible, given that Darwin himself had deployed typological language in some passages in the *Origin* and conceived the archetype concept in the light of common descent, albeit light-years distant from the evolutionary interpretation given by Haeckel to "Promorphologie" in the *Generelle Morphologie* (1:375–574) or by Gegenbaur in his Archipterygian theory ("Archipterygium").

Haeckelian typology is very far from that population thinking that represents the true turning point in Western thinking (Mayr, Introduction). This new way of thinking is, to follow Hodge, related to the British antiessentialist tradition derived from Berkeley and Hume, authors basically absent from Haeckel's culture, in which Lyell's geology had little room. However, in Darwin's published work this population thinking is not so clearly detached from the traditional typology of this time, as it is in the twentieth-century development of Darwinian evolutionism, the so-called New Synthesis. Some passages in the *Origin* in fact suggest that some form of developmentalism and even recapitulationism would not necessarily be inconsistent with Darwin's view (39–43). This justified Gegenbaur, Haeckel, and Huxley himself in thinking that Darwinism and their typology were not opposed, and allowed them to be called "Darwinists" (see di Gregorio, "A Wolf" and *T. H. Huxley's Place*). Only if we read those Darwin passages in the broader context of the whole Darwinian theory, and without the morphological bias of Gegenbaur, Haeckel, and Huxley, do we realize that Darwin's emphasis was in fact on the pervasiveness of variation, the role of the individual, and the diversity of nature, rather than the unity of type.

Bowler has included Haeckel and Huxley in his category of "pseudo-Darwinians"—superficially but not truly Darwinians. Perhaps a better term, at least for Haeckel and Gegenbaur, would be "semi-Darwinians," given that natural selection was necessary for them, although they never seriously inspected the scientific range of that concept and instead considered the products of evolutionary process set into action by selection, that is descent—descent as such, rather than the Darwinian "descent with modifications." Thus Haeckel and Gegenbaur claimed to be ardent supporters of natural selection, which Huxley in fact never did, making him even less Darwinian than Haeckel and Gegenbaur.

The Haeckel-Huxley relationship did not develop in one direction only, from Haeckel to Huxley; the English naturalist often provided his German colleague with important insights. In his Radiolaria monograph, the work that launched him as a first-class naturalist, Haeckel had deployed Huxley's studies on animal individuality and the alternation of generations in his discussion of organ and individual, later to become one of the central themes of the *Generelle Morphologie* (1:241–374). Huxley had defined the individual animal as "the sum of phenomena presented by a single life: in other words, it is all those animal forms which proceed from a single egg taken together" (*SM* 1:150). Huxley's view was seen by Haeckel as the forerunner of his own concept of genealogical individuality. Moreover, Huxley had studied the relationship between plants and animals, another topic treated by Haeckel and his friends and inspirers, Gegenbaur and Max Schultze.

Haeckel had sent Huxley a copy of his Radiolaria monograph, thus initiating a personal friendship and professional partnership that lasted until Huxley's death, in spite of the differences that existed or arose between the two. Huxley's copy of Haeckel's monograph has a number of annotations, mostly concerned with Schultze's protoplasmic version of cell theory as supported and used by Haeckel.

In 1853 Huxley had published "The Cell Theory," in which he had allowed only a secondary role to protoplasmic substance. By 1868 his view had been radically altered in favor of a protoplasmic interpretation of cell theory. Protoplasm was now defined as the formal (physical) basis of life. Huxley's new position seems clearly related to his readings of, and annotations to, Haeckel's monograph and shows the two-directional influence typical of Haeckel's career: Huxley provides Haeckel with useful material on invertebrate zoology, which Haeckel includes in his own work; he then provides Huxley with a broader theoretical framework for his research. In their partnership it was Haeckel, the younger man, who provided most of the theoretical insight, where Huxley had not much to offer.

Their friendship could have been jeopardized by the *Bathybius* affair, which at first was interpreted as a specimen of Haeckelian *Monera* (see *CE* 8:69–109; *SM* 3:330–39; McGraw; Rhebock; Rupke, "Bathybius"). As Huxley wrote to Haeckel:

> I have as yet received no separate copies of [a] paper (published in the *Quarterly Journal of Microscopical Science* for this month) which I read before the meeting of the British Association of Norwich.
>
> It is about a new "Moner" which lives at the bottom of the Atlantic to all appearance, and gives rise to some wonderful calcified bodies.
>
> I have christened it *Bathybius Haeckelii* and I hope you will not be ashamed of your godchild. I will send you some of the mud itself with the paper. (qtd. in Uschmann and Jahn 18)

Haeckel interpreted this discovery as the decisive proof of his philosophico-scientific view, as the foundation stone for his grand universal system, and triumphantly wrote to Huxley "Vivant Monera" (qtd. in Uschmann and Jahn 18). But when some results of the *Challenger* expedition became known in 1875, and *Bathybius* was shown to be just an organic precipitate, Huxley admitted his error and even joked about it, whereas Haeckel insisted that *Bathybius* was an ancestral animal form. However, no trace of argument between the two can be found in their surviving correspondence; they held their friendship in too high an esteem.

Huxley and Gegenbaur

Huxley's attitude to Gegenbaur was less direct and more difficult. With his reform of morphology, Gegenbaur had produced a scientifically strong framework for evolutionism that had no equal in his time and constituted the backbone of Haeckel's system (see Coleman). This notwithstanding, Gegenbaur never achieved (nor did he aim at) fame outside the circle of experts. That the roots of his evolutionary view were independent of Darwin's is clear if we consider the path he took to become an evolutionist, from comparative anatomy to an evolutionary reinterpretation of the vertebral theory of the skull. In some respects his scientific career had many resemblances to Huxley's: he started with invertebrate zoology and moved to vertebrate anatomy. The vertebral theory of the skull, originally formulated by Goethe, Oken, and Beneden, and logically reflecting *Naturphilosophie*'s assertion of the repetition and multiplication of parts within the organism, has been a long-running focus of morphological debate (see Russell 141–49). According to the theory, the skull is an expanded portion of the vertebral column. Savigny, Meckel, and Carus proposed various interpretations of the theory, while its supporters could not agree on the actual number of vertebrae comprising the skull. Even the great Johannes Müller referred to the theory (see Lohff). Along with such embryological typologists as von Baer and Huschke, Müller emphasized the role of the embryonic development of the skull as the key to the understanding of the vertebrate type.

One of the most ardent supporters of the vertebral theory was Richard Owen, who saw it as a prerequisite of his concept of homology and thus of his whole view. For him it made sense of the unity of structure of the vertebrate skull and therefore of the vertebrate type, underpinning his view of organic relationships as formal rather than temporal modifications of the Archetype. After publication of the *Origin,* Owen was targeted as the main scientific opposition to Darwin's views in Britain, but his duel with Huxley had started well before 1859 (see Desmond, *Huxley*). Huxley, aware of the indispensability of the vertebral theory to Owen's conception, in 1858 attacked his archenemy's view in "On the Theory of the Vertebrate Skull." Huxley—on his home territory, comparative embryology—studied the skulls of human, turtle, and carp, using the embryological type concept at a remove from its idealistic overtones. Huxley's conclusion was that all vertebrate skulls exhibit a common plan, but that this is not evidence for the vertebral theory because "it is no more true that the adult skull is a modified vertebral column, than it would be, to affirm that the vertebral column is a modified skull" (*SM* 1:585). However, in 1858 Huxley was unable to provide a rival justification of the unity of plan of vertebrate skulls.

Gegenbaur, on the other hand, had always been a keen supporter of the vertebral theory, from as far back as his idealistic, preevolutionary stage, and was sympathetic to Owen's work. After reworking his *Grundzüge* in an evolutionary light, Gegenbaur embarked on a thorough revision of the vertebral theory, convinced that the origin of the vertebral skull would be the key to the phylogenetic relationships of the vertebrates. The Goethe-Oken doctrine underwent a complete overhaul in Owen's system of osteology. With the establishment of an "Archetypus" of the entire skeleton, what Cuvier's school had established as comparisons within relatively narrowly defined departments, was consolidated into a harmonious whole. The vertebral theory could thereby be extended equally across "all (4) classes of vertebrates" whose cephalic skeleton could be demonstrated to consist of four vertebrae and appendages, with numerous modifications of parts (Gegenbaur, *Untersuchungen* 4).

But thanks to the *biogenetisches Grundgesetz,* which expressed the connection between ontogeny and phylogeny, it was possible to show the relationships of the vertebrae to the skull in a more general way than the anatomical manner of Owen: descent theory, as supported by comparative embryology, would turn Owen's formal archetypology into the record of a living system. In this new framework the vertebral theory could, and indeed had to, survive Huxley's attack. This seems to have strained things between Huxley and Gegenbaur.

Another possible reason for coolness between Gegenbaur and Huxley was the latter's friendship with Anton Dohrn. Dohrn, the founder of the Naples Zoological Station, had been Gegenbaur's and Haeckel's student at Jena, but had criticized both his teachers' reconstruction of vertebrate phylogeny and their morphologically inclined version of evolutionism. Dohrn insisted on a more physiological interpretation and had indeed proposed the introduction of the concept of change of function as of great relevance in evolutionary explanation. Moreover, he was highly critical of Gegenbaur's version of the vertebral skull, all of which made him persona non grata to Haeckel and Gegenbaur.

Both Huxley and Darwin were strong supporters of Dohrn's Naples Station, and Gegenbaur's and Haeckel's enmity toward Dohrn greatly damaged the cause of natural science and of evolutionism in Europe. Huxley, albeit so sympathetic to physiology as to have among his disciples the founder of the new physiological school in Britain, George Foster (see Geison), never practiced physiology seriously himself, while Darwin approved of Dohrn's change of function principle, but was rather dubious of his speculations on the origin of vertebrates, influenced as he confessed he was by Geoffroy's metaphysics.

Huxley and Haeckel Differ

Haeckel's great hope was to have his *Generelle Morphologie* translated into English, and for that reason he sought Darwin's and Huxley's support. Huxley could have been of special help, since he was directly involved with academic and editorial politics. Huxley required a drastic revision and shortening of Haeckel's text, with special attention to the elimination of the speculative parts, which indeed were many and frequently too crude for a British audience. The *Generelle Morphologie* was often (and often unnecessarily) aggressive, even toward living authors, and propounded a general view that was overtly anticlerical if not anti-Christian. This might pass in Germany, but it could be tactically harmful for the new scientific attitude that Huxley was propounding in England. This suggests why Huxley was so hesitant about a possible English translation of the *Generelle Morphologie,* which the Ray Society seemed to be willing to produce. He had demanded extensive cuts from Haeckel, and even if Haeckel did agree to these, no translation ever appeared. The philosophical parts would have been rejected by the English scientific public, especially the fourth book, entitled "Generelle Promorphologie oder allgemeine Grundformlehre der Organismen." In it Haeckel was attempting a descent-inclined reworking of idealistic morphology, an undoubtedly fundamental part of his work but one that would have attracted too much attention to the connection of Haeckel's views with *Naturphilosophie,* which Huxley did not consider politically correct in 1868. Although Huxley had been influenced by the *Descendenztheorie,* he was not prepared to accept the all-embracing ideology that went with it in the *Generelle Morphologie,* and that came to be called "Darwinismus" (see Moore). This disapproval of Darwinismus was visible in Huxley's review ("The Natural History of Creation") of Haeckel's *Natürliche Schöpfungsgeschichte,* a popularized version of the *Generelle Morphologie* based on Haeckel's lectures at Jena. The review was to say the least lukewarm; Huxley was steadily moving toward a view that he "extricated from ontological controversies and dissociated from hostile worldviews" (Moore 388).

Huxley's diffidence at teaming up with Haeckel on the ideology of Darwinismus is again detectable in his attitude toward the Haeckel-Virchow controversy. In 1877, in a speech delivered in Munich at the annual meeting of the German physicians and natural scientists, Haeckel not only emphasized the decisive role of *Entwickelungsgeschichte* [development history] for the various branches of science as a basis for a new evolutionary *Weltanschauung* but also campaigned for its introduction into the school curriculum (Krausse, *Ernst Haeckel* 92–95). Virchow had criticized Haeckel's stance, claiming that the theory of evolution was not sufficiently established, useful for research but not apt to be taught as the indisputable fact that Haeckel thought it was. Virchow's criticism was a severe blow to Haeckel, who took it as a real defection of the great pathologist from his monistic camp. Virchow had been one of Haeckel's most influential teachers at Würzburg and had inspired him to work on cell theory. Virchow was in the limelight not only as a leading man of science but also as a prominent member of the left liberal *Fortschrittpartei* (see Sheehan), and as such a staunch opponent of Bismarck who, by contrast, had grown into a hero to Haeckel after the Franco-Prussian war. Academic hostility was also involved. Gegenbaur endeavored (and, in Nyhart's opinion, largely failed) to make his reformed morphology into the core of research in German science institutes. Virchow, with DuBois Reymond and their Berlin friends, opposed that attempt. The feud between Haeckel and Virchow became a cause célèbre in European science and went on for some time, with Darwin, for one, clearly sympathetic toward Haeckel.

When Haeckel's reply to Virchow was translated into English, Huxley wrote a "prefatory note," which, rather than siding openly with Haeckel, was lukewarm and showed his intention to keep his distance from his friend's ideology. As Huxley wrote with respect to Virchow's view: "It is supposed to be a deadly blow to the doctrine of evolution; but, though I certainly hold by that doctrine with some tenacity, I am able, *ex animo,* to subscribe to every important general proposition which its author lays down" (Prefatory Note ix). Huxley agreed with Virchow on the incompleteness of human knowledge and on the decisive role of experimentation in establishing scientific truths. His whole writing was clearly against Haeckel's all-embracing ideology and in favor of his own, very different restrictive and agnostic attitude toward science. It should be noted that there is no sign in their surviving correspondence of Haeckel's having taken offense at Huxley's stance.

What really concerned Huxley was the surging evolutionary ethics in his time, which had in Britain its leading representative in Herbert Spencer. The controversy reflected on Huxley's political stance, since he was a staunch opponent of Spencer's laissez-faire (*CE* 1:251–89), while also opposing socialism (9:147–87). His view was that of an "enlightened conservative" (Hutton), of benevolent and somewhat patronizing interventionism of the state without renouncing the free market economy. Huxley criticized Spencer's concept of the relation between physical and social organisms, and launched his final all-out attack on naturalistic ethics in his celebrated lecture "Evolution and Ethics" of 1893. The analogies between Spencer's views and Haeckel's "Biologismus" are not difficult to see (see Weindling). Thus, if Huxley was prepared to accept Haeckel and his *Descendenztheorie* as a capital influence on his scientific work, he was not prepared to go the whole way with his colleague's views of the relationship between ethics and science, and rather sided with the traditional separation of the two endeavors in the manner of Hume and Kant.

Huxley's research was shaped by his close connection with the German tradition of science, and the key characters in understanding this fact are Karl Ernst von Baer and Ernst Haeckel. Von Baerian embryological typology molded Huxley's science to such an extent that even after his official endorsement and public defense of Darwin's views, only a form of evolutionism that could salvage those von Baerian roots would make sense to him. Haeckel's interpretation of evolution seemed to combine embryological typology with real evolution of forms in time. It is no surprise, then, that Huxley applied evolutionary reasonings to his scientific research only after reading Haeckel's main theoretical work. It was Haeckel's science that represented the decisive element that catalyzed Huxley's research toward evolution. But Haeckel's evolutionism was inclined toward descent and the drawing of phylogenies rather than natural selection. Thus, scientific "Haeckelismus" allowed Huxley to play down natural selection, for which he had no use. If this interpretation of the relationship between Huxley and the "German community" is correct, it follows that Huxley's own research is not the place to look for evidence of any direct impact, positive or negative, of the publication of the *Origin,* especially prior to the late 1860s, when systematologists in general began to adjust. The same line of thought also dispels the expectation-in-principle that the doctrine of natural selection ought to be featured somewhere in those works of Huxley that do apply the idea of evolution, for that expectation is countered by the observation of the substantial irrelevance of natural selection to the current concerns of systematics. Endorsing evolution and endorsing natural selection are not the same thing. Huxley's heavily qualified acceptance of natural selection and the survey of his research show that he became an evolutionist, but of the Haeckelian, not the Darwinian, kind.

To follow Haeckel all the way would have meant turning the theory of evolution from a successful scientific theory into a whole Weltanschauung. This Huxley was not prepared to do. Although Huxley would argue that the task of science was to provide the data on which to build our knowledge of the world, and therefore no sensible action could be possible without scientific knowledge, he would not accept that moral values could be directly extracted from scientific data. Against Spencer and Haeckel, he rejected naturalistic ethics and clearly distinguished what *is* from what *ought* to be. In this he sided with the philosophical tradition that included Hume and Kant, and separated ethics from science.

REFERENCES

Ackerknecht, Erwin H. *Rudolf Virchow: Doctor, Statesman, Anthropologist.* Madison: U of Wisconsin P, 1953.

Agassiz, Louis. *Recherches sur les Poissons Fossiles.* Neuchâtel: Institute Lithographique de H. Nicolet, 1833–43.

Barry, Martin. "On the Unity of Structure in the Animal Kingdom." *Edinburgh New Philosophical Journal* 22 (1836–37): 116–41, 345–64.

Bartholomew, Michael. "Huxley's Defence of Darwin." *Annals of Science* 32 (1975): 525–35.

Bowler, Peter J. *The Non-Darwinian Revolution: Reinterpreting a Historical Myth.* Baltimore: Johns Hopkins UP, 1988.

Carpenter, William B. *Principles of General and Comparative Physiology.* London: J. Churchill, 1839.

Coleman, William. "Morphology between Type Concept and Descent Theory." *Journal of the History of Medicine* 31 (1976): 149–75.

Darwin, Charles. *The Variation of Animals and Plants under Domestication.* 2 vols. London: J. Murray, 1868.

Desmond, Adrian. *Archetypes and Ancestors: Palaeontology in Victorian London, 1850–1875.* Chicago: U of Chicago P, 1984.

———. *Huxley: The Devil's Disciple.* London: Michael Joseph, 1994.

di Gregorio, Mario A. "A Wolf in Sheep's Clothing: Carl Gegenbaur, Ernst Haeckel, the Vertebral Theory of the Skull, and the Survival of Richard Owen." *Journal of the History of Biology* 28 (1995): 247–80.

———. "Order or Process of Nature: Huxley's and Darwin's Different Approaches to Natural Sciences." *Journal of the History and Philosophy of the Life Sciences* 2 (1981): 217–41.

———. *T. H. Huxley's Place in Natural Science.* New Haven: Yale UP, 1984.

Dohrn, Anton. *Der Ursprung der Wirbelthiere und das Princip des Functionswechels.* Leipzig: E. Engelmann, 1875.

Engelhardt, Dietrich von. *Historisches Bewusstsein in der Naturwissenschaft von der Aufklärung bis zum Positivismus.* Freiburg and München: Alber, 1979.

Gegenbaur, Carl. "Über das Archipterygium." *Jenaische Zeitschrift* 7 (1872): 131–41.
———. *Grundzüge der Vergleichenden Anatomie.* Leipzig: W. Engelmann, 1870.
———. *Untersuchungen zur Vergleichenden Anatomie der Wirbelthiere.* Vol 3. Leipzig: W. Engelmann, 1872.
Geison, Gerald L. *Michael Foster and the Cambridge School of Physiology.* Princeton: Princeton UP, 1978.
Gould, Stephen Jay. *Ontogeny and Phylogeny.* Cambridge: Harvard UP, 1977.
Haeckel, Ernst. *Generelle Morphologie der Organismen.* 2 vols. 1866. Berlin Walter de Gruyter, 1988.
———. *Natürliche Schöpfungsgeschichte.* Berlin: Georg Reimer, 1868.
———. *Die Radiolarien.* Berlin: Georg Reimer, 1862.
Hodge, Michael Jonathan S. *Origins and Species.* New York: Garland, 1991.
Hutton, Richard H. "The Great Agnostic." *Spectator,* July 6, 1895, pp. 10–11.
Huxley, Leonard. *Life and Letters of Thomas H. Huxley.* 2 vols. N.Y.: Appleton, 1901.
Huxley, Thomas Henry. *Collected Essays.* 9 vols. 1894. N.Y.: Greenwood Press, 1968.
———. "Fragments Relating to Philosophical Zoology, Selected from the Works of K. E. von Baer." *Taylor's Scientific Memoirs, Natural History* 3 (1853): 176–238.
———. "The Natural History of Creation—by Ernst Haeckel." *Academy* 1 (1869): 12–14, 40–43.
———. *The Oceanic Hydrozoa.* London: Ray Society, 1859.
———. Prefatory note. *Freedom in Science and Teaching.* By Ernst Haeckel. N.Y.: Appleton, 1879. v–xx.
———. *The Scientific Memoirs of Thomas Henry Huxley.* Ed. Michael Foster and E. Ray Lankester. 4 vols. plus supplement. London: Macmillan, 1898–1903.
———. *T. H. Huxley's Diary of the Voyage of the HMS* Rattlesnake. Ed. Julian Huxley. London: Chatto & Windus, 1935.
Krausse, Erika. *Ernst Haeckel.* Leipzig: D. G. Teubner, 1984.
———. "Johannes Müller und Ernst Haeckel." *Johannes Müller und die Philosophie.* Ed. Michael Hagner and Bettina Wahrig-Schmidt. Berlin: Akademie Verlag, 1992. 223–37.
Lenoir, Timothy. *The Strategy of Life: Teleology and Mechanics in Nineteenth-Century German Biology.* Chicago: U of Chicago P, 1989.
Lohff, Brigette. "Johannes Müller Rezeption der Zellenlehre in seinem *Handbuch der Physiologie des Menschen.*" *Medizin historisches Journal* 13 (1978): 247–58.
Mayr, Ernst. *The Growth of Biological Thought: Diversity, Evolution, and Inheritance.* Cambridge: Harvard UP, Belknap P, 1982.

———. Introduction. *On the Origin of Species* (facsimile of 1st ed.). By Charles Darwin. Cambridge: Harvard UP, 1964. vii–xxvii.

McGraw, Donald J. "Bye-Bye Bathybius: The Rise and Fall of a Marine Myth." *Bios* 45 (1974): 164–71.

Milne-Edwards, Henri. "Considérations sur Quelques Principes Relatifs à la Classification Naturelle des Animaux, et Plus Particulièrement sur la Distribution Méthodique des Mammifères." *Annales des Sciences Naturelles (Zoologie)* 1 (1844): 68–99.

Moore, James. "Deconstructing Darwinism: The Politics of Evolution in the 1860s." *Journal of the History of Biology* 24 (1991): 353–408.

Nyhart, Lynn K. *Biology Takes Form: Animal Morphology and the German Universities, 1800–1900.* Chicago: U of Chicago P, 1995.

———. "The Disciplinary Breakdown of German Morphology, 1870–1900." *Isis* 78 (1987): 365–89.

Ospovat, Dov. *The Development of Darwin's Theory: Natural History, Natural Theology, and Natural Selection.* Chicago: U of Chicago P, 1981.

———. "The Influence of Karl Ernst von Baer's Embryology, 1828–1859: A Reappraisal in the Light of Richard Owen's and William B. Carpenter's Paleontological Application of von Baer's Law." *Journal of the History of Biology* 9 (1976): 1–28.

Owen, Richard. *On the Archetype and Homologies of the Vertebrate Skeleton.* London: Van Voorst, 1848.

———. *On the Nature of Limbs.* London: Van Voorst, 1849.

Paradis, James. *T. H. Huxley:* Man's Place in Nature. Lincoln: U of Nebraska P, 1978.

Passmore, John. "Darwin's Impact on British Metaphysics." *Victorian Studies* 3 (1959): 41–54.

Ràdl, Emanuel. *Geschichte der Biologischen Theorien Seit dem Ende des Siebzehnten Jahrhunderts.* 2 vols. Leipzig: W. Engelmann, 1905.

Raikov, Boris. *Karl Ernst von Baer, 1792–1876: Sein Leben und Sein Werk.* Leipzig: J. A. Barth, 1968.

Rehbock, Philip. "Huxley, Haeckel, and the Oceanographers: The Case of Bathybius Haeckelii." *Isis* 66 (1975): 504–33.

Roger, Jacques. *Les Sciences de la Vie dans la Pensée Francaise au XVIIIe Siècle.* Paris: A. Colin, 1963.

Rupke, Nicolaas. "Bathybius Haeckelii and the Psychology of Scientific Discovery." *Studies in the History and Philosophy of Science* 7 (1976): 53–62.

———. *Richard Owen: Victorian Naturalist.* New Haven: Yale UP, 1994.

———. "Richard Owen's Vertebrate Archetype." *Isis* 84 (1993): 231–51.

Russell, Edward S. *Form and Function.* London: J. Murray, 1916.

Schultze, M. "Über Muskelkorperchen und das, Was Man Eine Zelle Zu Nennen Habe." *Archive für Anatomie, Physiologie, und Wissenschaftliche Medizin* (1861): 1–27.

Sheehan, James J. *German Liberalism in the Nineteenth Century.* Chicago: U of Chicago P, 1978.

Uschmann, Georg. *Geschichte der Zoologie und der Zoologischen Anstalten in Jena, 1779–1919.* Jena: G. Fischer, 1959.

Uschmann, Georg, and I. Jahn, eds. "Der Briefwechsel Zwischen Thomas Henry Huxley und Ernst Haeckel." *Wissenschaftliche Zeitschrift der Friedrich-Schiller-Universität Jena, Mathem.—Natürwissenschlaftliche Reihe* 9 (1959–60): 7–33.

Weindling, Paul. "Darwinism in Germany." *The Darwinian Heritage.* Ed. David Kohn. Princeton: Princeton UP, 1985. 685–98.

Winsor, Mary P. *Starfish, Jellyfish, and the Order of Life.* New Haven: Yale UP, 1976.

Thomas Henry Huxley and Nineteenth-Century Biology

ROBERT G. B. REID

THOMAS HENRY HUXLEY'S CAREER WAS DIVIDED BETWEEN original research and a variety of broadly educative activities, encompassing college teaching and curriculum design, presenting public lectures on biology, and writing zoological text books as well as essays on the liberal arts. He also acted occasionally as a government commissioner and civil servant. The image of the original scientist has, however, in the century since his death, become largely obscured by that of the writer and polemicist. Over fifty years ago, Aldous Huxley remarked of his grandfather, "As a literary man he is still a living force. His non-technical writings have the persistent contemporariness that is the quality of all good art" (50). More recently, when I asked a colleague a question about Darwinism, he unawaredly replied with a quotation from Huxley, believing that he was stating an original opinion in his own words. An aphoristic diamond had passed through several intellectual alimentary tracts and emerged bright and undigested. But the impact of Huxley's prose and the survival of such aphorisms should not surprise. As accumulated commentary about Huxley, particularly the studies of his rhetoric and other essays in this volume will confirm, Aldous was certainly right in recognizing the enduring quality of his art.

It is the other side of Huxley that I examine and another comment of his grandson's that I question. Aldous dismissed the importance of the original science: "As a scientific man, Huxley, like all his great contemporaries and predecessors, is now a mere historical figure" (50). The tendency to undervalue older science as more interesting to historians than to working scientists notwithstanding, Huxley's work as a scientist (for me, as a zoologist, in particular) is worth reexamining. His development as a biologist unfolds an almost mythic journey from pre-Darwinian foothills to the peaks of the evolutionary paradigm. I look at the course of that journey in biology, demonstrating links between his purely biological research and more general concepts, discussing the significance of his choices of subject matter—whether fortuitous or precognitive—and commenting on his subsequent impact on biology. Often, the subjects Huxley chose, and the questions he raised in the mid-nineteenth century, anticipated issues we are debating at the end of the twentieth.

How Huxley went about choosing his subject matter was a question of educational background, time, chance, and access to research material. Referring to himself in his 1890 autobiography as a "sort of mechanical engineer *in partibus infidelium*," he explains that family influences guided him instead into a medical program (*CE* 1:7). A mechanistically inclined medical student would naturally find much of interest in functional anatomy, which Huxley included under the general heading of "physiology." He noted, "the only part of my professional course which really and deeply interested me was physiology, which is the mechanical engineering of living machines" (1:7). The biological part of his undergraduate program was largely concerned with human anatomy and physiology, with a little comparative anatomy thrown in. In addition he read the French botanist De Candolle, attended botany lectures, and won a botanical medal in a competitive examination, but showed little subsequent participatory interest in it until 1886, when, during a prolonged illness, he went mountaineering in Switzerland to seek the "ur-gentian" (*Life* 2:146, 147). In 1888, at sixty-three, when most zoologists of his age would have been content to collect attractive specimens for their rockeries, he produced a comprehensive report on the systematics of the gentians.

Physiology, per se, did not form a major part of his original research, although he gave lectures and demonstrations in physiology and published a competent elementary textbook on the subject in 1866 that ran to five editions. His appointment as assistant surgeon to the HMS *Rattlesnake*, from 1846 to 1850, provided his founding experience as a field biologist but no opportunity to conduct experimental research. Here his first close encounters with exotic animals involved largely the Acaphelae, which are now called Cnidaria: the jellyfish, sea anemones, and siphonophores like *Physalia,* the Portuguese man-of-war. He also investigated the planktonic tunicates—salps, doliolids, and tadpolelike *Appendicularia*—and planktonic "cephalous gastropods" such as Pteropoda—the "sea butterflies." Reports on some of these organisms were forwarded for publication in England during the *Rattlesnake*'s voyage and had already established him as a biologist to be reckoned with before many even knew his face.

At the prompting of Richard Owen the list of topics he had initially intended to tackle included fish brain anatomy, fish ectoparasites, barnacles, marine worms, Acaphelae, including corals, as well as molluscs and tunicates. His *Rattlesnake* diary noted a pipe-dream of sorts: "I have a grand project floating through my head, of working up a regular monograph of the Mollusca, anatomy, physiology and histology, based on examination of at least one species of every genus. But I fear me much that, as the saying goes, my eyes are bigger than my belly" (*Diary* 24). At that time he was in South Africa, revising his priorities, but the subjects that had fortuitously appeared on the first leg of the voyage, to South America, remained at the top of the list. Ultimately, his ambition of producing a large, systematic study was achieved with his 1859 monograph *On the Oceanic Hydrozoa.*

Unlike Darwin, and unlike the official naturalist of the *Rattlesnake,* McGillivray, Huxley was not an avid collector nor taxonomist, and a psychological depression during the third cruise of the *Rattlesnake,* which visited parts of the Great Barrier Reef that teemed with interesting marine invertebrates, left him indifferent to all but the occasional jellyfish and mollusc that caught his fancy. His unremarkable 1849 report on the bivalve *Trigonia* is perhaps a reflection of a lack of any directing hypothesis for molluscan studies at that time. However, it is surprising that he made nothing of the corals and various species of giant clams that abounded on the Great Barrier Reef. Recovering his vitality on the fourth cruise he was by then more interested in the aboriginal human fauna of Australia. This established a foundation for a much later enthusiasm for ethnology.

Comparative anatomy and the application of the concept of unity of plan was Huxley's greatest early interest and the most productive area of his overall research career, constituting approximately one-third of his original titles in all fields. Unity of plan was an idea based on the phenomenon of homology: the underlying structural similarity in organisms that evolved from common ancestors, although some of its proponents, including Huxley, did not yet see it in the evolutionary context that provided its ultimate explanation. Although familiar with the evolutionary theories of Lamarck, Goethe, and Geoffroy de St. Hilaire, Huxley initially took the conservative side promulgated by Cuvier and his English disciple Richard Owen.

Cuvier, in 1812, proposed that there were functional and anatomical homologues, or "types," within the major *embranchements* of the animal kingdom. Thirty-five years later, Owen suggested that each of these groups had an identifiable original form, or "archetype." Cuvier believed that each organism was internally integrated and outwardly adapted to the environment, but had not evolved that way. Intermediate forms could not have the same degree of success in life as the existent adapted forms and therefore could never have survived. Owen, on the other hand, believed that once the archetype had been created it could have evolved into the diversity of extant forms, but he concealed this element of his thinking in the wake of the hostile reception accorded to the evolutionary thought of Robert Chambers's 1844 *Vestiges of Creation in Natural History* (Rev. R. S. Owen).

On board the *Rattlesnake,* Huxley, not yet an evolutionist, was content to accept Cuvier's and Owen's theories and to attempt to work them through in his studies of Cnidaria and Mollusca. One difficulty he faced was the relative chaos of zoological systematics at the phylum level. All the organisms in the Cuverian system that had some kind of spherical shape or radial symmetry were corralled in the subkingdom Radiata. The Mollusca included the Brachiopoda, which Huxley later cast adrift in the direction of the Polyzoa, and the annelid-arthropod group was confused. Although he professed no interest in systematics, he later helped to rationalize the phyla, for example, in an 1876 lecture entitled "On the Classification of the Animal Kingdom."

The Cnidaria (Acaphelae) were not just fortuitously the first subjects that came to his attention, they presented classic problems, pertaining to asexual reproduction and the so-called alternation of generations, which had already been studied by Steenstrup and Chamisso, with whose work Huxley was familiar. The "polyp question" raised by the asexual reproduction of *Hydra* had already been discussed by the eighteenth-century entomologist Réaumur, whose mechanistic view of biology had impressed Huxley (see Dawson). Huxley later described the Frenchman in an 1877 letter to Darwin as in some aspects "the biggest man between Aristotle and Darwin" (*Life* 1:514).

The conclusion of some of Huxley's biographers that his genius developed in part on the *Rattlesnake* because he had no library nor preconceptions and was forced to follow his own burgeoning philosophy is an oversimplification, although not to be bogged down in other people's minutiae is certainly an advantage. Serendipitously, in working with planktonic fauna, he had specimens that were transparent, and their cell layers, which he called their "foundation membranes" in his major 1849 report on the medusae, since he did not endorse the cell theory at that time, were visible to the naked eye. He had accepted the names "ectoderm" and "endoderm" for the foundation membranes by the time he published *On the Oceanic Hydrozoa* in 1859 (137). The transparency that made his material easy to study faded after a few hours and such details could not be seen in opaque preserved museum specimens. Furthermore, there were plenty of diverse forms for him to examine, all showing two foundation layers, despite their superficial structural differences. Here he elaborated and applied Owen's concept of homology.

> Now the organs of two animals or families of animals are homologous when their structure is identical, or when the differences between them may be accounted for by the simple laws of growth. When organs differ considerably, their homology may be determined in two ways, either—1, by tracing back the course of development of the two until we arrive by similar stages at the same point; or, 2, by interpolating between the two a series of forms derived from other animals allied to both, the difference between each term of the series, being . . . accounted for by the laws of growth. The latter method is that which has been generally employed under the name of *Comparative Anatomy,* the former having been hardly applicable to any but the lower class of animals. Both methods may be made use of in investigating the homologies of the Medusae. . . .
>
> The above definitions may be thought needless or trite but the establishment of affinities among animals has been so often a mere exercise of the imagination that I may be pardoned for pointing out the guiding principles which I have followed and by which I would wish to be judged. (*SM* 1:23)

The Royal Society editors apparently welcomed an elementary definition. This early publication not only demonstrates the clarity of Huxley's prose, it indicates an evolutionistic leaning. Huxley's theoretical approach extended beyond the Cnidaria, pointing out the similarity between their foundation layers and the serous and mucous layers (ectoderm and endoderm) of vertebrate embryos, picking up on the developmental observations of von Baer. A combination of meticulous anatomical description with the idea of the underlying unity of plan in the Cnidaria, together with his clarity of exposition, resulted in a model study that garnered for him a fellowship in the Royal Society in 1851. Huxley did not know that the report on medusae had been published until he returned to England, and this uncertainty may have contributed to his depression during the third cruise. "I made a good many observations during our cruise, and sent home several papers to the Linnaean and Royal Societies; but of these doves, or rather ravens, which left my ark, I had heard absolutely nothing up to the time of my return; and, save for the always kind and hearty encouragement of the celebrated William MacLeay, whenever our return to Sydney took me within reach of his hospitality, I know not whether I should have had the courage to continue labour which might, so far as I knew, be valueless" (*On the Oceanic Hydrozoa* viii).

Two 1851 *Rattlesnake* reports on tunicates further illustrate the combination of painstaking observation and general theory that mark the best of Huxley's biological publications. As he wrote to a colleague in physiology, "I have been working in all things with a reference to wide views of zoological philosophy" (*Life* 1:100). His wide views occasionally engendered a lyrical enthusiasm that is rarely found in modern scientific reports: "The sky was clear but moonless, and the sea calm, and a more beautiful sight can hardly be imagined than that presented from the decks of the ship as she drifted, hour after hour, through this shoal of miniature pillars of fire gleaming out of the dark sea, with an ever-waning, ever-brightening, soft bluish light, as far as the eye could reach on every side. The *Pyrosomata* floated deep, and it was with difficulty that some were procured for examination and placed in a bucket of sea water" (*SM* 1:53–54). Once the scene is set and the bucket full, he becomes all business, and follows with thirty-five detailed points of comparative anatomy and histology.

Huxley noted that all the organs of the salps have homologies in the other ascidians or tunicates. Then he went on, in an 1851 commentary on Müller's research on echinoderms, to consider the great issue of alternation of generations, which had "bid fair to unite all the aberrant generative processes of the Invertebrata . . . under its conditions" (*SM* 1:115). Suspicious of such allure, Huxley subsequently concluded, in "Upon Animal Individuality," that "there is no such thing as a true case of the 'Alternation of generations' in the animal kingdom; there is only an alternation of true generation with the totally distinct process of Gemmation or Fission [i.e., asexual reproduction]" (1:49). It was better to describe animals that had distinct life cycle phases, some of which might generate asexually before they came to a sexual phase, as a series of zooids: distinct forms of the same individual. An echinoderm's life consisted not of alternating generations but of the larval zooid followed by the adult echinoderm zooid. He also considered here the meaning of individuality in "colonial" animals: "A whole tree of Sertularia, a Pennatula, a Pyrosoma, a mass of Botrylli, must no longer be considered as an aggregation of individuals, but as an individual developed into many zooids" (1:118). These arguments were further developed in a series of papers in which the work of earlier authors were elaborated. Although Huxley had been somewhat limited initially in his *Rattlesnake* work by the lack of a proper library, he was able to catch up at the Sydney libraries while the ship was in port and by the use of facilities provided by the naturalist W. S. MacLeay (*Life* 1:39).

The *Rattlesnake* findings continued to dominate Huxley's publications in the early 1850s. In 1853 the Royal Society published his paper on the cephalous molluscs in its Philosophical Transactions. Here he endeavored to identify the molluscan archetype, a search complicated by the great plasticity of the phylum. He proposed an archetype for the opisthobranchs and made perspicacious comments about general molluscan features, such as the possession of the radula and the particular prosobranch gastropod phenomenon of "flexure," now called "torsion." Although the bivalve molluscs have no radula, he pointed out that their typical gastric crystalline style is also found in some gastropods. As evidence of the enduring impact of Huxley on this subject, his disciple Lankester took up the concept of an archetypal, "schematic" mollusc in the late nineteenth century, and the twentieth-century malacologist C. M. Yonge elaborated a new model of the hypothetical primitive, or archetypal, mollusc in 1947. A former student of Yonge's, I have recently written on the archetypal bivalve, which places me in an almost direct line of intellectual descent, along with innumerable other zoologists and paleontologists of a certain age. However, the rest of us did not have to break all new ground, and we enjoyed the conceptual advantage of an evolutionary paradigm. Extrapolating from the cephalous molluscs, Huxley was all but forced into that context. If, he speculated, "all Cephalopoda, Gasteropoda, and Lamellibranchiata, be only modifications by excess or defect of the parts of the definite archetype, then, I think, it follows as a necessary consequence . . . [that] there is no progression from a lower to a higher type, but merely a more or less complete evolution of one type" (*SM* 1:192). Now we know that he did not mean evolution in the modern sense, but was actually proposing embryonic saltations in different directions, which essentially comes down to the same thing. Only, he did not pose the question how. Nor did he ask whence came the diversification within those major classes. Though poised to be an evolutionist, his own saltation did not occur until six years later.

In 1854 Huxley obtained the appointment to the Royal School of Mines that guided his professional path to paleontology. However, during the 1850s he continued to publish on a variety of topics, many of them arising from his *Rattlesnake* fieldwork, along with others that began to draw him away from the zooplankton. Points of more general interest concerning the latter continued to appear. In a study of *Appendicula,* an aberrant organism that its discoverer, Chamisso, had guessed might be a ctenophore, Huxley established its tunicate affinities, pointing out that it was a sexually mature larval form. This was to become a key to the relationship between protochordates and vertebrates.

His new interests included parasitology, tooth development, blood circulation in annelids, the biology of tubeworms, brachiopods, and Protozoa, sensory organs, and nerve function. The appointment at the School of Mines was to be a temporary double post including the lectureships in paleontology and natural history. In "Autobiography" he notes, "I refused the former point blank, and accepted the latter only provisionally, telling Sir Henry [the director] that I did not care for fossils, and that I should give up Natural History as soon as I could get a physiological post" (*CE* 1:15). Pedagogical necessity soon developed into a profound interest in paleontology, and publications in this field began to appear regularly from 1856 onward, ultimately constituting about a quarter of Huxley's total literary output.

Huxley's early scientific publications are marked by a scrupulous attribution of priority to earlier workers and by a generous assessment of their contributions to the field. After this first flush of collegiality, he gradually began to go overboard in his condemnation of those who offended his perception of the rules of the game, particularly those who theorized on the basis of little or no evidence, or who misreported earlier works. Scurrilous anonymity also infuriated him.

He first tasted blood in an 1854 review of the tenth edition of Chambers's *Vestiges of Creation in Natural History*. The hostile reception of the first edition in 1844 from religious fundamentalists is supposed to have diverted Owen entirely from embracing evolutionism and to have delayed Darwin from publishing on evolutionary theory for ten years. Now, at the point where Darwin was finishing up his work on barnacles and contemplating the reins of the evolutionary bandwagon again, here was one of the most respected young scientists of the time hammering at the theory from the scientific side. Huxley's fundamental complaint was that Chambers was a progressionist who assumed that there is an inexorable complexification of life with time. This would then require that existent types all be more sophisticated than their fossil relatives, and such did not appear to be the case. Fossil fish were just as anatomically complex as modern fish, and the extinct trilobites were more complex than some living crustaceans. Huxley's arguments against that kind of progressionism were logically valid. The modern, conventional view is that while progress from the simple to the complex has occurred overall, it may be arrested in some lines of evolution. An inevitable complexification is not the product of an inherent drive, as the progressionists once believed. Huxley knew nothing of the impact that hostility to *Vestiges* had had on Owen and Darwin. By the tenth edition he regretted that "this once attractive and still notorious work of fiction" had flourished like a weed and needed the attention of gardener Thomas Henry (*SM* supplement:1).

It was not just the anonymous "Vestigiarian" who came under fire. The work of certain professional biologists cited by Chambers also met with Huxley's withering polemic, for example, "Agassiz, the great investigator of [the Paleozoic fishes], whose lively fancy has done at least as much harm to natural science as his genius has assisted its progress" (*SM* supplement: 10). This was two years before his own grievance with Richard Owen developed, so when Chambers referred to the latter's work Huxley hastened to find that it had been taken out of context rather than to criticize Owen. Huxley's attack on *Vestiges* even sees him drawing approvingly upon the opinion of the Cambridge professor of geology, Sedgwick: "Surely we have waded far enough through this lumber-room of second-hand scientific furniture, this attempt to build a tower of Babel heaven-high with half-burnt bricks; at any rate, far enough to convince the reader that however the Vestigiarian may wince under the remark, Professor Sedgwick was quite justified in asserting that he is 'not only unacquainted with the severe lessons of inductive knowledge, but possesses a mind apparently incapable of comprehending them' " (supplement:17).

Sedgwick, who had trained Darwin in field research, and may even have influenced his evolutionary thinking, was soon to turn his invective against his prize pupil, only to find himself dismissed as one of Cambridge's "old fogies." Darwin looked to up-and-coming young naturalists such as Huxley for support. It was slow in coming, and did not arrive with any degree of enthusiasm until Huxley had had the opportunity to read *Origin.*

Huxley's taste for blood found a new victim in Richard Owen, the director of the British Museum. The origin of the grievance is obscure. Leonard Huxley, the son and biographer of T. H. Huxley, believed that it began in 1857, when Owen gave a series of lectures at the School of Mines and styled himself as "Professor of Paleontology," which one would think a minor presumption considering Huxley's rejection of the lectureship in paleontology and his declared lack of interest in fossils three years earlier. Leonard also believed that his father already felt himself Owen's superior in his own areas of research and "had the choice of submitting to arbitrary dictation or securing himself from further aggressions by dealing a blow which would weaken the authority of the aggressor" (*Life* 1:153). Whatever the reasons, Huxley had found in Owen's writings ammunition in the form of careless reading of other authors, slapdash interpretations of his own observations, and the development of empty theories based on inadequate objective evidence.

In 1858 the Royal Society Croonian Lecture provided Huxley with the opportunity to confront Owen regarding his theory of the vertebrate skull. Following the lead of the nature philosopher Oken, Owen had theorized that the skull was derived from vertebrae. Huxley, on a careful examination of recent embryological evidence, demonstrated that the putative homologies did not exist and that "the spinal column and the skull start from the same primitive condition, . . . whence they immediately begin to diverge. . . . Thus it may be right to say, that there is a primitive identity of structure between the spinal or vertebral column and the skull; but it is no more true that the adult skull is a modified vertebral column, than it would be to affirm that the vertebral column is a true modified skull" (*SM* 1:585). A mere four years earlier Huxley had excused Geoffroy's outrageous attempt to derive a fish from a squid as a "true inspiration," although the method of working it out was wrong. In the same 1854 lecture, "On the Common Plan of Animal Forms," Huxley had given special credit to Owen's contribution to the interpretation of unity of plan in the vertebrates. So this was not simply a matter of disputing Owen's derivation of the skull; it was a U-turn from respectful praise to calculated attack. Moreover, it was to become a prolonged and one-sided vendetta.

In the same year Huxley published his only significant entomological study, on the greenfly *Aphis*. He explained that he had deviated from zooplankton and fossils to search for a type species of insect for general study. It was also a subject of opportunity, since the pest had infested his geranium plant and he had the live material in hand. Moreover, aphids presented the classic problem of parthenogenesis (reproduction in this case by the female insect without the cooperation of a male). A subject widely discussed by others, including Bonnet, Trembley, and Réaumur a century before, it had drawn the attention of Owen in 1843. Thus there was ample justification for Huxley to have taken it up, but it was also an opportunity to attack Owen once again for erroneous observations and for misreporting his source material. Several pages are devoted to the demolition of Owen's hypothesis of a "spermatic force" that might have persisted from a previous sexual encounter and contributed to the mysterious parthenogenesis.

Ten years later Huxley had even more reason to return to the fray, since by that time the pendulum had swung to evolutionism, and Owen had turned obstructionist, coaching Bishop Wilberforce for his attempt to "smash Darwinism" at the Oxford meetings of the British Association in 1860 and being suspected of writing a destructive anonymous review of *Origin* (Hull 214). At the Oxford meeting Owen himself gave an early presentation against Darwinian evolution, basing his arguments on the putative dissimilarities between human and gorilla brains. Huxley confined his response to a flat contradiction. The manner in which he deflated Wilberforce in public debate has entered into the biological mythos, to the degree that Hooker, who gave the scientific defense of Darwinism, is ignored (Løvtrup, "Hooker"). Huxley's contribution was brief and polemical. Interestingly, it was Robert Chambers, the anonymous author of *Vestiges,* who persuaded Huxley to risk "episcopal pounding" and remain at Oxford for Wilberforce's presentation (*Life* 1:193). He apparently bore no grudge for the attack on *Vestiges* that Huxley himself finally admitted had been "needless savagery" (1:180).

The next occasion for the renewal of Huxley's feud with Owen was his 1868 report on *Archaeopteryx.* The central issue was the possible relationship of reptiles and birds, specifically whether this feathered fossil was more reptile than bird. Huxley had discovered that the interpretation of the orientation of the only known, headless fossil was crucial. Owen had confused the right and left legs and read the ventral aspects of the vertebrae as dorsal, bringing him to determine it a reptile. Huxley concluded that "*Archaeopteryx* is more remote from the boundary-line between birds and reptiles than some living *Ratitae* [flightless birds] are." Moreover, if, as subsequently happened, "when the head of *Archaeopteryx* is discovered, its jaws contain teeth, it will not the more, to my mind, cease to be a bird, than turtles cease to be reptiles because they have beaks" (*SM* 3:345). I will not be so naive as to enquire why Huxley or any budding scholar should feel obliged constantly to defer to some elder statesman whose authority is as much a matter of time served and patronage as quality of scientific work. Owen's own scientific output was prodigious, but he was frequently in too much of a hurry to pay attention to the details that Huxley would fasten upon, and Huxley set his own standards so high that he was invulnerable to replies in kind. Huxley's attacks on Owen, though obsessive, were consistent with an antiauthoritarianism that lasted throughout his career. He remarked in 1866:

> The improver of natural knowledge absolutely refuses to acknowledge authority, as such. For him, scepticism is the highest of duties; blind faith the one unpardonable sin. And it cannot be otherwise, for every great advance in natural knowledge has involved the absolute rejection of authority, the cherishing of the keenest scepticism, the annihilation of the spirit of blind faith; and the most ardent votary of science holds his firmest convictions, not be-

> cause the men he most venerates hold them . . . but because . . . whenever he thinks fit to test them by appealing to experiment and to observation—Nature will confirm them. (*CE* 1:40–41)

In 1880, just after the twentieth anniversary of the publication of *Origin,* he even cautioned against such blind acceptance of Darwinism:

> History warns us, however, that it is the customary fate of new truths to begin as heresies and to end as superstitions; and, as matters now stand, it is hardly rash to anticipate that, in another twenty years, the new generation, educated under the influences of the present day, will be in danger of accepting the main doctrines of the "Origin of Species," with as little reflection, and it may be with as little justification, as so many of our contemporaries, twenty years ago, rejected them. (*CE* 2:229)

Would that Huxley's general advice on authoritarianism had consigned this worst bane of biology to mere history; would that his warning regarding blind faith in Darwinism had been effective by the twenty-year limit. Alas, time's arrow points to antiauthoritarians becoming authorities, with destructive powers beyond their own reckoning. St. George Jackson Mivart, one of Huxley's protégés, had begun to establish a respectable career as a primatologist. He confessed to Huxley that a conflict was developing between his Roman Catholic religious beliefs and his interpretations of primate evolution. Huxley the materialist showed him no mercy and set out to give his writings the Owen treatment. He even checked and criticized Mivart for the misuse of religious sources in the latter's 1871 *Genesis of Species,* which is the most comprehensive critique of Darwinism of the nineteenth century. "Only fancy my vindicating Catholic orthodoxy against the papishes themselves," he congratulated himself in a letter to Hooker (qtd. in Gruber 87). He allowed his "agnostic" zeal to cloud his tacit agreement with some of the biological opinions presented by Mivart's book, such as those pertaining to saltations and the genetic predispositions of parallel evolutionary lines.

Imagine Huxley's chagrin when Owen's grandson asked him to contribute an essay on Owen's position in the history of anatomical science to a posthumous biography of his grandfather. This recalled Cuvier's being invited to deliver the eulogy on Lamarck, a parallel that cannot have escaped Huxley when, in the essay, he was discussing Cuvier, Lamarck, and the nature philosophers: "Here we see the men, over whose minds the coming events of the world of biology cast their shadows, doing their best to spoil their case by stating it; while the man [Cuvier] who represented sound scientific method is doing his best to stay the inevitable progress of thought and bolster up antiquated traditions" (*SM* 4:670). Huxley managed to deliver a fairer assessment of Owen than Cuvier had of Lamarck, but was he ruefully thinking of himself in his contemplation of Cuvier? Not only had Huxley persistently rejected evolutionism until the late 1850s, there was another major paradigm that he had denounced: the cell theory.

The cell theory was largely formulated independently, in the late 1830s, by the botanist Schleiden and the zoologist Schwann at the University of Jena. It was a burning issue when Huxley was a medical student. The theory stated simply that all organisms were cellular in composition, that all cells came from other cells, and that consequently the cell was the basic functional-morphological unit of the organism. Any microscopist could have visual evidence of the existence of plant cells, since they are bounded by thick, cellulose cell walls. It was not obvious that animals were likewise constructed, since their cell membranes could not easily be seen through a microscope. A cell boundary could only be discerned in embryos and cartilage; this was enough for Schwann but not for Huxley. In an 1853 lecture to the Royal Institution he asserted that

> the existence of separate cells is purely imaginary, and . . . the possibility of breaking up the tissue of a plant into such bodies, depends simply upon the mode in which certain chemical and physical differences have arisen in the primarily homogenous matrix [the "periplast"]. . . .
>
> In all young animal tissues the structure is essentially the same, consisting of a homogeneous periplastic substance with imbedded Endoplasts. . . .
>
> *In all animal tissues the so-called nucleus* (Endoplast) *is the homologue of the primordial utricle (with nucleus and contents) . . . of the Plant, the other histological elements being invariably modifications of the periplastic substance.* (*SM* 1:217–18)

Even plant cell walls were epiphenomena of the local physico-chemical conditions rather than integral parts of the microanatomy. Huxley downplayed the significance of the nuclei because earlier biologists had attributed an organizing vital force to them—although he smuggled in "general determining Laws of the organism" as a substitute.

The cell theory is now so well established that it is only mentioned to undergraduates as a historical footnote. However, in the late nineteenth century, the concept of the cell as the fundamental anatomical and physiological building block of the organism was to be a necessary paradigm for research into embryology, genetics, and biochemistry. Huxley had egotistically rejected a productive working hypothesis in favor of an ill-conceived notion of his own. By insisting on the continuity of the protoplasm as the fundamental stuff of organisms, he fell, in 1868, into his most embarrassing error: *Bathybius.* Deep-water mud samples contained films of gelatinous matter with a variety of inclusions. Subsequently, it was discovered that this was simply an artifact of preservation of the samples, but Huxley no sooner sounded his usual caution than he disregarded it:

> it seems to me to be in the highest degree important to keep the questions of fact and the questions of interpretation well apart.
>
> I conceive that the granule-heaps and the transparent gelatinous matter in which they are embedded represent masses of protoplasm. . . . "Urschleim," which must, I think, be regarded as a new form of those simple animated beings which have recently been so well described by Haeckel in his "Monographie der Moneren." I propose to confer upon this new "Moner" the generic name of *Bathybius,* and to call it after the eminent Professor of Zoology in the University of Jena, *B. Haeckelii.* (*SM* 3:337)

He was indifferent to the absence of nuclei and regarded the coccolith skeletons that were caught up in the ooze as elaborations of "Urschleim." This he finally renounced in 1879 at the Liverpool British Association meetings (*Life* 2:5). Huxley's 1866 text book, *Lessons in Elementary Physiology,* accepted cells as anatomical units, but treated tissues and organs as the functional units. At the level of biochemical knowledge then available it did not really matter much. By the time of his outstanding 1879 monograph, *The Crayfish,* he had finally accepted the cell as a functional unit and enthusiastically elaborated on its archetypal role in homology.

Apologists such as Lankester, an editor of the *Scientific Memoirs* (*Life* 1:152), and Ashforth (22) have tried to infer that Huxley's critique of the cell theory was justified by subsequent developments, approvingly citing a metaphor from one of Huxley's reviews "that cells are not the cause but the result of organisation—in fact, are . . . to the tide of life what the line of shells and weeds on the sea-shore is to the tide of the living sea" (*Life*

1:152), even suggesting that informed opinion had swung once more toward Huxley's point of view. It is true that the pendulum eventually swung too far in favor of cellular reductionism, but not enough to justify the view of the cell as an insignificant epiphenomenon of the whole organism.

The theory of evolution was another major paradigm in the making to which Huxley, during the 1850s, showed not simply detachment but incisive criticism. I have already noted his legitimate complaints regarding progressionism in *Vestiges.* In his 1855 lecture on the hypothesis of the progressive development of animal life in time, he admitted that "the further we go back in time, the more different are the forms of life from those which now inhabit the globe," but within the major subkingdoms, then identified as the Protozoa, Radiata, Annulosa, Mollusca, and Vertebrata, the major classes appeared to have always existed, and the differences between the past and the present occurred at the level of the orders or lesser taxons (*SM* 1:300). Even extinct ordinal types were few in number, most of these reptiles. There was

> strong evidence in favour of the belief that a general uniformity has prevailed among the operations of Nature. . . .
>
> That there has been a succession of living forms in time, in fact, is admitted by all; but to the inquiry—What is the law of that succession? different answers are given. . . .
>
> According to the affirmative doctrine, commonly called the theory of Progressive Development, the history of life, as a whole, in the past, is analogous to the history of each individual life in the present; and as the law of progress of every living creature now is from a less perfect to a more perfect, from a less complex to a more complex state—so the law of progress of living nature in the past was of the same nature; and the earlier forms of life were less complex, more embryonic than the later. (*SM* 1:301)

This was essentially the evolutionary teaching of von Baer, whose work had influenced Huxley from his student days and for whom he had the greatest respect, fulsomely disclaiming in this case "the slightest intention of desiring to depreciate the brilliant services which its original propounder had rendered to science" (*SM* 1:304). But fossil types were not embryonic versions of modern types, so this kind of evidence for evolution was inadequate or unacceptable.

The same lecture went on to attack Cuvier's correlation principle, whereby the anatomy and habit of a fossil organism could be derived from only fossil fragments. Accepting that some anatomical extrapolations were valid, he was highly critical of the view that "physiological," that is, functional, conclusions could be drawn. In an 1894 retrospective assessment of the correlation principle as applied by Cuvier and Owen, Huxley inferred with some justification that their success was more the result of their extensive experience with fossil forms than the simple application of the principle. Cuvier, although not an evolutionist, was certainly a believer in the adaptive fit between an organism's form and way of life, which Huxley also attacked as a general principle, in structuralistic terms. This shotgun blast drew rebukes from various quarters, including one from Darwin, who was finding his evolutionary mettle once more.

In the late 1850s, while his research was dominated by paleontology, Huxley was aware of the need for a hypothesis that would encompass the mechanism of diversification of homologous structures and rationalize unity of plan. However, his first positive evaluation of evolutionism came only in 1859, after the Linnean Society had published the Darwin-Wallace papers in late 1858.

> If . . . we view "Persistent Types," in relation to that hypothesis which supposes the species of living beings living at any time to be the result of the gradual modification of pre-existing species—a hypothesis which, though unproven and sadly damaged by some of its supporters, is yet the only one to which physiology lends any countenance—their existence would seem to show, that the amount of modification which living beings have undergone during geological time is but very small in relation to the whole series of changes which they have suffered. In fact, palaeontology and physical geology are in perfect harmony, and coincide in indicating that all we know of the conditions in our world during geological time, is but the last term of a vast and, so far as our present knowledge reaches, unrecorded progression. (*SM* 2:92–93)

Conversations with Darwin on the gist of the idea were unconvincing, and the Darwin-Wallace papers were not quite enough, but *Origin of Species* provided the persuasive power of a critical mass of evidence adequate for a working hypothesis. Huxley's review of *Origin* appeared in the *Times* (London) on December 26, 1859. The penny had dropped. "My reflection, when I first made myself master of the central idea of the 'Origin' was, 'How extremely stupid not to have thought of that!' " (qtd. in F. Darwin 2:197). Instead of an inherent, universal, organic progress there was a natural mechanism that could select or reject, according to the prevailing conditions of life. Some lines could persist unchanged while others evolved. A lecture to the Royal Institution was in order and was delivered in February 1860 (*SM* 2:388–93). He illustrated Darwinism with examples of the artificial selection of pigeons and the natural selection of horses. But because there was no proof that selection produced mutually infertile species, it fell short of being a "satisfactory theory," although he had by this time no doubt "that one of these days either Mr Darwin's hypothesis, or some other, will get up and walk, and that vigorously; and so save us the trouble of any further discussion of this objection" (2:392). (This problem was finally to be solved when the role of the chromosomes in sexual reproduction was realized.)

Huxley was quick to fill in one of Darwin's blanks, the question of where humans came in, with "On the Zoological Relations of Man with the Lower Animals" in 1861. Noting that Linnaeus had perceptively included *Homo* with the apes, he concluded, again contradicting Owen, that the human genus was on the same level as the other "Quadrumana" rather than in an order by itself. He followed through in 1863 with one of his most enduring books, *Evidence as to Man's Place in Nature.* Still regarded as a useful historical introduction to anthropology, the material, some historical, some up-to-the-moment at the time of publication, is woven into a synthesis that would still be difficult to surpass. It deals with the natural history of the apes, human and ape embryology, and similarities between human and gorilla brains. A consideration of the then fragmentary remains of Neanderthals led him to conclude that these were not intermediate between apes and humans but a divergent evolution from a common stock. The extant apes were not the ancestors of humans, but even more diversified products of that common ancestry.

In the modern context of scientific disciplines it is perhaps stretching "nineteenth-century biology" to include anthropology and paleontology. Huxley himself, as he said in 1877 in "On the Study of Biology," regarded any human endeavor as within the realm of biology and had no qualms about spreading the mantle of his own biological research to cover ethnology and psychology. He was also aware that the Greek word *bios* pertained to human life and affairs (*CE* 3:269), while allowing that for practical reasons the study of plants and animals gave large enough scope for biology. "But I should like you to recollect that that is a sacrifice, and that you should not be surprised if it occasionally happens that you see a biologist trespassing upon questions of philosophy or politics; or meddling with human education; because, after all, that is a part of his kingdom which he has only voluntarily forsaken" (3:271). Huxley's paleontological studies cannot be omitted here since they were so important to his understanding of evolution. They helped to build the foundations of the subject and to provide evidence that he felt at the time of publication of *Origin* was sadly lacking. Initially, he seized on whatever paleontological material came to hand like an explorer in a dark cave, only the vendetta with Owen providing a thread of continuity. His subjects included: crustaceans, belemnites (extinct squid-like molluscs), fish, amphibians, crocodilians, pterodactyls and other extinct flying reptiles, plesiosaurs, dinosaurs, *Archaeopteryx,* and fossil whales. The greatest focus was on the labyrinthodont amphibians, crocodilians, and dinosaurs. He generalized that the latter were not only birdlike but close to the line of bird evolution. His 1869 study of the dinosaur *Megalosaurus,* another opportunity to take a dig at Owen, includes this entertaining table talk:

> I find that the tibia and the astragalus of a Dorking fowl remain readily separable at the time at which these birds are usually brought to table. The cnemial epiphysis is also easily detached at this time. If the tibia without that epiphysis and the astragalus were found in the fossil state, I do not know by what test they could be distinguished from the bones of a Dinosaurian. And if the whole hind quarters, from the ilium to the toes, of a half-hatched chicken could be suddenly enlarged, ossified and fossilized as they are, they would furnish us with the last step of the transition between Birds and Reptiles; for there would be nothing in their characters to prevent us from referring them to the *Dinosauria.* (*SM* 3:485–86)

One can imagine him waving the debris of his dinner at an impromptu family lecture.

His zoological research continued during this concentrated effort on paleontology and included the comparative anatomy of fish, mammals, and birds, observations on sensory organs, and preparative work on crayfish for his monograph. He also made a brief foray into microbiology, repeating some of Pasteur's experiments, and published on yeast, *Penicillium,* spontaneous generation, and the history of biology.

Huxley assessed the value of paleontology to Darwinism in his 1880 lecture to the Royal Institution, "On the Coming of Age of the *Origin of Species*," noting great progress in filling the gaps that had yawned twenty years before: the relationship between birds and reptiles, vertebrates and invertebrates, flowering and nonflowering plants. He could very well have claimed much responsibility for those advances. Those he had not touched with original study he had helped to popularize. At the level of firsthand research his greatest contribution to nineteenth-century biology in the early years of Darwinian evolutionism was in paleontology. This is not to downplay the importance of his pre-*Origin* work on the zooplankton, which alone would have assured his reputation as a great naturalist of the last century.

Teaching was not in Huxley's original professional program. The physician, turned naval surgeon, turned marine zoologist, initially had no responsibilities except to his patients and peers. But necessity, and his appointment to the Royal School of Mines, brought him to one of the most successful and enduring applications of his talents. Biology, a term that was coined independently by Lamarck and Treviranus in 1801, had not yet become a logos. Nowhere was it taught as such, although natural history had been sporadically taught and was regularly tacked onto medical programs, along with subdisciplines such as comparative anatomy and botany—for its significance to materia medica. Huxley therefore assumed not only the role of a teacher of the natural historical sciences but also that of a propagandist, urging the value of their dissemination, especially during early schooling. In 1854, even before he had a permanent teaching position, he noted the "utter ignorance as to the simplest laws of their own animal life, which prevails among even the most highly educated persons in this country" (qtd. in Mitchell 164). Even trained physicians could hardly explain the physiology of respiration, he added.

The steps he took to remedy the situation were numerous and the progress of his campaign was sustained for the rest of his career. In addition to his own professional teaching, he delivered regular, evening, research seminars at the Royal Institution, in language that was remarkably free of jargon. His respect for rhetoric was paramount; the printed transcripts of his lectures reveal a powerful orator. A peripheral interest in etymology also emerges, for example, in his discussion of the scientific and common names of Crustacea in *The Crayfish*. His pedagogical effectiveness comes out even more strongly in his series of lectures to working men at the Jermyn St. Museum and the Working Men's College. "I am sick of the dilettante middle class, and mean to try what I can do with these hard-handed fellows who live among facts" (*Life* 1:149). His lecture "On a Piece of Chalk" was first given as a working man's lecture at the Norwich meetings of the British Association in 1868. Opening with a lyrical passage on the European chalk beds, it comes down to earth with the production of a common chalk marker, such as might come out of the pocket of any carpenter or mason. Chemistry, mineralogy, marine biology, hydrography, paleontology, ethnology, evolution, and archaeology are all interwoven into the interdisciplinary presentation, in a way that is sadly lacking in modern curricula composed of splintered subdisciplines.

In 1870 Huxley joined the London School Board, which catalyzed him to practice what he preached and to put educational philosophy to work. In 1871 he inaugurated a practical course in biology for high school teachers at the South Kensington Natural History Museum, which he then adapted for use at the Royal College of Science, as the Royal School of Mines had been renamed when it moved to South Kensington. Teaching laboratories had not existed at the old institute. In 1880, feeling a need to remedy the dearth of suitable biological textbooks, Huxley joined the physicist Balfour Stewart and the chemist Roscoe in editing a series of science primers for Macmillan, writing an *Introductory Science Primer,* having already contributed a text on earth sciences, *Physiography.* By then he also had under his belt the text books *Elementary Lessons in Physiology,* aimed at the secondary school level, *An Introduction to the Classification of Animals, Manual of the Anatomy of Vertebrated Animals, A Course of Practical Instruction in Elementary Biology, Anatomy of Invertebrated Animals,* and *The Crayfish.* The last is a unique combination of original investigation and general discourse on the Crustacea, taking the British crayfish as the type species.

The expression "type" can be confusing, although Huxley tried to be explicit in his definitions: "In every group there is some average form; some form which occupies a sort of central place, around which the rest seem to arrange themselves . . . the nearest actual embodiment of the common plan. Such a form is commonly called the TYPE" (*SM* 3:134). In his earlier, pre-*Origin* writing, "type" had the more fundamental meaning of the original or archetypal form from which others had diversified. However, the usage came to signify a representative form that was convenient for practical teaching purposes. Detailed examination of the type specimen could be extrapolated to acquire a general knowledge of the group. Another teaching method that was already in vogue was a cumbersome examination of many predissected specimens; but Huxley demanded hands-on training as well as a fundamental understanding of the phylogenetic relationships.

In his teaching laboratories at the Royal College of Science, the type specimens used were *Amoeba* and *Vorticella* for the Protista, *Hydra* for the Cnidaria, a starfish for the echinoderms, a snail, a squid and a freshwater mussel for the molluscs, a lobster and a crayfish for the crustaceans, a beetle or cockroach (not the greenfly!) for the insects, skate and cod for the fish, and frog, tortoise, pigeon and rabbit for the other vertebrate classes. In some cases the practicality of keeping specimens alive in the laboratory, and the economy and ease of procurement were just as important as settling on a truly representative form. The freshwater mussel is actually a rather peculiar type, but marine animals are harder to keep alive for long periods. Needless to say, the lobster has long since vanished from the type species menu.

While a full discussion of the implementation of Huxley's ideas on education is beyond my present scope, it is necessary to point out that he not only constructed the foundations of secondary schooling in biology, he also was responsible for some reforms of university curricula, including medical training. His influence on the teaching of comparative anatomy and physiology is still felt, although the relative importance of the former has dwindled as that of molecular biology has expanded.

This raises the question of what Huxley would have thought of modern biology as it heads into the twenty-first century. There exist in his writings some observations that have cautionary or prophetic value, although, like all futurists, he was much better at analyzing the present than at accurate prediction. In the pedagogical arena he would have had mixed feelings about the emphasis on molecular biology relative to organismic studies. His own emphasis was always holistic, having been formed by the dual influences of nature philosophy, qua unity of plan, and the study of physiology, which ever emphasizes the need to put organisms back together again once their bits have been analyzed. Nevertheless, he would have approved of what this new discipline can do to confirm and add to evolutionism and materialism in general.

In the study of type specimens in the nineteenth century, it was possible for students to acquire a fundamental, firsthand knowledge of the most up-to-date comparative anatomy with the simplest of instruments, supplemented by microscopic studies. It is not possible for students to obtain the molecular equivalent of such original information at firsthand, without the use of equipment whose expense is completely out of proportion to that required in the past. The result is the emergence of small elitist groups of the kind that Huxley scorned, with the rest having to content themselves with scratching the experimental surface and with book-learning.

Another recent development is the intrusion of political correctness into physiology laboratories. Not only Huxley, but all the great, historical teachers of physiology, taught that progress could only be made through practical experience. Nowadays, even amongst academics, there is a growing feeling that the use of animals for teaching purposes is unethical, and what remain of comparative anatomy and physiology laboratories are under constant pressure to find computer programs or biochemical systems to substitute for organisms. During Huxley's career vivisection was a hot issue that took much of his attention, since he was falsely accused by Lord Shaftesbury, during the debate on the Vivisection Bill in 1876, of having recommended vivisection for high school laboratories (*Life* 1:460). Huxley was uncomfortable both with the vivisection law and with vivisection:

> I have always felt it my duty to defend those physiologists who, like Brown-Séquard, by making experiments on living animals, have added immensely not only to scientific physiology, but to the means of alleviating human suffering, against the often ignorant and sometimes malicious clamour which has been raised against them.
>
> But personally, indeed I may say constitutionally, the performance of experiments upon living and conscious animals is extremely disagreeable to me, and I have never followed any line of investigation in which such experiments are required. (*Life* 1:463)

He went on to emphasize that he required his laboratory demonstrators to use only animals that had been anaesthetized or had had their brains pithed, such as in the classic demonstration of sciatic nerve function in frogs. However, despite the recommendations of the Commission on Vivisection, of which Huxley was a member, the law came down against such demonstrations, approving only experiments that involved original research, leaving Huxley to lament that the mild discomfort given to a live frog in demonstrating the circulation of the blood in its foot webs was now liable to the same legal punishment as the amphibian's willful torture, and that England would have to import her physiology from Germany and France as she imported hock and claret (*Life* 2:467, 474). Regarding the progress of science in 1887, he optimistically said, "Unless the fanaticism of philozoic sentiment overpowers the voice of philanthropy, and the love of dogs and cats supersedes that of one's neighbour, the progress of experimental physiology and pathology will, indubitably, in course of time, place medicine and hygiene upon a rational basis" (*CE* 1:123). What might he have made of current philozoic firebombs in laboratories and death threats uttered against physiologists?

In a lighter vein, in "Oysters and the Oyster Question," Huxley remarked:

> For the comfort of those lovers of oysters who entertain strong views upon the vivisection question, and who may be perturbed by the reflection that they not only vivisect their favourites, but swallow them alive, it may be well to remark that the nervous system of the oysters is more poorly developed than that of any of their allies among lamellibranchiate molluscs. . . . Of course I do not suggest that this consideration can have any value in the eyes of those who maintain that the infliction of suffering on animals for the benefit of mankind is absolutely unjustifiable. Rigorists of this class are bound to denounce the brutal deglutination of oysters, no less than the cruel crushing of fleas, and the infamous poisoning and drowning of flies and cockroaches. (*SM* 4:581)

Huxley had no long-term vision regarding the overexploitation of natural resources, although overfishing of herrings and oysters was already being debated. He had previously, with Sir James Caird and Mr. Shaw-Lefèvre, formed a commission to inquire into the sea fisheries of the United Kingdom. It made its report in 1866. As Inspector of Fisheries, succeeding Buckland to the post in 1881, Huxley reported on the general biology of the herring and the oyster. He concluded, "the best thing for Governments to do in relation to the herring fisheries, is to let them alone. . . . Let people fish how they like, and when they like. . . . It will be time to meddle, when any satisfactory evidence that mischief is being done is produced" (*SM* 4:492). This was based on his calculation of the enormous potential productivity of the herring and the relatively slight impact of the fishery relative to natural predation. He even excused heavy ground fishing on the spawning beds, since egg-eating fish were being removed. Unfortunately, inadequate calculations and authoritarian opinions of this kind have dogged fisheries to the present time, and evidence of mischief, far less effective governmental meddling, has barely preceded the almost universal collapse of fisheries.

On the oyster question Huxley took a similarly libertarian view. There was no point in regulating the native stocks in public beds since he knew at that time of no bed that had been permanently wiped out by overfishing. But the combination of population expansion, domestic and industrial pollution, and overfishing have removed the majority of productive public beds from the present scene. Huxley did correctly see the future in the enhancement of private culture and the development of reliable oyster seed production; globally these are responsible for most commercial oyster production at the present.

On the question of evolutionary theory some issues raised by Huxley have reemerged in the last twenty years. However, later in his career as Darwin's bulldog, his views became polarized in favor of gradualistic Darwinism. His opinion of catastrophism is a case in point. As late as 1869, in his presidential address to the Geological Society, he noted that while evolutionism was destined to "swallow up" catastrophism, and uniformitarianism, a new synthesis was possible. Catastrophes, different and more powerful forces than "those we now know," were a necessary part of evolution, although uniformitarianism "has compelled us to exhaust known causes before flying to the unknown." For the synthesis he used a clock analogy: the striking of the clock was catastrophic and the ticking gradual. "Nevertheless, all these irregular and apparently lawless catastrophes would be the result of an absolutely uniformitarian action; and we might have two schools of clock-theorists, one studying the hammer and the other the pendulum" (*SM* 4:418–19).

Huxley also had difficulty with Darwinian interpretations of plant and animal distribution, given the modern map of the terrestrial world. Although he might have been aware of the speculations of Francis Bacon and von Humboldt regarding the curious topographical match of the continents of Africa and South America, he does not seem to have known of the opinions of Antonio Snider, who in 1858 proposed, on the basis of fossil evidence, that the continents had once constituted a single land mass. Addressing the Geological Society in 1870, still as its president, and not one to mince words, even as a relative outsider, he asserted that a vast alteration of the physical geography of the globe was necessary to understand biogeography and proposed massive upheavals and depressions of the ocean beds to make it comprehensible (*SM* 3:539). By 1880, however, in "The Coming of Age of the 'Origin of Species,'" he misrepresented Cuvier's catastrophism as "wholesale creations and extinctions of living beings" and condemned "the wild speculations of the catastrophists, to which we all listened with respect a quarter century ago" (*CE* 2:231).

Nowadays the idea that frequent catastrophes have wiped out large portions of the plant and animal kingdoms in a relatively short time is almost conventional wisdom. These were by no means "the result of an absolutely uniformitarian action," but were asteroid impacts. The theory of continental drift, based on the concept of plate tectonics, while resisted with all the vigor a dying paradigm could muster until almost twenty years ago, is now also an integral part of biological teaching. Darwin could literally stand on the side of an erupting volcano and ignore it as a geobiological force, soothing himself with Lyell's gradualism. But one might have expected more from Huxley, who never quite abandoned saltatory evolution, that is, evolutionary leaps. On the other hand, he did stand fast on the close relationship between dinosaurs and birds. Although dismissed by subsequent generations of paleontologists, it became a cause célèbre in 1975, when Robert Bakker reopened the controversy with his contention that not only were the dinosaurs the direct progenitors of the birds, they were also warm-blooded. This continues to be a matter of heated debate.

Before Huxley embraced Darwinian evolution, he had accepted that archetypes had somehow to diversify, and so he suggested an epigenetic saltation that would produce new forms by an embryological change in the archetype. This was not something he hotly pursued, nor was it a new concept. Geoffroy had suggested such radical changes, and Cope, from the 1860s on, was speculating on the effects of relative acceleration and retardation in embryos as an element of putative saltation (see Cope). Even the despised "Vestigiarian" was somewhat of an evolutionary epigenesist (Løvtrup, "Robert Chambers"). On saltation per se Huxley had remarked, "we have always thought that Mr. Darwin has unnecessarily hampered himself by adhering so strictly to his favourite 'Natura non facit saltum'. We greatly suspect that she does make considerable jumps in the way of variation now and then, and these saltations give rise to some of the gaps which appear to exist in the series of known forms" (*CE* 2:97).

At the present time the possibility of epigenetic saltations is being pressed by numerous biologists from a variety of areas of research (see Reid, *Evolutionary Theory* ch. 13), and from orthodox neo-Darwinism itself has arisen the suggestion that the gait of evolution may not be uniformly slow but, *qua punctuated equilibria,* "herky jerky," as Eldredge and Gould suggest.

Another recent school of thought that criticizes the "adaptationism" of neo-Darwinism must also find comfort in Huxley's earlier thinking (Gould and Lewontin). In 1856 he challenged:

> Regard a case of birds, or of butterflies, or examine the shell of an echinus, or a group of foramenifera, sifted out of the first handful of sea sand. Is it to be supposed for a moment that the beauty of outline and colour of the first, the geometrical regularity of the second, or the extreme variety and elegance of the third, are any *good* to the animals? that they perform any of the actions of their lives more easily and better for being bright and graceful, rather than if they were dull or plain? So, to go deeper, is it conceivable that the harmonious variation of a common plan which we find everywhere in nature serves any utilitarian purpose? that the innumerable varieties of antelopes, of frogs, of clupeoid fishes, of beetles and bivalve mollusks, of polyzoa, of actinozoa, and hydrozoa are adaptations to as many different kinds of life, and consequently varying physiological necessities? Such a supposition with regard to the three last, would be absurd. . . . Who has ever dreamed of finding an utilitarian purpose in the forms and colours of flowers, in the sculpture of pollen-grains, in the varied figures of the frond of ferns? What "purpose" is served by the strange numerical relations the parts of plants, the threes and fives of the monocotyledons and dicotyledons? (*SM* 1:311)

These rhetorical questions raised a number of issues that are currently being debated (see Reid, *Evolutionary Theory* ch. 10). It is unfortunate that he did not pursue them, after he became a convinced evolutionist. Here he was criticizing Cuvier's correlation principle, which was founded on a non-evolutionary adaptationism—every aspect of the anatomy and actions of an organism had to be correlated with its habit and habitat.

Adaptationism has provided the bread and butter for generations of neo-Darwinists. Example after example of what the early commentaries set forth as being impossible to explain by natural selective advantage has been scrutinized and a plausible hypothesis proffered, although the exhaustion generated by the search for just one hypothesis usually ends the quest just at that point. Darwin thought that these questions could be answered by another kind of correlation principle: if the manifestation of a nonadvantageous trait were genetically correlated with an advantageous one it would continue to appear in a breeding population (See Darwin, *Origin* 108–10). This agrees with one of the saner versions of the modern "selfish gene" concept. Darwin's logical explanation and legitimate examples do not, however, explain the "harmonious variation of a common plan." This is the realm of modern biological structuralism and evolutionary epigenetics, in which questions of adaptation to environment of particular traits are no longer central (see Webster and Goodwin; Hughes and Lambert). More than matters of ontogeny and phylogeny, they are virtually matters of ontology, and as such they require the attention of the kind of metaphysician personified by Huxley himself.

In T. H. Huxley we find a man who through his personal journey shaped nineteenth-century biology, ushering it from the pre-Darwinist era to evolutionism. It is remarkable how, before his acceptance of evolutionary theory, the concept of unity of plan, not so much a working hypothesis as an intuitive sense of relatedness, gave his zoological research the drive that led to the establishment of his career. But ultimately his epistemology was too mechanical, suffering from a self-imposed curbing, if not failure, of creative imagination, and he refrained from the kind of intuitive leaps that distinguished other great theoreticians. Darwin may have considered himself Huxley's intellectual inferior, but he did allow himself to suspend disbelief and speculate widely, despite the lip-service he paid to Baconian methodology. Huxley, affecting to despise Baconian epistemology as "humbug," suffered from the very limitations of its mechanical inductive method. As a pure scientist he was sometimes an inspired builder, but not an architect, and much of his success can be put down to sheer industry and output. Some of the research was mundane—adequate for the time, but opportunistic and goalless. Some was ill-inspired by a meaningless vendetta. But much was central to the development of key areas of invertebrate zoology, comparative anatomy, and paleontology, and contributed to grander syntheses than he had the literary and oratorical talents to propound to audiences much wider than only his peers and students. This ability, together with his design and construction of a formal biological education, has made his influence felt throughout the twentieth century, and as a scientist, far from being a "mere historical figure," he raised issues that will be relevant in the twenty-first.

REFERENCES

Ashforth, Albert. *Thomas Henry Huxley.* N.Y.: Twayne, 1969.

Bakker, Robert. "Dinosaur Renaissance." *Scientific American* 232 (1975): 58–78.

Chambers, Robert. *Vestiges of Creation in Natural History.* 1844. N.Y.: Humanities Press, 1969.

Cope, Edward, D. *The Origin of the Fittest.* Chicago: Open Court, 1887.

Cuvier, George. *Recherches sur les ossemens fossiles des quadrupèdes.* 4 vols. Paris: Déterville, 1812.

Darwin, Charles. *The Autobiography of Charles Darwin and Selected Letters.* Ed. Francis Darwin. 3 vols. 1892. N.Y.: Dover, 1958.

———. *On the Origin of Species.* 1872. 6th ed. London: Watts, 1929.

Darwin, Francis. *The Life and Letters of Charles Darwin.* London: Murray, 1887.

Dawson, Virginia P. *Nature's Enigma: The Problem of the Polyp in the Letters of Bonnet, Trembley, and Réaumur.* Philadelphia: American Philosophical Society, 1987.

Eldredge, Niles, and Stephen J. Gould. "Punctuated Equilibria: An Alternative to Phyletic Gradualism." *Models in Palaeobiology.* Ed. T. J. M. Schopf. San Francisco: Freeman Cooper, 1972. 82–115.

Gould, Stephen J., and Richard Lewontin. "The Spandrels of San Marco and the Panglossian Paradigm: A Critique of the Adaptionist Programme." *Proceedings of the Royal Society,* B. Biological Science 205 (1979): 581–90.

Gruber, James W. *A Conscience in Conflict.* N.Y.: Columbia UP, 1960.

Hughes, Arthur J., and David M. Lambert. "Functionalism, Structuralism, and 'Ways of Seeing.'" *Journal of Theoretical Biology* 111 (1984): 787–800.

Hull, David L. *Darwin and His Critics.* Cambridge: Harvard UP, 1973.

Huxley, Aldous. *The Olive Tree and Other Essays.* London: Chatto & Windus, 1936.

Huxley, Leonard. *Life and Letters of Thomas H. Huxley*. 2 vols. N.Y.: Appleton, 1901.

Huxley, Thomas Henry. *Collected Essays*. 9 vols. 1894. N.Y.: Greenwood Press, 1968.

———. *The Crayfish: An Introduction to the Study of Zoology*. London: Kegan Paul, Trench, Trubner, 1879.

———. "The Darwinian Hypothesis." Review of *The Origin of Species*. *Times* (London), December 26, 1859.

———. *An Introduction to the Classification of Animals*. London: J. Churchill & Sons, 1869.

———. *Introductory Science Primer*. London: Macmillan, 1880.

———. *Lay Sermons, Addresses, and Reviews*. London: Macmillan, 1870.

———. *Lessons in Elementary Physiology*. London: Clay & Sons, 1866.

———. *A Manual of the Anatomy of Invertebrated Animals*. London: J. & A. Churchill, 1877.

———. *A Manual of the Anatomy of Vertebrated Animals*. London: J. Churchill & Sons, 1871.

———. *On the Oceanic Hydrozoa*. London: Ray Society, 1859.

———. *Physiography*. London: Macmillan, 1877.

———. "The Reception of the 'Origin of Species.'" *Life and Letters of Charles Darwin*. Vol. 2. Ed. F. Darwin. London: Murray, 1887. 179–204.

———. *The Scientific Memoirs of Thomas Henry Huxley*. Ed. Michael Foster and E. Ray Lankester. 4 vols. plus supplement. London: Macmillan, 1898–1903.

———. *T. H. Huxley's Diary of the Voyage of the HMS* Rattlesnake. Ed. Julian Huxley. London: Chatto & Windus, 1935.

Huxley, Thomas Henry, and H. N. Martin. *A Course of Practical Instruction in Elementary Biology*. London: Howes & Scott, 1875.

Lankester, Edward Ray. "Mollusca." *Encyclopaedia Britannica*. 9th ed. 16:632–95. London: British Dictionary, 1883.

Løvtrup, Soren. "Robert Chambers, Charles Darwin, and the Mechanism of Evolution." *Bulletin Biologique de la France et de la Belgique* 112 (1978): 113–26.

———. "Hooker, Huxley, and Bishop Wilberforce." *Rivista di Biologia* 77 (1984): 417–27.

Mitchell, P. Chalmers. *Thomas Henry Huxley*. London: G. P. Putnam's Sons, 1901.

Mivart, St. George J. *The Genesis of Species*. London: Macmillan, 1871.

Owen, Richard. *Lectures on the Comparative Anatomy and Physiology of the Invertebrate Animals*. Hunterian Lectures. Vol. 1. London, 1843.

———. "Report on the Archetype and Homologies of the Vertebrate Skeleton." *Report of the British Association for the Advancement of Science* (1847): 169–340.

Owen, Rev. Richard S. *The Life of Richard Owen.* London: Murray, 1894.

Reid, Robert G. B. *Evolutionary Theory: The Unfinished Synthesis.* Ithaca: Cornell UP, 1985.

———. "Pedal Feeding and the Evolution of Bivalve Mollusks." *McGraw Hill Yearbook of Science and Technology.* Ed. Sylvia Parker. N.Y.: McGraw-Hill, 1994. 56–58.

Snider, Antonio. *La Création et ses mystères dévoilés.* 1858. Cited by P. M. Hurley. "The Confirmation of Continental Drift." *Continents Adrift.* San Francisco: Freeman, 1972. 57–67.

Webster, G., and Brian C. Goodwin. "The Origin of Species: A Structuralist Approach." *Journal of the Society of Biological Structure* 5 (1982): 15–47.

Yonge, Charles Maurice. "The Pallial Organs in the Aspidobranch Gastropoda and Their Evolution throughout the Mollusca." *Philosophical Transactions of the Royal Society* B.232 (1947): 443–518.

Genius in Public and Private

PAUL WHITE

> I have nearly traversed half the globe and have found only error and discord till I came to your cottage, where truth and happiness reside.—Bernardin de St. Pierre, *The Indian Cottage*

> Thus is true moral genius . . . "ever a secret to itself."—Carlyle, "Characteristics"

TWO FIGURES HAVE PREDOMINATED IN CAREER HISTORIES OF Victorian men of science: the heroic genius and, more recently, the rising professional. In such narratives—that of innate brilliance overcoming social obstacles to reach its natural height, and that of a division of intellectual labor proceeding from an expanding middle class—women have had little place except as supporters of men or preys to exclusion. Accounts of Thomas Huxley's life, including the recent and lively biography by Adrian Desmond, have largely reproduced these romantic and bourgeois sagas. Henrietta Heathorn, his wife of thirty years, has appeared as his constant "help and stay," or as the occasional, if spry, victim of his irascible nature and emotional despotism. These portraits owe much to mainstream Victorian studies, which represent the period as an age of separate public and private spheres, and an era in which gender roles were likewise rigidly distinguished.[1]

Rather than interpret the lives of Huxley and Heathorn according to these categories, I use their histories to critique these categories. How did Huxley conceive of his masculinity? Why were representations of women, of the feminine, and of domesticity essential to his self-understanding and his work? How did Heathorn, in fashioning herself, shape scientific identity? And how did she contend with a man whose cultural role was conflated with hers, who was often represented (and denigrated) as feminine, and whose economic position made the existence of a home and thus of her own moral purpose uncertain? To address these questions is to show how the contours of public and private life, and of gender itself, were shaped by continuous negotiations between the sexes. It is to suggest ways of revising accounts in which women's involvement in the sciences—as practitioners, assistants, popularizers, and audiences—have been viewed either as proto-professional or as passive and subordinate, while the gender roles of men have appeared fixed and undifferentiated.[2]

The letters and diaries of Heathorn and Huxley are particularly valuable sources for examining the intricate relationship of science and domesticity during the mid-Victorian period. In this essay I draw most heavily on the documents they composed during their prolonged engagement from 1847 to 1855. Over this eight-year period, in which she resided in Australia and he on board a surveying vessel and then in London, they exchanged several hundred lengthy letters and kept separate journals for each other to read during the long intervals of separation. While he struggled for social acceptance and a living among scientific elites, and she strove to manage households that were not her own, their correspondence played a fundamental role not only in sustaining their relationship, but in constructing their own identities. The form that this self-fashioning took—reading and writing—was more than a product of the peculiar circumstances of their separation. Both Huxley and Heathorn drew considerably on literary figures—in some cases the same figures—for models of character. Through writing, their lives were crafted according to these models. Literary activity was perhaps the chief means by which she, an aspiring gentlewoman and poet, and he, an aspiring man of science, received and reworked the social norms of their time and station.

The literature on which Huxley and Heathorn drew was hardly a stable code of behavior, however. Because literature during the period was a particularly controversial locus for transgressing (or reinforcing) the boundaries between public and private and between masculine and feminine, the documents they produced were a volatile medium for establishing the role that each would have in the other's life. An activity in which women displayed their mental capacities respectfully and built a sympathetic community in private, writing was also a central forum through which women, who were increasingly prolific not just as novelists and poets, but as essayists, biographers, historians, and reviewers, might advance subversive claims to moral and intellectual power.[3] Likewise, written work was still the finished form of science, the preeminent manifestation of its practitioners' genius, and the most important means, therefore, of constructing its practitioners' cultural authority. It is no coincidence that the explicitly gendered referent "man of science" became current at the same time that authors like Carlyle invented the heroic "man of letters." By constructing such personas, learned men tried—with only mixed success—to exclude women from debates about the meaning of high culture and to define and defend their polite activities as masculine and worthy of respect at a time when the gender identity of such activities was highly ambiguous.

My essay thus explores how these cultural constructs—the Victorian man of science and the Victorian wife, scientific work, and home—emerged together and constituted each other. I argue that scientific identity was shaped considerably by ideals of domesticity and feminine virtue, as well as by other traditionally masculine models of work. The values Huxley believed to inhere in the domestic setting and the values of his calling were the same. The virtues of scientific genius—its perfect transparency, its honesty and integrity, its purity of purpose—these were the virtues of the happy home. Science and domesticity were aligned in this way because each was defined in opposition to the same fallen world of work ruled by self-interest. This alignment was highly unstable, however, in part because Huxley was compelled to assume a position in the male-dominated world of professions and in part because the associations of scientific work with domesticity, and of scientific men with feminine virtue, were not easily managed on his own terms. The meaning of science could only be established through engagement with others whose interests and ideals differed drastically from his, most crucially his fiancée. This process of gendering was not always reciprocal and mutually fulfilling. But amid a great deal of conflict and misunderstanding, what they were able to give each other was surprising. Henrietta transformed Thomas's science into work in the world. Thomas transformed Henrietta's home into a fairyland of science.

Imperial and Sentimental

When Thomas Huxley met Henrietta Heathorn, his official title was assistant surgeon, a low-ranking officer in the Royal Navy. Neither medical practice nor naval service were ideal careers to him, but there were few options open to a young man of lower-middle-class family who was drawn to comparative anatomy. His formal education had ceased at age ten when the school where his father taught declined and the family moved from Ealing to Coventry in difficult straits. At thirteen he began a sequence of apprenticeships with his two brothers-in-law, both medics, that was to continue for six years. In his spare time, he read in the library of the Royal College of Surgeons and attended lectures at Sydenham College and later at Charing Cross Hospital where he and his elder brother obtained scholarships. By the age of twenty, he had completed his first examinations for a University of London degree but was considerably in debt and too young to be certified by the Royal College of Surgeons. Needing a livelihood, he wrote to Sir William Burnett, physician-general of the navy, in July of 1846: "I take the liberty of addressing myself directly to you as the Head of the Department to inquire whether (supposing I can produce the requisite qualifications) any application from myself simply will have any chance of success. I hope that you will consider the importance of your judgment to me as sufficient cause for what would otherwise appear an uncalled for intrusion on my part" (HP 11:194). He submitted a summary of his medical training and his prizes in botany, chemistry, anatomy, and physiology, offering to read French and German, to draw and, if need be, to dissect on the spot whatever Sir William might suggest.

Poised between humility and impertinence, Huxley displayed in this letter the kind of mixed manners men of science would have to acquire to enter circles where patronage operated in tandem with meritocracy, but where the criteria of merit were not firmly established. Like the other young or aspiring men of science he would later befriend—William Carpenter, John Tyndall, Edward Forbes—Huxley was in need of gainful employment and posed himself as hard working, self-made, and averse to the entrenched privileges of aristocracy. Like them, however, he also saw himself as the bearer of a high-cultural tradition and as the possessor of innate and lofty powers that raised him above other commercial and professional men. With few scientific positions available in universities at midcentury, he would derive much of his financial support and popular prestige by associating himself with trade, industry, and material progress—associations that he could seldom make with good conscience. What he wished was to be honored as an aristocrat of the mind while being rewarded as a useful producer.

Carlyle's remarks on genius, copied in the diary of a young Huxley in 1842, captured this persona nicely: an intellectual force, hardly embodied, sweeping the individual along as it carried the age, inscrutable even to its visionary self. Conceived by eighteenth-century writers as an innate and effortless capacity, genius lingered on in the Victorian period, coming to describe a variety of self-made man fraught with contradictions: the genius-at-work whose peculiar labor was original rather than repetitious or mechanical, moral rather than base. Much of the effort that men of science expended, effort that exhausted them and in some cases ruined them, went toward securing a place for themselves to work that was isolated from the world and its wicked ways, from the bustle of commercial and political interests, from the routine of manual and middling pursuits (qtd. in HP 31:169).[4]

Thus Huxley, while struggling for recognition, comported himself as a prodigy. Still needing friends in high places, he assumed an air of self-reliance. With Burnett's support, he obtained a naval appointment and, by persistent activity in the laboratory at Haslar Museum where he was first posted, was able to convince others that he was not chiefly a ship's surgeon. Through the influence of Sir John Richardson, inspector of hospitals and one-time Arctic explorer, he got leave to depart on a long voyage to the South Seas on HMS *Rattlesnake*. The ship's captain, Owen Stanley, provided him with introductions to leading metropolitan naturalists, Forbes, John Gray, and Richard Owen: "From all these men," he wrote to his sister, "much is to be learnt which becomes peculiarly my own, and can of course only be used and applied by me" (*Life* 1:29).

At sea, Huxley hoped to enter the scientific community of his dreams. Just before sailing he wrote, "Surveying ships are totally different from the ordinary run of men-of-war. . . . From the men who are appointed having more or less scientific turns, they have more respect for one another than that given by mere position in the service, and hence that position is less taken advantage of. They are brought more into contact, and hence . . . almost proverbially stick by one another. . . . I may read, draw, or microscopise at pleasure, and as to books, I have *carte blanche* from the Captain to take as many as I please" (*Life* 1:28–29). It seemed to Huxley that in a place where men respected scientific learning, they would also respect each other; and they would do so not for considerations of rank, but for matters of character and accomplishment. Here among men he thought were his scientific fellows, he looked to find the integrity, loyalty, and intimacy not to be had in the wider world ruled by self-interest.

The surveying vessel also gave him the opportunity to mingle with other sorts of men and to graft his budding scientific identity onto more proven manly stock. The *Rattlesnake*'s official mission was to survey the Torres straits to make safe passage for British merchants and emigrant steamers, and to assess sites in North Australia for new colonies. The genius that explored dark interiors of marine life could now venture into uncharted waters and unseen isles. Tasks like reading and microscope gazing, which might otherwise have seemed sedentary and pedantic, could become part of the heroic saga of empire building. Local flora and fauna were fields to be conquered and, if he was to earn his reputation as a naturalist, made his own. While the ship was anchored off the coasts of New Guinea and northern Australia, Huxley frequently accompanied the crew ashore and spent much time observing the local people. Only captain's orders prevented him from traveling overland with another expedition led by Edmund Kennedy who, along with most of his party, was killed in an encounter with natives. In his diary, Huxley explicitly linked scientific discovery and the sacrifices it entailed with colonial enterprise. "There lies before us a vast continent—shut out from intercourse with the civilized world . . . and rich . . . in things rare and strange. The wild and noble rivers open wide their mouths inviting us to enter. All that is required is coolness, judgment, perseverance, to reap a rich harvest of knowledge and perhaps of more material profit. . . . A little risk is also needful" (*Diary* 166).

To sustain the manliness of his conquering self, Huxley had to manufacture complementary representations of his objects of discovery. On the coasts of New Guinea, he sketched a fertile landscape, rich with resources unused by supine natives who were, in turn, "diminutive," "perfectly modest," and of "primitive simplicity" (*Diary* 218–28). The colonial land and its local inhabitants were opulent, civil in character if not in accomplishment—indeed effeminate. This discourse, now familiar to us as orientalism, by which colonial others—their bodies, language, religion, ways of seeing and knowing—were made feminine, full, and ripe was at midcentury just acquiring its scientific ethos. It enabled a ship's surgeon to play the part of a regal imperialist, to enter the primitive domains of imagination, feeling, and tradition unschooled by reason, to liberate the innocent and ignorant from the jealous rule of priestly missionaries, of superstition and myth, and to establish an enlightened paternity—the tutelage of science.[5]

This emerging discourse was not a firm resource, however, for he soon began to question the integrity of the empire's manly mission. In part, this was a reaction to the ship's crew who, he quickly discovered, were not the scientific fellows he had imagined. The same men who were responsible for "the dreary business of charting" were also responsible for what he came to see as the more brutal business of discovery. The same men who threw his laborious dissections overboard as waste and made fun of his books, were the men who killed out of fear and who plotted the eventual destruction of peaceful communities. He described how the crew would respond if the local peoples behaved with the duplicity of European explorers and their historians: "Suppose we landed at a village, the natives received us with every appearance of friendship, took one or two of our number up to their houses, and then as we were shoving off treated us to a shower of spears? What should we say? When should we have held forth sufficiently about their treachery? And yet some day or other a big book will appear with a statement to the effect that 'every effort was made to conciliate the natives and treat them kindly' " (*Diary* 184).

In private, Huxley parried the emasculating ridicule of his mates by mocking their gallantry. How courageous was the "brave Captain" who would have all native men bound on the beach to satisfy him of the security of "his little body," or who might be scared off by "old women, if at all shrewishly inclined?" By renouncing violence, Huxley could be more manly than men of war who refused to sacrifice themselves, even while they sacrificed others, for progress: "What is the advancement of knowledge and the opening of wide fields for future commerce to my comfort, my precious life, I should like to know?" (*Diary* 166, 183–84, 202–3). This was a peculiar form of manliness, that abjured force, that had sympathy for intimate communities of the weak and the simplicity, honesty, and peacefulness that made them so—a manliness that preferred studying natives to putting them to work. Huxley found resources here in another literary tradition, sentimental fiction.

Among the books that his coarse companions mocked were romantic tales about cultured men of feeling whose mission was to beautify the world with truth. From Goethe's *Werther,* in which a young genius was more appreciated by women and children than by cool and composed men of affairs, Huxley drew the moral that men who spent their energies in the pursuit of wealth and power sacrificed their "greater soul" and that the public who wept for men of science as departed from the right path were a "judicature of fools" (HH 58). Whenever "rebellious or discontented," he turned to Carlyle's lives of Heine or Jean Paul. Here he could read how the son of a Chemnitz weaver braved poverty, contempt, and third-rate schooling to become an academic "son of Genius," or how the mind of a writer—piercing and grasping, yet meek—might win a universe for itself and make all those who used to cavil at it into stargazers (Carlyle; *Diary* 70). Such literary models, mingling rugged intellectualism with warm sympathy and the investigation of nature with sweet domestic delights, were invaluable for a would-be man of science who wished to partake of the heroism, while abhorring the violence, of empire. It was for the man of culture to refine not only the savagery of primitive peoples but also the brutish ways of conquerors.

Huxley displayed his attraction to sentimental fiction and its men of feeling in a long description of Mauritius, the scene of Bernardin de St. Pierre's Rousseauist fable *Paul and Virginia.* In a letter to his mother he confessed his "raptures" over the island, its luxuriant countryside, and its people of all colors and varieties of costume hastening to market. "In truth, it is a complete paradise, and if I had nothing better to do, I should pick up some pretty French Eve (and there are plenty) and turn Adam. N.B. There are no serpents in the island." Instead, he visited the tombs of two storybook lovers, whose tale he believed to be "a fiction founded on fact." "Paul and Virginia were at one time flesh and blood, and . . . their veritable dust was buried at Pamplemousses in a spot . . . visited as classic ground." The resting place was a garden wilderness, behind the home of an English mechanist who built steam engines for the island's sugar mills. The lovers' ashes lay in two funeral urns, each raised on a pedestal whose ornaments had been dilapidated by the "merciless ravages" of English relic hunters. "I had a full sense of the abomination of such practices and contented myself with plucking a couple of roses, which now scent my desk" (*Diary* 28–30). Huxley made a sketch of the scene and was prompted to remark on another of Bernardin de St. Pierre's tales, *The Indian Cottage:* "I have not forgotten the lively feelings it excited in me, though I fear that such feelings would not return by a reperusal; intercourse with the world the flesh and the devil having extinguished any latent spark of sentimentalism . . . that may have existed in my bosom." The visit, the twelve-mile walk it entailed, the rich descriptions it occasioned, the censure of ruthless Englishmen, the sketch, and finally the plucking of sweet flowers were all undertaken, as he wrote his mother, "for old acquaintance sake. . . . I never was greatly given to the tender and sentimental, and have not had any tendencies that way greatly increased by the elegancies and courtesies of a midshipman's berth" (*Life* 1: 37–38).

Paul and Virginia and *The Indian Cottage* were among the more widely read tales in a sentimental genre about civilization redeemed by the truths and affections that issued from nature, where nature was epitomized by the happy home. Men of learning were thus mandated to perform through culture what women performed through feeling: the domestication of rough and ultimately destructive masculinity. Bernardin de St. Pierre prefaced the 1806 edition of *Paul and Virginia* with this eulogy to women: "You are the flowers of life. . . . You civilize the human race. . . . You are the Queens of our beliefs and of our moral order" (qtd. in Jordanova, *Sexual Visions* 33–34).[6] *The Indian Cottage* closed with a man of learning, who had traversed the earth in search of truth and happiness, narrating these revelations from a country cottage: "Truth must be sought with a pure and simple heart. It is only to be found in nature. It should only be communicated to good men. . . . No one can be happy without a good wife" (Bernardin 288). Men of culture were to reform the world by extending the virtues of domesticity beyond the household, to refine the manners of rude imperialists, and to make the colonies a happy home, a place where English (or French) industrialists might build shrines to learning and reside in verdant green at peace with people of color.

If Huxley could not make himself an empire builder without remorse, neither could he be a sentimental culture-hero without incurring other guilty associations. Carlyle's great men of letters often verged on impetuosity. They might fret and storm to no useful purpose, their powers of discernment mistaken for the artificial delicacy of modern times. Goethe's Werther, too sensitive to do anything but weep and take his life, might reveal the hazards of refinement, the degree to which artistry and learning incapacitated one for manly action. Such readings were encouraged, for the qualities that distinguished men of culture in works by Bernardin de St. Pierre, Goethe, and Carlyle were qualities that in Victorian society were ascribed much more exclusively to women's nature. By midcentury in England, the pure and regenerative ethos of the home had been reconstructed by several generations of evangelical writers with the expressed object of forming women's character. Men's work, too, was refashioned as a wilderness of trial and strife, the necessary complement and practical support to domestic bliss. Within this ideology of separate spheres, the activities of cultured men might be denigrated as ornamental, leisurely—in short, womanly. It is not surprising that Huxley's reaction to the commercial and military barbarism of empire, though in keeping with the dictums of many evangelical writers on slavery and factory production, was mute in comparison. His criticism of the ship's crew, its merciless, irrational violence and its travesty of true manliness, were completely at odds with the colonial venture in which he was engaged, and with his own emerging orientalist role, which pitted masculine culture against feminine nature and viewed progress as a conquest of the latter. His professional future and sense of manhood were utterly dependent on the imperial enterprise and its assumptions of the femininity of the other.

While at sea, Huxley verged between the imperial and the sentimental, trying to devote himself to scientific work and to the cultivation of warm and worthy companions. Just a few months after departure, he described in his diary a retreat into silence from messmates too coarse to associate with men of science. He pined for the "social ease" and "friendly influences of a home circle." Above all, he longed for fellowship with his sister Lizzie, her cultivated mind and taste, her "tenderest heart," and her "more than man's firmness and courage" (*Diary* 15, 71). With his sister, now emigrated to the United States, and Edward Forbes, the well-appointed and genteel London naturalist, he located the trust, intimacy, and intellectual sympathy he missed on board. With them he began to share his work. He wrote to Lizzie of his investigations on the Physalia (the Portuguese man-of-war), the ideas of classification to which they had given rise, the papers that he was writing, and the scientific ends he hoped to achieve: the reduction of three apparently separate and incongruous groups into modifications of a single type. "There!" he concluded, "Think yourself lucky you have only got that to read instead of the slight abstract of all three papers with which I had some intention of favouring you" (*Life* 1:36).

Early in the voyage, Huxley had begun reflections that he believed would eventually lead to a reorganization of Cuvier's class of Radiata into several new groups. But this was only one of his narrower concerns. As his notes and correspondence with Forbes indicated, Huxley's research program was guided by a classification scheme known in contemporary zoological circles as Quinarianism. It was associated most closely with the work of William MacLeay, a member of the Linnean Society, who had left England in 1826 for Australia, where he and Huxley eventually met and had many conversations. The system was widely debated by naturalists and had gained some renown (or infamy) through its incorporation by William Swainson in his volume for Lardner's *Cabinet Cyclopedia* ("Treatise on the Geography and Classification of Animals," 1835) and its praise in the anonymous *Vestiges of the Natural History of Creation.* The system divided the animal kingdom into five groups, each of which contained five members in turn (and so forth). Each group, and the subgroups within it, were related linearly by "affinity." The last member of each series was related, in turn, to the first, forming a circle. Finally, each member of a circle was related to corresponding members of other circles by "analogy." "This uniformity," Huxley wrote in his notes during the voyage, "is surely in favour of the arrangement and tends to strengthen the conviction that it is really natural and well founded." He considered the "Circular System" to be the "Keplers Laws" of animal form. He saw his work, which took him into embryology, as part of a search for the laws behind these laws: "The laws of the similarity and variation of development of Animal form are yet required to explain the circular theory—they are the true centripetal and centrifugal forces in Zoology" (HP 37:45, 40:149).[7]

Huxley's search for the zoological law of gravity would lead him to address what he called the more "philosophical" questions of taxonomy. For example, Huxley's first public lecture, delivered at the Royal Institution in 1852, addressed the topic of "animal individuality" (see *SM* 1:146–51). He clearly found in this abstract, descriptive, and disciplined search for order with microscope and ink a sanctuary from the confusion and conflict that the public world of men posed to him, as well as a haven from fiercer acts of discovery that created order with force. While anchored off Mauritius on his twenty-second birthday, he described in his diary what a score of years in the world had made him "a bundle of glorious and inglorious contradictions as men call a man. 'Ich kann nicht anders! Gott hilfe mir!' Morals and religion are one wild whirl to me—of them the less said the better. In the region of the intellect alone can I find free and innocent play for such faculties as I possess. And it is well for me that my way of life allows me to get rid of the 'malady of thought' in a course of action so suitable to my tastes" (*Diary* 26). The region of intellect, the order it constructed, and the company it kept were to Huxley a refuge and, as he would soon confess, a home away from the clash and din of work in the British empire. The domestic sphere and scientific community he lacked on ship were being reconstituted in nature and correspondence.

While Huxley was brooding over the meaning of his science at sea, it was his unusual fate to have women as his chief respondents. Shortly after the *Rattlesnake* reached Australia in 1848, he met the woman who would eventually be his wife of thirty years and who became, together with his sister and mother, his chief confidant. The letters he exchanged with Henrietta Heathorn were perhaps the most important medium through which his still tenuous identity as a man of science, and hers as a woman and wife, were shaped.

Woman's Work

The life of Henrietta Heathorn did not begin when she met her future husband and was graced, like England's colonial subjects, with high culture and liberated from the frailties of the feminine condition. It is not easy, however, to construct an account that portrays her as other than a happy wife and hostess to the great, and that presents her long engagement to Thomas as more than a testament to their enduring love and deep sympathy. Almost all her surviving documents were written for his eyes. Duly edited, they tell a fairy story of romance. In a fragment written toward the end of her life, Henrietta described first meeting "the young officer" at a dance, the interest they found in their talk, his calling later on a swift horse, paralyzing her with a fixed gaze, and offering to remove all hindrances from her path in life (HP 62:37). The poetry she wrote during their engagement and the letters and journal entries from the period re-create such chivalric bliss again and again: "I always knew some day the Prince would come for me. He blew the horn and stormed the gates and slew the giants in his way" (Huxley,

Poems 14). As they sat together on the rocks beside the sea near Sydney, talking to her heart's content of their home and all that they would make it, she felt lucky to find one with whose tastes and thoughts she could agree, one who answered the ideal her heart told her she would find, even before she knew him. "Bless you my own dear Henry . . . the spring of happy thoughts . . . the source from whence in every grief or vexation I draw sweet consolation—the fount of all good feelings—the incentive to right resolves and their fulfillment" (*Diary* 234, 235, 238).

But these fairy stories were not conceived in complacency. Like Thomas's sagas of imperial discovery and airy zoological theories, they were often constructed as places of order and beauty where there seemed to be none. While Thomas was sailing around the South Seas sketching the natural world, trying to make space for himself as a man of science, Henrietta lived in the home of her brother-in-law and half-sister outside Sydney, where she worked managing their household and helping raise their two children. She had emigrated from England in her early teens, though not directly. Together with her sister she spent two years in Germany, where she studied literature, before joining her parents in Australia. The details are sketchy, but it is clear that her father, who owned a brewing business, was often insolvent. Most likely, it was to ease his financial burden that Henrietta, at the age of sixteen or seventeen, came to live with her brother-in-law William Fanning, a successful merchant, and to assume considerable responsibilities there. Henrietta planned to marry Thomas on his promotion to full surgeon. But this promotion was not forthcoming and never really sought. Thomas wished instead to gain a position that would enable him to devote most of his time to what he considered the more lofty pursuit of science. It was this ambition—which was shared and reconstructed by her—that prolonged their engagement for eight years and that occasioned much doubt and reflection by both of them on the nature of his work and their future home.

Home for them was thus not a place for which they patiently waited while he obtained a scientific living. Precisely what made the home a pure and stable center of value was a matter of intense debate. Their different conceptions of home and its social and moral foundations became a source of extreme tension that repeatedly called into question his manhood and her womanhood. In order to examine this process of mutual definition, Henrietta's writing must be interpreted not as simple description or fantasy, but as narrated accounts of the kind of woman she wished to be and of the kind of man she wished him to be. Such writing was her best, and often only, means of manifesting her will, of influencing him, and of overcoming the enormous disagreement between their ideals: "I have promised to keep a journal and this promise made to one inexpressibly dear shall be faithfully kept—a journal not only of daily occurrences but thoughts which bad or good shall be registered, even tho' intended for his perusal, for should he not see me as I am? I will hide nothing from him" (*Diary* 211).

In the journal that Henrietta began the day of Thomas's departure for his last survey expedition, she described the activities that gave her greatest satisfaction. Chief among these was household management. Though she never listed accounting among her occupations, it is certain that she made the bulk of the purchases for the home and commanded its servant staff, which included a cook, a butler, a maid, a nurse, and a gardener. Her activities were typical of those recovered by historians of women as essential to the Victorian economy.[8] They were also activities that women themselves, despite much middle-class moralizing to the contrary, regarded as "work." Henrietta defended this position at the very outset of her relationship with Thomas. Not only did she make frequent mention of her household responsibilities as labor, she did so against "gentlemen's assertions" that women were creatures of leisure who lived to shop and against Thomas's own urging that she spend more time "improving" herself: "The day passed as Mondays generally do, very busy all the morning in household matters and all the afternoon at work. . . . It is absurd of Hal to bid me read and practise regularly—what with making my own things in the house I have full employment" (*Diary* 218, 214). An aspiring poet and devotee of German literature, Henrietta too had visions of the house as a home for polite culture. But she had a very different notion of its place there and of what made a home harmonious.

Henrietta's domestic activity held much of its significance for her as a form of religious devotion. If, as social historians have argued, religious institutions were places where women could exercise authority and a range of expression that was not available in other spheres of public life, the most important religious institution for middle-class women was the household. The representation of the home as sanctuary and heaven, and of the wife as innocent angel and nurturing madonna, were among the more famous contributions to English identity by the so-called Evangelical movement. Much of this ideology—endowing women with religiosity because their yielding nature made them superior spiritual vessels—reflected the social and political anxieties of ruling classes of men and was continuous with other constructs of a passive, frail femininity: illness, excitability, silence, and (asexual) reproduction. But many middle-class women found in religious beliefs a moral meaning for their work in the home, as well as a moral foundation for their authority over others in it. Henrietta's own religious life, while including regular attendance at church, radiated from the place where she led the family in hymns at the piano, prayed for strength for her daily tasks, and cared for her sister's children with what she described as "holy love." She read Carlyle's life of Jean Paul—but not, like Thomas, as consolation for a life of scholarly discontents, rather as an aid to the religious fulfillment that she proudly displayed to her fiancé. "I rose a better creature, more cheerful and happy. He struggled thro' deepest poverty and pain. Mind conquered the infirmities of the body and the evils of life. . . . And shall not I whose troubles are but faint and miniature shadows of his, strive against and subdue them? Henceforth I will. . . . All sorrow is selfish. I will become better and God help me in my intentions that they be deeds not words" (*Diary* 286).[9]

This religious recipe for mastery was essential to Henrietta's self-command and her command of the household. In a long passage she told how, returning home one evening to find the staff all "tipsy," the horse escaped, the kitchen on fire, and the nurse hurling abuse, she astonished a friend by her forceful manner of redressing the servants. In a measured tone, she quelled the inferno, restored order to chaos, and admonished the insolent creatures for intemperance. "I sent for the gardener to sleep in the house and shut up immediately that I might get them off to bed and after very quiet and firm measures restored the house to peace. . . . They being penitent . . . I forgave them after very serious lectures" (*Diary* 276–78). If the moral authority conferred to women by religion was exercised most frequently over servants, it was also available as a resource to address men in private who did not measure up and as a basis for entry into the sphere of letters and (more subversively) politics. If Henrietta's religious code prescribed subordination to a more awesome and imposing man of learning in the house, it also gave her strength to intervene in matters where she felt Thomas was wanting or she was wronged.

Henrietta's life, though centered physically and spiritually on the home, was not confined to its four walls. She exhibited her proximity to nature by describing her daily swims, long riding trips, and rowing expeditions. Drawing perhaps on Gothic sagas to construct her feminine persona, she was part intrepid adventurer, as suggested by an account in which she was out walking with some friends and killed a poisonous snake, while the others screamed and leapt away. "Mr. Hamilton said I was very brave and I looked upon myself as a miniature heroine" (*Diary* 281). Together with the Fannings she attended parties and balls, where she mingled with the upper ranks of Sydney society and where, as she repeatedly told Thomas after their engagement, she had to dissuade prospective suitors among leading professionals and men of business. She appears to have had a close circle of women friends whom she often visited, especially a neighbor Alice Radford, who had also become recently engaged. She occasionally remarked on the "excessive simplicity" of some of these women whose lives did not seem attached to the highest causes. Her education in Germany and her knowledge of its literature would have been exceptional in England and was even more so in Australia. Like all women, however, her professional prospects were virtually nonexistent. These acquirements and conditions may have had much to do with her attraction to a would-be man of science, who held out possibilities of a residence in London, company with persons of education, and a life of great spiritual gravity. This attraction was not always supported by her family and friends, however, who urged her not to waste her youth on so unpromising a lover. Nor was it always encouraged by Thomas's behavior, which she at times found incomprehensible and domineering.

> The afternoon brought you, my own dear one . . . but you were in such a strange mood that I felt you did not make me glad. You were capricious; if I talked you would have me silent and if I laughed it grated on your ear and only at length, when I looked as sad as I felt and suggested I would try crying, did you utter any kind words or fold me lovingly to you. . . . Before dinner time your fitfulness had returned and a little nasty spirit possessed me to tease you—till you warmly told me I had better no. I then of course felt more wishful than ever to do so and returned again and again . . . half in sport and half in earnest—till I went up to dress for dinner and told Alice part of my imaginary grievance. She had guessed something from your altered manner . . . [and] remarked you were dull and not joyous as usual. I reasoned

> myself into no very amiable mood much to Alice's amusement and so went down to dinner with a white dress but a naughty dark heart, punishing myself for your supposed harshness instead of you. . . . Alice to whom I told all would not have me wretched all the evening and whispered you I was unhappy—and then dearest you talked and kissed away all my evil fancies. . . . How happily the rest of the evening passed, peace stole into my heart and abode there and when he had gone and I laid my head on my pillow I resolved that I would never more torment myself and him again. Love will tarnish if 'tis always petted. (*Diary* 240–42)

The editor of Henrietta's journal, her grandson Julian Huxley, has suggested that this passage be read as evidence of her emotional volubility: under the strain of her ardent love for Thomas and his imminent departure, she retreats to her woman's world of taunting and tears, where facts are indistinguishable from flights of fancy. But Henrietta's story was of course a retrospective one written for a man whom she considered capricious. Her tears, teasing, and woman friend were her resources in what she believed to be a contest of wills. In the tale, her friend's intervention resolved the conflict. But she constructed a deeper resolution through her writing. In this scene and many others, it was always she who succumbed, she who could not make her needs understood. But with the closing moral and a change of person—here from "you" to "he" when referring to Thomas—she moved into a highly fictional voice and wrote a happy ending to the distressful evening. Such narrative shifts, apparently unnoticed, gained her a distance from Thomas. By placing him in her literary framework, she could shape the meaning of his actions, and in so doing, make the whole ethics and politics of their relationship explicit, both to herself and to him. Through journal writing, she drew together her various models of womanhood and manhood to develop the strategies needed to cope with their differences and to modify them. It was the same process that Thomas used on board ship to reconcile the life of a romantic genius with that of an imperial and economic man. But given the very different opportunities open to her and the apparent dynamics of their face-to-face relations, this writing was perhaps the only medium through which she could actively shape her world and theirs.

Such lively portraits of the young couple during their last few months together in Australia were drawn over and over again, all with the same dizzy mixture of playfulness and pain. On one occasion, he tried to steal a purse she had made for his birthday before the date arrived. "We had great fun all the evening about it . . . until his old habit of trying the strength of his power over me came upon him." If she refused him such a trifle, surely she would do likewise in greater matters? With this prerogative, he declined to call the next day. "My heart ached. . . . I could no longer keep in my tears and would not at first bid him goodnight believing he would not come on the morrow, when he was so kind and loving and good to me. . . . I always punish myself most by any request I deny him" (*Diary* 245). In another incident, he demanded that she give up her only photograph of him which she had treasured for three years and which she slept with under her pillow. After promising to return it, he then said she would never have it, and that she was not to even ask for it.

> I felt he was asking almost more than I could perform. . . . Surely he had done it on purpose to vex me—he was tyrannizing—he knew I could refuse him nothing and he asked so much—slowly and with difficulty I quelled the agony that filled my heart and promised not to ask again. . . . For a moment I felt he was unkind to try me—only for a moment—for I remembered how few ways I had to shew my love for him. I had yielded my will to his, even more a secondary affection; and I was, if not happy, at peace, for he would see that I would rather pluck the dearest object from my heart than offend him. (*Diary* 239)

In these passages Henrietta demonstrated the virtue that used to be demanded of men in public but that was still expected of women in private: sacrifice. By placing herself in the traditional role of the self-effacing woman, she could both explain to herself why she was forced to accept his demands and redeem the painful compromising of her will. Though Thomas returned his photograph in a locket that she could wear around her neck, though he called on her the next day and was kind and good even though she did not give him his birthday purse—still he could not convey his love for her without displaying his power over her. In these narrated encounters, the structure of their most intimate feelings was being worked out. Henrietta accepted that theirs would be a relationship in which his affection would be expressed through assertion and hers through yielding, a relationship in which his gifts would require that she give up something, while hers would not be received on her own terms. She accepted her place in a world in which women had very restricted means of expressing their worth, although the men they loved offered fresh occasions for abnegation. But to write of such acceptance was also to make this virtue clear to one who gave no indication that the terms of this relationship were troubling, who showed no signs of knowing the intricate ways in which her happiness depended on the movement of her will, and who, after denying her other means of manifesting her devotion to him, might have failed to recognize even this—her self-denial. If an air of mystery and caprice, self-command and a command of others were the characteristics of genius, Henrietta showed Thomas their resemblance to the coarse masculinity that he had condemned in the British empire. "You draw out my thoughts and feelings—and appropriate them most tyrannically—and yet 'tis perhaps one of the things that has bound me with stronger love to you. You are a tyrant still conquering by strength where influence fails, indeed you have tonight acted very meanly . . . and I have half—only half a mind, remember—to give you up as Will was constantly advising" (*Diary* 242).

Henrietta's sacrificial displays were thus not passive concessions to Thomas's might. They were contracts that established the terms of their relationship according to her principles. These accounts made it clear that she would only offer herself to him on the condition that he live up to certain ideals of which he was continually falling short. If Thomas, despite his learning, vocation, and invention, was still an unrefined and imperfect man, perhaps Henrietta could complete him. But according to whose models of manhood? While Henrietta was martyring herself on his behalf, Thomas was setting his own conditions through his writing to her, requiring that she fulfill various supportive roles if he was to succeed as a man of science.

Improvement by Domestication

In one of his first letters to his fiancée after their engagement, Thomas wrote how her confiding, tender love had awakened him to a nobler life and a purer course of action. His pledge to work for her and to provide a place for their love opened a space for her power to operate on him.

> Man is and must be influenced by the woman he loves. . . . She is that living ideal of goodness before his eyes. . . . When one is sick of the world, of its petty intrigues, its lesser and greater selfishness and dirt eating, when one is disposed to think that earnestness and truth and firm kind goodness have utterly disappeared from the earth, how great a blessing it is to feel assured that there is yet one in whom all these qualities live and so verily form a part of everyday life—out of the storybooks. . . . Man is as clay in the hands of the father—woman. (HH 7–8)

In this passage, Thomas described his entry into a symbolic order of Victorian work and home in which the former was a place of atonement and trial undergone individually and without regard for the feelings of others, and the latter a place of salvation ruled by sympathy and affection. This Victorian home and the values therein functioned to complement, facilitate, and redeem his (still imagined) public life. Engagement with Henrietta enabled him to mobilize the virtues of this domestic sphere to anchor his drifting self. From Henrietta's loving base he could enter a world of struggle and intrigue, pursue a livelihood, vent ambition, and yet maintain a sense that he was working for something pure. He could also order his own affections, making love a condition of work, a consolation amid struggles in the world, a source of satisfaction for desires that, unsated, would fester and detract from other (greater) things. In his letters to her, Thomas evoked their mutual mentor Carlyle to tell how the logic-mill of men's minds might be upset by "the most unspeakable affections and sympathies," and thought become a weariness, and existence a "foolish wandering" (HH 2–3). This ideology of separate spheres was ubiquitous. If Huxley had known him at the time, he might have turned for advice to Charles Darwin, who, among his own reasons for marriage, had numbered the joys of feminine charm and constant companionship, the pure happiness of the hearth and a wife beside it, and domestic relief from a "whole life, like a neuter bee, working, working, and nothing after all" (Burkhardt and Smith 443–45). Huxley's circumstances were more pressing, however, because the comforts of a home remained very tenuous and intangible for him. They had to be furnished entirely by a fiancée's love and its literary rendering. This may, in part, explain the tone of his letters in which Henrietta's powers, though natural, were virtually dictated to her. But his correspondence was also a continuation of their courtship dynamic in Sydney: even while confessing his utter dependence on her, Thomas had to make himself commanding. "The thought that it is my duty to discipline myself for her sake . . . nerves my better feelings—and often her image is my good genius, banishing evil from my thoughts and actions. . . . You have purified and sweetened the very springs of my being which were before but waters of Marah, dark and bitter were they. And strangely enough, too, not merely is your influence powerful over my heart, but my intellect is stronger, my thoughts more rapid, my energy less exhaustible, I never could acquire more rapidly or reason more clearly" (*Diary* 65–66).

The asymmetry of the lovers' accounts was appropriate within an ideology that gave women considerable but very circumscribed powers and that tended to equate their domestic strength with political and professional weakness. Thus Henrietta's influence on Thomas, unlike his on her, was never "tyrannical." Nor was the genius that he claimed to draw from her one that she would ever be able to display herself. For men of science, however, this ideology was difficult to wield because what they drew from domesticity was not just consolation and nourishment for their public lives, but the very virtues that defined their public persona. They could not readily leave work for home, for what gave their work its special authority were qualities they deemed domestic. After his engagement, Thomas began to write of Henrietta's love much as Whewell wrote of mathematics and Mill of sensation. It was "fixed and unchangeable," the "firm ground" and "foundation" for his "belief that truth and sacredness do exist in this lower world." Where research without her had brought only doubt, love brought dawn and a rosy promise of enlightenment: "I begin to fancy that when the bright day comes many things now dark may become clear to me" (HH 23).

The feminine muse, like his sentimental fiction, was not always so inspiring, however. Huxley repeatedly confessed his own tendency to suffer like women from the very virtues that distinguished them. The qualities of his genius—its volatility and excess, its intuition, its power of passion superior to will—these were also the qualities of feminine weakness. In his journal and letters to Henrietta, he described his "restlessness" and "instability of temper." His thoughts were like his "strides up and down this quaking deck." His intellect was "acute and quick rather than grasping and deep" (*Life* 1:103–4; HH 2–3, 36). "I am," he later wrote, "the son of my mother." From her he inherited the "emotional and energetic temperament" and the "rapidity of thought" that were always both a boon and a burden. "I have a woman's element in me," he wrote his sister, "I hate the incessant struggle and toil to cut one another's throat among us men, and I long to be able to meet with some one in whom I can place implicit confidence, whose judgment I can respect, and yet who will not laugh at my most foolish weaknesses, and in whose love I can forget all care" (*CE* 1:3–4; *Life* 1:65–69). While at sea, uncertain of a career, with only a small network of correspondents, most of them women, with whom to share his work, and with the conditions of his imagined scientific community fulfilled in a fiancée's love, Huxley was particularly prone to these conceptions. If he could find in Henrietta and their love a moral and epistemological ground of scientific work, if he could derive from the home an ideal of scientific community and a sanctuary from the corrupt world of public men, still his authority as the man of the house was suspect if he never left the house, if his rootless invention found discipline in a woman's love. Unable to rely on the established boundaries of a home and workplace should his genius appear feminine, he had to evoke Henrietta's "strong natural intelligence" and her "firmness of a man" to explain her influence (*Life* 1:67).

Such anxieties over effeminacy, though not new to men of learning, were especially acute by the mid-Victorian period. Eighteenth-century writers had considered that many of the qualities that distinguished them from their rough brethren were typically feminine—their delicate, sensitive constitution, their lively imagination, their refined taste. They had tried to make their intellects firm, grasping, and reaching. They had insisted that the severity of study and the drudgery of contemplation demanded virile strength. But they did not have to represent their activity as work. By the time of the Great Exhibition, the polite practices of learned men formed the subject of a highly gendered utilitarian critique. The same ideology that judged women incompetent for public occupations also challenged the fitness of men of science, juxtaposing their leisured life of mind with muscular exploits of economic man, profiting by toil, strife, and stoic endurance (Collini; Hilton). Even the donnish likes of Whewell, Herschel, and Babbage had flirted with the cult of entrepreneurship and were compelled to ascend podiums to defend the practicality of theoretical sciences and of a mysterious, ineffable genius that could not be taught. Huxley had to read his sentimental fiction against a flood of Smilesian literature in which Britain's real heroes were engineers, in which originality was regulated by the needs of commerce, in which Watt and Boulton were geniuses, and Newton a brilliant mechanic (Smiles 1:xvi–xvii, 4:292–307). Because of the prominent place that industry had come to occupy in British national culture, and because of their own dependence on industry for professional opportunities, this was not a perspective that men of science could easily scorn. To do so was to risk being associated with femininity in a most damaging way.[10]

The same ideology that refused to acknowledge either domestic or scientific activities as work did accord each of them dignity as moral endeavors (domestic orderliness, natural law-giving), as forms of religious devotion (the care of children, natural theology), or as polite culture. In the case of women, it was clear where these respectable activities were supposed to reside: the home. Even the writing of proper lady authors was conceived as a domestic pastime or recreation, for their leisure was an important sign of their husband's prosperity. For men who worked, the home was there too: a source of private pleasures and nourishments to redeem the hardships and crassness of their public lives. But what of men whose activities were not considered work? Clearly the "home" was a locus where the meaning of science, of womanhood, and of manhood were being constructed, but in different and conflicting ways.

At times, Huxley tried to distance himself from the domestic practice of natural history by insisting on his disregard for collecting and by invoking a mechanical image of nature that enabled him to claim that what he was really doing with microscope, sketch pad, and Circular theories was engineering. But except for an occasional row with his shipmates or a rumination on empire, he had no way of settling the relationship of his science to other men's work except through his relationship with Henrietta. One of the ways in which he tried to convince her of his adequacies as a man was by associating his science with the practice that was perhaps most central to her activity in the home: religion. Huxley at first tried to avoid the subject because of their extreme differences and the tension it aroused. But religious discussions eventually occupied a substantial part of their correspondence and became an important forum in which he constructed the manly mission of science. In these discussions he tried to turn the focus away from theology or the substance of belief, and toward the method of arriving at it. He began to articulate a position that he would eventually put to great use in debates with clergyman: scientific inquiry was a form of religious devotion.

> I have thought much of our afternoon's conversation, and I am ill at ease as to the impression I may have left on your mind regarding my sentiments. If there be one fact in a man's character rather than another, which may be taken as a key to the whole—it is the tendency of his religious speculations. Not by any means, is the absolute nature of his opinions in themselves a matter of so much consequence, as the temper and tone of mind which he brings to the inquiry. . . . May his fellow men then form no judgment upon the point? Surely they must . . . but let them . . . inquire whether he be truthful and earnest—or vain and talkative—whether he be one of those who would spend years of silent investigation in the faint hope of at length finding the truth—or one who conscious of capability would rather gratify a selfish ambition by adopting and defending the first fashionable view suited to his purpose. . . . On grounds of this kind only can a judgment be justly formed. On these my own dear only must you form your judgment of me. (HH 3)

While defending his unorthodoxy, and challenging the authority of others, including Henrietta, to judge him on their terms, Thomas implied that belief was itself the product of work—a work to which his profession was especially devoted, the production of truth. Thomas was now the earnest truth-seeker laboring in silence, not the orthodox believer who acquired his positions idly and selfishly. In other letters, he suggested where this concerted practice of inquiry fit in the world of other men's labors. He described to her the Crystal Palace as "the great Temple of England" where fifty thousand faithless people worshiped every day while "their professed teachers" disputed about prevenient grace and the Thirty-Nine Articles. He remarked how the "profound ragings" of Carlyle's *Latter Day Pamphlets* awakened in him a vision of "that union of deep religious feeling with clear knowledge" by which alone men could live in peace. He hoped that "his efforts and those of such men as he—earnest resolute and singleminded men" would "do something towards producing a thorough change in our public men and public acts" (HH 112). In the process of defending his faith (or lack of it) to Henrietta, Thomas began to represent science as part of a high-cultural religion whose leaders would save civilization from the blind worship of commerce and industry, power and mammon. It was a heroism close to one preached at this time by liberal Anglicans like Thomas Arnold, Frederick Maurice, and Charles Kingsley—famous critics of the churches for failing to work in the world and leading architects of Christian manliness.[11] Combining the characteristics of the religious devout with the man of action—determination, courage, strength—inquiring men of science could appear as prophetic figures, razing outworn traditions and helping to build new moral foundations.

> Of one thing I am more and more convinced—that however painful for oneself this destruction of things that have been holy may be—it is the only hope for a new state of belief. . . . That a new belief—through which the faith and practice of men shall once more work—is possible and will exist—I cannot doubt. Whether it will arise in my time or not is another matter. At any rate, what am I, that I should not be content even by negation to help in the "Forderung der Tag" as the great poet has it?
>
> Think well of these things Menen. I do not say think as I think—but think in my way—Fear no shadows—least of all that great spectre of personal unhappiness which binds half the world to orthodoxy. They say "how shocking, how miserable to be without this or that belief!" Surely this is little better than cowardice and a form of selfishness. The Intellectual perception of truth and the acting up to it—is so far as I know the only meaning of the phrase "oneness with God." So long as we attain that end does it matter much whether our small selves are happy or miserable? (HH 165–66)

If Thomas, who wrote this letter from England during the fourth year of their engagement, was not yet Henrietta's temporal provider, he might at least be her spiritual master, the man who would lead her to God. Like his other, more muscular model of manly activity—the conquering imperialist—this one of the secular prophet could not be consistently maintained, however. It failed because as yet Thomas felt his religion of science to be both morally empty and theoretically disabling and because religion for him entailed an indifference to temporal things that was completely at odds with Henrietta's own religious experience, which was centered on the everyday tasks of her home. The alleged virtue of the truth-seeker—his scientific method—was hollow because, according to him, it had produced no truth. But it had also failed to produce or to respect the kind of household in which Henrietta could exercise her faith. His anxiousness for the scientific certainty that would make him a spiritual provider involved him in doubts about his worthiness as a man of action able either to meet his lofty goals or to furnish the things of most importance to her.

> Have you not that which of all things I most covet, a firm and clear belief received equally by your reason and your heart? . . . And shall I above all others, I who love you seek to withdraw this from you? Had I any fixed system of belief . . . I should seek to win you over. . . . But this state of doubt from which I cannot extricate myself is yet I am fully aware bad and evil, and as in itself negative can lead to empty denial only, not to working fruitful truth. . . . Depend upon it man was not made for doubt but for belief. The end of man is to act and belief is the source of action. Doubt leads to little better than moral paralysis. (HP 31:66)

Behind his concession that those like Henrietta with "simple and confiding faith" had a happiness, a confidence, and an ability to act in the world that he lacked and desperately wanted, may have been anxieties about the reliance of his genius on temporal provision and Smilesian models of man. But even if the epistemological foundations of Thomas's science had a material basis that he could not consciously acknowledge, it was not merely adventitious that his failure thus far to furnish the basis for a marriage and a family coincided with his inability to undermine her faith. How could he successfully apply his scientific canons of evidence to the unhappy bonds of orthodoxy without a position of distinction and means? How could he ask her to leave her home for his when he had none of any substance to offer? The home of Thomas's floating and beleaguered life of science was not, it would seem, the home that Henrietta knew.

Pressing Points of Economy

After Thomas's last stay in Sydney in 1850, the couple did not see each other for four years. In London the difficulties of finding a place for his science in the world of work were more concrete than they had been at sea. Shortly before reaching England, he wrote Henrietta that he no longer expected a promotion to full surgeon on the basis of his scientific work but hoped for a shore appointment and a grant to publish his research. He took up temporary lodgings with his brother in London, but soon moved to a place where he could leave his books about, read and write in solitude, and greet the "great world only when necessary." "I have drawn the sword, but whether I am in truth to beat the giants and deliver my princess from the enchanted castle is yet to be seen." Within three months, he wrote that any attempt to live by a scientific pursuit was a farce, that he could earn distinction but not bread, and that he would sacrifice it all to be with her, away from the "buzz of the world" in a "quiet cottage" with only the prattle of children "about our knee" (HH 135, 140).

Such dire sketches were intermittent, however. He was enormously affected by the distinction that he began to acquire among his new scientific peers. Royal Society membership had been quick, and its gold medal soon followed. Thus encouraged, he persisted in representing his vocation as lying outside the sphere of practical work where it could remain free of the self-interest, jealousy, and ambition that characterized his social climb—corruption that he could then blame for impeding him. "Why persevere in this hopeless course?" he asked, anticipating Henrietta's dismay. A "hidden force" impelled him, a "sense of power and growing oneness with the great spirit of abstract truth." "I cannot help myself." (HH 146, 172, 229).

Thomas's letters asked Henrietta to accept that his lofty vocation was unvalued in the world of livings and professions, that her own decisions and attachments had been just as impractical, and—through his stoic injunctions to "fear not" lest their small selves be miserable—that temporal provisions were of little importance. The point was put to her rather differently by her brother-in-law, the successful businessman, who often voiced his lack of confidence in the prospects of her fiancé, whom he affectionately called the "spider stuffer." In her journal and letters, she referred Thomas to the repeated advice she received from Will to give him up, "You will never be married, Nettie" (*Diary* 236). If a home with Thomas was uncertain, her own domestic role was rendered more difficult and precarious when the Fannings moved back to England shortly after Thomas's departure. She returned to the house of her father, whose money problems began to worsen as he speculated, with unfortunate results, in the boom-and-bust business of gold. These pressures combined to instill in her a fear of domestic ruin that Thomas, a scientific knight-errant, himself in debt, with a moral aversion to the practical, could not abate. Was he really the "earthly ark" from which she would never wander?

In her journal she described an afternoon in 1849 when, in her room amid gathering clouds, wild wind, and rain, she imagined that a stroke of lightning rendered her blind. She ran to her sister in a panic: "I shall never see again, I can never be his wife. . . . My utter helplessness and uselessness came before me as I thought of the many services I could never more render even to myself. I was a burden, a useless being in this world, and almost believing the picture I had sketched I cried bitterly" (*Diary* 236). Such feelings of impotence bespoke the absence of possibilities for women to be useful outside marriage, but they arose more particularly from a fear that Thomas could not provide the kind of home in which she could demonstrate her worth. Her waking dream was a glimpse at what Thomas's anxieties about his own identity, and his very different models of manhood and domesticity, meant for her—namely, her own failure as a woman, brought on by his failure to provide the support that she expected of a man.

Not surprisingly, these mixed feelings of inadequacy emerged in her religious life. The correspondence they shared over religious matters was in some ways typical of that between men of science and their often more orthodox wives or fiancées. Henrietta, like Charles Darwin's fiancée Emma Wedgwood, for example, defended the religious beliefs that were central to her sense of purpose and authority, gathering evidence for scriptural accounts of creation and pointing out the uncertain foundations of all knowledge, including that of science (Burkhardt and Smith 2:122–23, 126, 169). She credited Thomas, as Emma did Charles, with an earnest disposition, but found their religious differences a painful void, a bar to intimacy, and a threat to domestic harmony. Henrietta's position was especially difficult, however, because she had no other family with whom to share her faith. Without her father or mother for spiritual support, she looked to Thomas, but found him instead a formidable obstacle to reaching higher things: "I fondly hoped he would have been the guide and instructor unto more perfect ways—but here my hopes have borne bitter fruit. Something has come over me of late; I cannot pray as fervently as I did" (*Diary* 228–29).

It is uncertain if Henrietta expected Thomas to conform to a particular model of pious Victorian manhood—the husband and father who led the family in prayer, Bible reading, or religious instruction by the fireside, who took his place as head of the household in the parish church. But his unorthodoxy clearly rendered the important contribution she hoped he would make to the sacred center of her home suspect. Such spiritual shortcoming was more grave because of the great pretense he himself had made of being a religious figure—one whose special calling was the pursuit of truth and goodness. But contrary to Thomas's account in which, out of kindness, he refused to make her share his skepticism, Henrietta's described her active resistance to his method of doubt and persistently placed herself and Thomas, not in the custody of his genius, but of her quite different God.[12] Poems and reminiscences written in later life suggest that she retained her religious beliefs and remained, as she was during their engagement, eager to accept his science as a form of natural theology. It was thus on her religious terms, drawn from an enduring English tradition of Christian learning, that she offered him a worshipful adoration that he feared he might not deserve. Shortly before his final departure from Sydney, she described how he called on her one afternoon to repeat her vows:

> He pictured his hopes frustrated thro' having over-rated his ability, even after years of well-directed energy, and asked me not if I would still love him—he knew that I still would—but if I could esteem him as before—for said he to be loved from compassion would be unbearable. The proud creature, that he should ever imagine my love could mix with such a feeling for him!—I told him that I reverenced as much as I loved him and believed that success or non-success was no criterion of merit, but he replied he would not give anything for the ability that could not compel fortune. (*Diary* 244)

Henrietta respected the detachment of material motives from material gain. She was quite happy to see Thomas in the role he fashioned for himself as high-culture priest or prophet. But according to her Providence, professional success was at least the reward of worth, if not the measure of it. Her own religious life depended on a middle-class home and the income to support it in crucial ways that she, unlike Thomas, could not ignore. Because of her own material circumstances and practical experience—vastly superior to his—in managing the needs of a household, she could not be consoled by his high-cultural castles (or cottages) in the air. As the engagement drew on, and he failed to procure a position while persistently devaluing temporal ends, she began to press him with points of economy he would sooner forget.

Henrietta resisted the efforts of Thomas to distance himself from practical concerns. When government positions opened in Australia during the gold rush, she cautiously suggested he pursue a career that would hasten their togetherness. But more frequently, she urged on his scientific person a more professional comportment. She did not accept his otherworldly concept of vocation, nor his hasty and ill-conceived resolves—which always came after his failure to get a particular scientific post—to take up brewing with her father or shopkeeping. After he missed getting a professorship at the University of Toronto in 1852 she wrote, "You talk of coming out—but *not to practice your profession*—will you let me say that I think I know you well—that you are unfitted by your previous habits from becoming a . . . storekeeper—your vocation does not lie in business matters . . . you have been so long employed in a particular and less practical branch of the profession. I mean less practical in a particular sense. That you might require to turn your attention especially to the requirements of practice before you commence on your own account" (HH 211).

Such revelations from Henrietta—his station in an impractical part of the scientific profession, his inattention to the requirements of practice—were perhaps less bruising to Thomas's ego than her persistent attribution of his genius to acquired cultivation. While Thomas, to sustain his confidence, had continued to cite his whole life history as testimony that secret powers had intended him for science, Henrietta suggested instead that what had kept him apart from the ordinary influences of men, in isolation, poverty, and lack of sympathy, were simply his "habits" and past "employments." What determined him to a life of esoteric studies and her to a life of fiscal concern were his present "tastes and pursuits" (HH 183). By contending with his genius—the mysterious force that he believed had fitted him for science and had fitted him to her—she was, in her own way, calling on him to work, but in a manner that he had always resisted: "I . . . only think it necessary to possess a competency that your own hopes and pursuits may neither be crushed nor impeded. . . . Dearly as I love you and you know how well that is, it cannot blind me to the fact that many an otherwise happy home has been beggared in peace from the perpetual anxiety for the future" (HH 211).

During his years of separation from Henrietta in London, Thomas produced sixteen articles for various metropolitan journals based on his *Rattlesnake* research. He spent much time in a failed effort to acquire funds to publish an extended monograph presenting his new ideas on marine classification to full advantage. To this effect, and to his own exasperation, he twice obtained written recommendations from the Royal Society and British Association for a government grant and personally called on the dukes and earls of the Admiralty and Treasury Departments. He also applied without success for the four natural history positions that fell open (at Toronto, King's College, Aberdeen, and Cork), foiled in his opinion "by petty intrigues" (*Life* 1:83–84). On leave from the navy at half-pay, he supported himself through translations of German zoology texts, museum work, a quarterly science column, and other miscellaneous journalism for the *Westminster Review.* After three years of this he wrote to Henrietta that his youth was gone, that work was no longer enjoyable, and that his proud Pegasus was now broken: "a carthorse pulling its load at full speed." Nor did he hold out any hope for future happiness. Life seemed "to afford few pleasures, but a most sandy desert of . . . drudgery" (HH 259).

With such words Thomas asked Henrietta to witness the conversion of his happy domestic activity into toil and of himself into a weary scientific piece-worker. He was now sacrificing himself for her, conforming to the punishing ethic of work that she had pressed him into serving. By presenting him with accounts of his scientific genius as a poor remunerative pursuit and of himself as one inexperienced and disinclined toward practical affairs—in short, as a man ill-equipped to support a wife and family—Henrietta had done something almost unforgivable. She had manifested to him certain material conditions of domesticity and social determinants and duties of the self-made man (whether a genius or not) that were fundamental to contemporary middle-class mores, but that he refused to accept. Rejecting his disheartened claims that his work had no value in the world of professions, she had instead suggested to him the ways in which science was in fact valued in that world if he would only pursue it *as* a profession, that is—according to rules and procedures that injured his pride. Because of their very different notions of work and home, and because her role was to support him in his battles with the world, her gentle appeals were interpreted by him as slights to the gravity of his calling and as challenges to his manhood. As early as February 1852, he began to question her capacity to understand the moral motives of his vocation and to rank her with those men of the world who had no appreciation for "earnestness," men for whom work was just a matter of selfish "likes and dislikes. . . . So far as I can judge, you do not seem to think that there is one course of life I ought to pursue rather than another—your own generous heart would have me follow that which I like best, but only because I like it best. . . . My brothers understand ambition and profit. Fanning understands ambition. None, not even you, . . . seem to comprehend the noble love of intellectual labour for its own sake" (HH 186).

Having adopted a discipline of improvement, and having applied himself in every direction to procure a living on which they could marry, Thomas continued to insist on the purity of his work from worldly ends. But he could no longer do so without also removing himself from Henrietta's company and sympathy. Thus, while his love for her and his desire to make a home with her did not dissipate, these domestic attachments came to be contrasted with his life of science in a way they had not been previously. The moral ethos of his science—its juxtaposition to the corrupt world of work—could only be sustained now by relocating the home and the women in it in that corrupt world, by redefining the home as a material possession, and by representing women as consumers of leisure and luxury, as pre-Raphaelite Madonnas of sensual delight. In this view—which of course was an enduring tradition among monkish scholars—the home and the affections therein were not bastions of purity. They were "earthly temptations" distracting men of science from their sacred calling. Once intertwined with his lofty vocation, the love he had for Henrietta now appeared to lead him away from the highest aims of his nature. "There are times when I cannot bear to think of leaving my present pursuits . . . and there come intervals . . . darling, when I would give truth and science and all hopes and all aims to be folded in your arms—to have your heart beating against mine and your lips upon my hot brow. . . . I know which course is right, but I never know which I may follow; help me darling against myself" (HH 218). This struggle to detach science from the passions of private life may also have been conditioned by the disarray in Thomas's own family during the period. His mother, to whom he was close, died in April 1852. This left his father, now paralyzed and a mental invalid, in the care of his elder brothers who were apparently little better off than Thomas. Though almost completely reticent about his brothers in the past, he began to describe their households to Henrietta in the bitterest of terms. They were places devoted to petty quarrels, usually over money, from which he did his best to stay away.

Faced with these acrid and accusing accounts, Henrietta asserted her respect for Thomas's special calling while reminding him again that the domestic foundation on which his science rested was unstable. She faced the prospect of a marriage in which her feelings and desires could not be expressed without questioning his prerogatives as a man and betraying her commitments as a woman. Not only would the values that defined the home be dictated to her, but the importance of the home and of her role in it would be diminished to dignify his work. Thomas had boasted of powers to tame chance, to discover the laws and solve the great problems of nature. She knew him as one who, despite his return to London, was still embarked on a global enterprise exploring dark continents, oceans, and life forms, opening paths to commerce and civilization. She accepted that their place in the world and their powers to shape it were different. While he was writing lectures and addressing large audiences on comparative anatomy, she remained in Australia, now working in the household of her parents, offering advice on matters of economy to her father, and writing. But in the margin of the letter in which Thomas had accused her of not loving him and of not understanding the value of intellectual labor, she inscribed the words, "he wronged me." In ensuing correspondence, she evoked his own cherished transparency, threatening to draw a veil between them if he would not let her express her feelings. She recalled for him that their culture and circumstances had placed them in different though overlapping spheres:

> Will you make no allowances for me remembering how different is my life to yours? mine a round of simple duties without one engrossing thought but you—yours a hard struggle I know dear Hal but still full of keen and pleasurable excitement, the hope of winning a name and a position. . . . Whenever I have considered the years, the life time that were to make me your companion I have been with the pleasurable conviction that however far beyond mine in cultivation our tastes were naturally the same—and whatever sorrows from without we might have to encounter I yet believed our home would possess the attraction of unity and be bright with that sunshine of the heart where fullest sympathy and closely blended interests intermingle with a deep and earnest love such as I know mine is. (HH 250)

Thomas's new taxonomy, in which scientific work and home, men of science and women, were discretely opposed, was not maintained for long.

It was only one scheme among others that he adopted as the circumstances and his mood demanded. As his separation from Henrietta deepened, he described for her how the life of science—full of tepid theories and pallid abstractions—left "a void in his heart" and "failed to satisfy his soul." He feared that his noble pursuit of beauty and truth was just a personal ambition, that it was no different from other men's aims, except perhaps in being less rewarded. In his letters, science and domesticity oscillated between purity and depravity. "I think of all my dreams and aspirations, and of the path which I know lies before me if I can only bide my time, and it seems a sin and a shameful thing to allow my resolve to be turned; and then comes the mocking suspicion, is this fine abstract duty of yours anything but a subtlety of your own selfishness? Have you not other more imperative duties?" (HH 177–78, 181, 218). If Thomas could not forsake his fiancée and marry his work, this was not only because the pursuit of science by itself was unfulfilling. Rather, what made science satisfying and significant lost its value if the Victorian home was just another tainted place in a fallen world. Other sources of moral authority and social distinction could be cultivated, such as religion and letters. But domesticity was too compelling to be renounced. As Henrietta's own experience revealed to him, the integrity of these other spiritual and high-cultural endeavors was strongly invested in the home. In the models of manhood and womanhood that eventually came to prevail in his life, science and domesticity were closely aligned.

> Women often forget that men are essentially different from themselves, that a man's actions cannot and ought not to be exhausted within the circle of his affections as their own may rightly be. . . . No woman who knows her true interests will ever begrudge the time her husband gives to . . . Science or Art. They are her best allies, for they all require earnestness and faith and fixity of purpose for their successful cultivation. In this pure sphere, the soul sickened and sceptical from intercourse with men meets truth face to face. . . . It returns to the world purified and thence fitter to recognize the good in all shapes, fitter therefore to love, for that means to recognize purity and goodness. (HH 164)

Thomas could assure Henrietta that his scientific affairs were not adulterous, because domestic values and those of science (and other learned practices) were the same. He was like other men in needing the nourishment of the domestic sphere, while extending his aims and purposes beyond it. But his work was not like other men's—base and self-gratifying. When men of science left their loving wives, they entered a moral sanctum like that of their own home—a "pure sphere" where souls communed with truth. Domestic affection and its restorative virtues reproduced and reinforced their own creative powers and moral endeavors: their role as discoverers and educators, their duty to civilize savage men and redeem civil society, to make the world a home, ordering the relations of classes, nations, and races as they ordered plants and animals, as their own households were ordered. In order to sustain these notions, however, Thomas had to gain entry into the polite society of scientific men, and to procure a living enabling Henrietta to join him, thus allowing their different conceptions of science and the home to coexist.

Fairy-lands of Science

Within a year of his return to London, Thomas had written to Henrietta and his sister of his new attachments with "immensely civil" men of the metropolitan elite who supported his work and character, and whom he had come to call his "scientific friends" (HH 177; *Life* 1:65–69). Owen would do anything in his power for him. Forbes would move heaven and earth. In the recommendations he gathered in application for a professorship at the University of Toronto, the most distinguished in Britain's natural history world—Thomas Bell, William Sharpey, John Gray, Charles Darwin—all testified to his industry, his power of intellect and expression, and his readiness to lead others. William Carpenter wrote, "he has displayed a rare combination of the best qualities of the accurate pains-taking and sagacious observer, with those of the profound and philosophic reasoner. . . . The energy and enthusiasm of his own character must exercise a most beneficial influence in promoting the diligence and zeal of those who may become his pupils" (HP 31:68ff). On receiving the Royal Society gold medal in 1852, Huxley praised the community that had made a place for his work to reside after its long journey in the wilderness. He confessed his dependence on a body of fellows who were so open, transparent, and free from the motives of personal interest that they could receive and reward men whose efforts were truly original and whose conclusions departed significantly from their own.

> The memoir for which you have done me the honor to award the Royal Medal today was printed by this Society during my absence in a remote corner of the world . . . far away from all sources of knowledge as to what was going on here. . . . I sent my memoir away to you with a doubtful mind—I questioned whether the dove thus sent forth from my ark would find rest for the sole of its feet. But it has this day returned, and with . . . an olive branch and with a twig of the bay and a fruit from the garden. . . . I trust I shall never forget the kindness and the aid I have received upon all hands from the men of science of our country. (HP 31:139)

Though he had yet to obtain a steady position, Thomas was giving occasional Royal Institution lectures by this time, organizing collections at the British Museum and sessions at the British Association for the Advancement of Science, and passing the evenings at boisterous dining clubs. Returning one night from the Red Lions—the fraternal supper order of the British Association—he wrote to Henrietta, "I have at last tasted what it is to mingle with my fellows—to take my place in that society for which nature has fitted me. . . . I can no longer rest where I once could have rested" (HH 172).

Among the men whom he now found to revere and trust implicitly was his "old hero," Sir John Richardson, a "man of men" to whom he was "mainly indebted" for his *Rattlesnake* appointment (*Life* 1:102). He also admired Joseph Hooker, sure to succeed his father as director of Kew and seated at evening banquets beside his bride-to-be, the daughter of Henslow (HH 156). Another of his "warmest friends" was John Tyndall, whose own career-path in the physical sciences in many ways paralleled his. When Huxley was appointed in 1855 to lecture regularly at the Royal Institution where Tyndall also taught, the latter would write, "we are now colleagues at home, and I can claim you as my scientific brother" (1:137). Chief among his new cohorts and paragons of purity was Edward Forbes, lecturer on natural history and paleontology at the Royal School of Mines and paleontologist to the Geological Survey. Except for Richardson, to whom he wrote a long letter shortly before his arrival to London, Forbes had been the only member of the London establishment with whom Thomas had corresponded during the voyage. After returning to London, Thomas called on Forbes frequently at the Museum of Practical Geology and the two became friends. He wrote to his sister in 1851, describing how the latter had supervised papers published in his absence, provided many important introductions, helped him in getting membership to the Royal Society, and penned a favorable review of his work—"all done in such a way as not to oppress one or give one any feelings of patronage. . . . I can reverence such a man and yet respect myself" (1:103–4, 73). Thomas's mentor was also one whose character and relations with others embodied the ideals of scientific community. Forbes, he confided to MacLeay and to his sister, was his "great ally, a first-rate man, thoroughly in earnest and disinterested, and ready to give his time and influence—which is great—to help any man who is working for the cause. . . . A man of letters and an artist, he has not merged the *man* in the man of science—he has sympathies for all, and an earnest, truth-seeking, thoroughly genial disposition which win for him your affection as well as your respect" (1:103, 102). According to Thomas, these moral qualities, rather than written work, gave Forbes distinction and influence in the scientific community. He made this point by drawing a contrast with Richard Owen, who could not gain a following "if he write from now till Doomsday" (1:102).

Early in 1855 Henrietta set sail for England despite poor health, and the couple were soon married. "I feel I have got home at last." In June of the previous year, Thomas had obtained a lectureship at the Royal School of Mines, and a position with the Geological Survey—both jobs unexpectedly vacated by Forbes who left London for a chair at Edinburgh. On her honeymoon she stayed with two of her husband's scientific friends, Frederick Dyster and George Busk, and their wives at Tenby. Still too weak to stand, she was waited on by Thomas and carried daily by him to the seaside. On the Dysters' veranda, she made drawings and wrote descriptions as her husband dissected fauna gathered in a day's dredging with Busk. "I like it," she told Dr. Dyster, "I am proud to be associated with his work" (HP 31:82). Late in life, she would reminisce, as she did of their romantic engagement, of her place in what now seemed to her the "fairy land of science":

> When we married I had not the least idea of the true meaning of Science. . . . I had only the dimmest idea, if any, of how a description of a marine creature should win for him fame—or help in any way to bring about his obtaining a position that would enable us to marry. . . . He did not enlighten me.

> But once we were married I began to learn. The whole atmosphere was scientific—his occupation, his friends, his books, the public lectures he gave that I attended. Little by little I saw and understood the great problems that underlay the dissection of even a fish or plant . . . of the interchange of force and heat, of the wonders of chemical changes. It was a revelation that ennobled the world I lived in . . . leading to undreamt of possibilities. . . . I lived in a new world, full of strange facts, but facts that were like fairy tales. (HP 62:2)

Henrietta's experience was in some ways representative. It was not uncommon for a Victorian wife to work in the home as her husband's scientific assistant. It is clear from other cases that divisions of labor were enacted, even celebrated, in laboratories and fieldwork. Class and gender hierarchies were part of scientific practice, inscribed in its most interior spaces (Abir-Am and Outram). But precisely what this work meant to her, to Thomas, and to a wider Victorian public is difficult to determine. A history of the working association of the Victorian husband and wife of science has yet to be written. If the relationship of Henrietta and Thomas is at all exemplary, the meaning of science in the home and the flow of influence and authority there were not matters that scientific men of the house could easily control.

In her reminiscences and in poems written over the course of her marriage, Henrietta suggested that her association with science was uplifting. Its moral principles, embodied in her husband, raised her, she thought, to "a plane far higher than that of most people" (HP 62:1). Such remarks not only imply an acceptance of the models of work and home that men of science were constructing for their own much-needed fulfillment and assurance, they rest on the elaborate model of acceptance that had framed Henrietta's relationship with Thomas from the beginning. Every point in their engagement where she contended with him and found dignity through sacrifice depended on attitudes of women's work as reproductive rather than creative and social relations in which women were properly instructors of children and the poor, popularizers rather than discoverers of knowledge, and dutiful attendants or silent muses to men of genius. Clearly, the establishment of a stable home with Henrietta, together with a secure place of work, enabled Thomas to become the empire builder of his dreams without much further vexation. Consoled by his now invisible wife (seen perhaps only on Sunday mornings at church) he went on to father, along with eight legitimate children, views that would widen the space between public and private, and reaffirm the notion of science as a product of sturdy, masculine reason. The sciences that fixed and ranked all living things on highly moral scales of being and development were running arguments about who could participate in constructing these scales and in defining these moral terms. By locating inferiority in women's bodies, minds, and daily tasks, the natural history, political economy, and—just emerging—psychology, biology, and anthropology that he and other men made all helped to constitute a social order in which women played a subordinate role together with other lesser classes, an order in which scientific men had happy homes for their feelings during the day of hard intellectual work.[13]

But if Henrietta came to revere her husband as one whose soul glowed with purest fire, who "grasped the torch of God's own flame," the price of this adoration was the fusion of elements Thomas would spend much of his career trying to disentangle: the conversion of science into natural theology, into romance, even into part of a woman's work at home (Huxley, *Poems* 11–12, 158). The terms in which scientific identity was settled by Henrietta, while raising Thomas to the status of a poet and prophet, also opened the discourse and the authority of science to other social groups—notably men of letters, clergymen, and women. The scientific writing and practice of these groups, though denigrated as "amateur" or "popular" by aspiring "professionals," often reached much larger audiences than that of elites and transformed the meaning of science in ways that did not readily support the assertions and interests of elites. Because men of science were not alone in claiming the virtues of the home, debates about gender during the period were interwoven with those about high culture, and high-cultural mediums such as writing were major scenes of resistance to the hegemony of men of science. At the same time, Henrietta's insistence that Thomas practice his vocation as remunerative work manifested to him the utter dependence of his science on a wider world of commerce and industry. The identity of science for her—as a useful profession, as a support for domestic economy, as a form of religious devotion, and as a literary art—was not easily resisted, for, far more than Thomas's agnostic and plain-speaking image, her view expressed the meaning of science for its various Victorian publics.

The efforts made by Huxley, and others with whom he engaged, to articulate and control these representations of science—as religion, as literature, as labor—are part of a larger story of the career that followed these early years at sea and in London. By examining the years when his relationship with his fiancée intertwined with those that he was trying to establish with natural objects and with other men in building his career, I have tried to suggest that, despite their long hymns to solitary genius, scientific work had no significance for men of science without a domestic community—a realm of sociability, affection, and sympathy that was in many ways an affront to their faith in reason and meritocracy. The contradictions in their self-image were managed through a series of overlapping and dichotomous spheres—home and work, private and public, genius and society, nature and culture—with a great flow of traffic between them that men of science tried, and often failed, to control. Thus the interrelations between genius and nature, and those between men of science themselves, were steeped in the affections that typified the home, while the home allegedly provided a sanctum for these emotions away from the practice of serious scientific work. The same must be said of Henrietta when she was Thomas's community, his partner against corrupt men who were governed not by truth but by personal interests—a relation that went unacknowledged when this work was affirmed by that sordid world, when its men seemed themselves to be the embodiment of truth and goodness and welcomed Huxley into their happy home.

NOTES

A different version of this essay appeared in the March 1996 issue of *History of Science.*

1. First an exemplar of the genius (see, e.g., Bibby), Huxley has become an epitome (virtually a caricature) of the rising professional. See, for example, Turner; Desmond, *Archetypes, Huxley,* and *Politics of Evolution;* and Rupke. For assessments of Heathorn, see *Life* 1:38–39; *Diary* 268–69; Jensen 23; and Desmond, *Huxley* 134–35, 166. For feminist critiques of such readings with an emphasis on women's exclusion from the professions, see Richards, "Darwin" and "Huxley"; and Russett.

Among recent versions of the separate spheres thesis, the most influential and subtle is that of Davidoff and Hall. Though rendered highly problematic by such works as Peterson's *Family, Love, and Work,* and Poovey's *Uneven Developments,* such arguments are still a commonplace in the new cultural histories of art and private life. See, for example, Wolff and Seed; and Perrot. For the most recent critiques of this thesis, see Vickery; and Wahrman. Much of the large literature on English masculinity takes the divisions of public and private for granted and fails to discuss femininity or the power relations that operate in conjunction with these gender identities at all. An exception is Roper and Tosh.

2. The important collection on women in science edited by Abir-Am and Outram emphasizes the juxtaposition of science (objective, public) to domestic life (subjective, private) as the former became professionalized. These interpretations have been complemented and confirmed by Schiebinger's studies of the eighteenth century. On the topic of gender and the sciences, I have found the work of Jordanova, Outram, and Daston, most helpful.

3. The central place of a variety of literary activities in the lives of gentlewomen has been shown by Peterson. On the controversies surrounding women's writing in this period, see Poovey, *The Proper Lady;* David; and Clarke, *Ambitious Heights* and "Strenuous Idleness."

4. Though focused on an earlier period, Schaffer's "Genius" and Shapin's "'The Mind' " have been most suggestive. Despite a large literature on the professionalization of the sciences, little is known, apart from the often hagiographic accounts of biographers, about what it was like to be a "rising professional." For a suggestion of how calamitous this process could be, see Secord; and Shortland.

5. Huxley's identification with empire building was quite typical for British naturalists on expeditions during the period, as a growing literature on natural history and empire shows (see Browne; Smith; and Pratt). Such studies have not focused on what I take to be the important ambivalence of this relationship, however. For their imperial roles, English men (Huxley included) would also revive medieval models of the virtuous conqueror and rescuer of women (see Girouard). For gendered accounts of orientalism, see especially Wee; and Inden.

6. As part of a broad examination of enduring associations of women with nature and men with culture, Jordanova draws on *Paul and Virginia* to illustrate one version of this pairing that gave the moral advantage to women. Huxley's own reading of this tale supports her important claim that such conceptions could coexist and contend in the very same person with a discourse of enlightenment (or what I have called here, with specific reference to the colonial context, orientalism). In contrast, Desmond interprets Huxley's difficulties (when they are not caused by climate) as class difficulties and his ship-board conflicts as class conflicts (*Huxley* 40–41, 54, 72, 85).

7. MacLeay's work, its reception among zoologists, and Huxley's debt to it are carefully examined in Winsor 81–97.

8. Reclaiming the important role of women in the Victorian middle-class economy is one of the central aims of Davidoff and Hall. For an account of the strenuous occupation of household management in Britain, see Branca.

9. The religion-saturated home of the evangelical middle class is a commonplace in British historiography; though most studies, when not focused on the confining effects of the angel-in-the-house ideology, have been concerned with how Christian piety gave women mobility and authority outside the home (see, e.g., Thompson; Prochaska; Owen; and Davidoff and Hall).

For a survey and critique of Victorian models of feminine weakness, see Vicinus.

10. For this gendered characterization of early-modern men of learning I draw heavily on Barker-Benfield. Though focused on France, recent works by Outram *(The Body)* and Vincent-Buffault provide highly relevant accounts of such shifts in manly comportment and the expression of the emotions.

On the anguished sympathies of Whewell, Herschel, and Babbage for middle-class values of work and utility, see Alborn. Just how compelling capitalist ethics and modes of production were to theoreticians and practitioners of the physical sciences has been shown by Schaffer's "Astronomers Mark Time"; and Wise and Smith.

11. On Christian manliness, see Hilton; Vance; and Newsome.

12. Her published poems suggest quite strongly that through the end of her life she believed in natural theology (Huxley, *Poems* 11–12, 75–76, 155–56) and a Christian sort of heaven (99–101).

13. For extended accounts of the gender and racial constitution of the Victorian sciences, see Richards, "Darwin"; and Russett.

REFERENCES

Abir-Am, Pnina, and Dorinda Outram, eds. *Uneasy Careers and Intimate Lives: Women in Science, 1789–1978.* New Brunswick: Rutgers UP, 1987.

Alborn, Timothy L. "Idols of the Factory: Genius and Industry in the Language of British Scientific Reform, 1820–1840." *History of Science,* 34 (1996): 223–75.

Barker-Benfield, G. J. *The Culture of Sensibility: Sex and Society in Eighteenth-Century Britain.* Chicago: U of Chicago P, 1992.

Bernardin de St. Pierre, Jacques Henri. *Paul and Virginia* and *The Indian Cottage.* London: J. F. Dove, 1828.

Bibby, Cyril. *T. H. Huxley: Scientist, Humanist, Educator.* London: Watts, 1959.

Branca, Patricia. "Image and Reality: The Myth of the Idle Victorian Woman."

Clio's Consciousness Raised: New Perspectives on the History of Women. Ed. Mary S. Hartman and Lois W. Banner. N.Y.: Harper & Row, 1974. 179–81.

Browne, Janet. "A Science of Empire: British Biogeography before Darwin." *Revue de l'histoire des sciences modernes et contemporaines* 45 (1992): 453–75.

Burkhardt, Frederick, and Sydney Smith, eds. *The Correspondence of Charles Darwin.* Vol 2. Cambridge: Cambridge UP, 1985.

Carlyle, Thomas. "Characteristics." *Edinburgh Review* (1827).

———. "Jean Paul Friedrich Richter." *Edinburgh Review* (1827).

———. "The Life of Heine." *Foreign Review* (1828).

Clarke, Norma. *Ambitious Heights, Writing, Friendship, Love: The Jewsbury Sisters, Felicia Hemans and Jane Carlyle.* London: Routledge, 1990.

———. "Strenuous Idleness: Thomas Carlyle and the Man of Letters as Hero." *Manful Assertions: Masculinities in Britain since 1800.* Ed. Michael Roper and John Tosh. London: Routledge, 1991. 25–43.

Collini, Stefan. "Manly Fellows: Fawcett, Stephen, and the Liberal Temper." *The Blind Victorian: Henry Fawcett and British Liberalism.* Ed. Lawrence Goldman. Cambridge: Cambridge UP, 1989. 41–59.

Daston, Lorraine. "The Naturalized Female Intellect." *Science in Context* 5 (1992): 209–35.

David, Deirdre. *Intellectual Women and Victorian Patriarchy.* London: Macmillan, 1987.

Davidoff, Leonore, and Catherine Hall. *Family Fortunes: Men and Women of the English Middle Class, 1780–1850.* Chicago: U of Chicago P, 1987.

Desmond, Adrian. *Archetypes and Ancestors: Palaeontology in Victorian London, 1850–1875.* Chicago: U of Chicago P, 1984.

———. *Huxley: The Devil's Disciple.* London: Michael Joseph, 1994.

———. *The Politics of Evolution: Morphology, Medicine, and Reform in Radical London.* Chicago: U of Chicago P, 1989.

Girouard, Mark. *The Return to Camelot: Chivalry and the English Gentleman.* New Haven: Yale UP, 1981.

Hilton, Boyd. "Manliness, Masculinity, and the Mid-Victorian Temperament." *The Blind Victorian: Henry Fawcett and British Liberalism.* Ed. Lawrence Goldman. Cambridge: Cambridge UP, 1989. 60–70.

Huxley, Henrietta A. *Poems of Henrietta A. Huxley with Three of Thomas H. Huxley.* London: Duckworth, 1913.

Huxley, Leonard. *Life and Letters of Thomas H. Huxley.* 2 vols. N.Y.: Appleton, 1901.

Huxley, Thomas Henry. *Collected Essays.* 9 vols. 1894. N.Y.: Greenwood Press, 1968.

———. *The Scientific Memoirs of Thomas Henry Huxley*. Ed. Michael Foster and E. Ray Lankester. 4 vols. plus supplement. London: Macmillan, 1898–1903.

———. *T. H. Huxley's Diary of the Voyage of HMS* Rattlesnake. Ed. Julian Huxley. London: Chatto & Windus, 1935.

Huxley Papers. Imperial College of Science, Technology, and Medicine Archives, London.

———. Huxley-Heathorn Correspondence. Imperial College of Science, Technology, and Medicine Archives, London. Cited as HH.

Inden, Ronald. *Imagining India*. London: Blackwell, 1990.

Jensen, J. Vernon. *Thomas Henry Huxley: Communicating for Science*. Newark: U of Delaware P, 1991.

Jordanova, Ludmilla. "Gender and the Historiography of Science." *British Journal for the History of Science* 26 (1993): 469–83.

———. "Nature's Powers: A Reading of Lamarck's Distinction between Creation and Production." *History, Humanity, and Evolution: Essays for John C. Greene*. Ed. James R. Moore. Cambridge: Cambridge UP, 1989. 71–98.

———. *Sexual Visions: Images of Gender in Science and Medicine between the Eighteenth and Twentieth Centuries*. London: Harvester, 1989.

Newsome, David. *Godliness and Good Learning: Four Studies on a Victorian Ideal*. London: Cassell, 1961.

Outram, Dorinda. "Before Objectivity: Wives, Patronage, and Cultural Reproduction in Early Nineteenth-Century French Science." *Uneasy Careers and Intimate Lives: Women in Science, 1789–1978*. Ed. Pnina G. Abir-Am and Dorinda Outram. New Brunswick: Rutgers UP, 1987. 19–30.

———. *The Body and the French Revolution: Sex, Class, and Political Culture*. New Haven: Yale UP, 1989.

Owen, Alex. "Women and Nineteenth-Century Spiritualism: Strategies in the Subversion of Femininity." *Disciplines of Faith: Studies in Religion, Politics, and Patriarchy*. Ed. Jim Obelkevich, Lyndal Roper, and Raphael Samuel. London: Routledge and Kegan Paul, 1987. 130–53.

Perrot, Michelle, ed. *A History of Private Life*. Vol 4. Cambridge: Harvard UP, 1990.

Peterson, M. Jeanne. *Family, Love, and Work in the Lives of Victorian Gentlewomen*. Bloomington: Indiana UP, 1989.

Poovey, Mary. *The Proper Lady and the Woman Writer*. Chicago: U of Chicago P, 1984.

———. *Uneven Developments: The Ideological Work of Gender in Mid-Victorian England*. Chicago: U of Chicago P, 1988.

Pratt, Mary Louise. *Imperial Eyes: Travel Writing and Transculturation.* London: Routledge, 1992.

Prochaska, F. K. *Women and Philanthropy in Nineteenth-Century England.* Oxford: Clarendon Press, 1980.

Richards, Evelleen. "Darwin and the Descent of Woman." *The Wider Domain of Evolutionary Thought.* Ed. D. E. Oldroyd and Ian Langham. Dordrecht: D. Reidel, 1983. 57–111.

———. "Huxley and Women's Place in Science: The 'Woman Question' and the Control of Victorian Anthropology." *History, Humanity, and Evolution: Essays for John C. Greene.* Ed. James R. Moore. Cambridge: Cambridge UP, 1989. 253–84.

Roper, Michael, and John Tosh, eds. *Manful Assertions: Masculinities in Britain since 1800.* London: Routledge, 1991.

Rupke, Nicolaas. *Richard Owen: Victorian Naturalist.* New Haven: Yale UP, 1994.

Russett, Cynthia. *Sexual Science: The Victorian Construction of Womanhood.* Cambridge: Harvard UP, 1989.

Schaffer, Simon. "Astronomers Mark Time: Discipline and the Personal Equation." *Science in Context* 2 (1988): 115–45.

———. "Genius in Romantic Natural Philosophy." *Romanticism and the Sciences.* Ed. Andrew Cunningham and Nicholas Jardine. Cambridge: Cambridge UP, 1990. 82–98.

Schiebinger, Londa. *The Mind Has No Sex? Women in the Origins of Modern Science.* Cambridge: Harvard UP, 1989.

Secord, James A. "John W. Salter: The Rise and Fall of a Victorian Palaeontological Career." *From Linnaeus to Darwin: Commentaries on the History of Biology and Geology.* Ed. A. Wheeler and James Price. London: Society for the History of Natural History, 1985. 61–75.

Shapin, Steven. "'The Mind Is Its Own Place': Science and Solitude in Seventeenth-Century England." *Isis* 79 (1988): 372–404.

Shortland, Michael. "Darkness Visible: Underground Culture in the Golden Age of Geology." *History of Science* 32 (1994): 1–61.

Smiles, Samuel. *Lives of Engineers.* 5 vols. London: Murray, 1874–77.

Smith, Bernard. *European Vision and the South Pacific.* New Haven: Yale UP, 1985.

Thompson, F. M. L. *The Rise of Respectable Society.* Harvard: Harvard UP, 1988.

Tosh, John. "Domesticity and Manliness in the Victorian Middle Class." *Manful Assertions: Masculinities in Britain since 1800.* Ed. Michael Roper and J. Tosh. London: Routledge, 1991. 44–73.

Turner, Frank. *Contesting Cultural Authority: Essays in Victorian Intellectual Life.* Cambridge: Cambridge UP, 1993.

Vance, Norman. *The Sinews of the Spirit: The Ideal of Christian Manliness in Victorian Literature and Religious Thought.* Cambridge: Cambridge UP, 1985.

Vicinus, Martha, ed. *Suffer and Be Still: Women in the Victorian Age.* Bloomington: Indiana UP, 1972.

Vickery, Amanda. "Golden Age to Separate Spheres? A Review of the Categories and Chronology of English Women's History." *Historical Journal* 36 (1993): 383–414.

Vincent-Buffault, Anne. *The History of Tears: Sensibility and Sentimentality in France.* London: Macmillan, 1991.

Wahrman, Dror. "'Middle-Class' Domesticity Goes Public: Gender, Class, and Politics from Queen Caroline to Queen Victoria." *Journal of British Studies* 32 (1993): 396–432.

Wee, C. J. W-L. "Christian Manliness and National Identity: The Problematic Construction of the Racially 'Pure' Nation." *Muscular Christianity: Embodying the Victorian Age.* Ed. Donald E. Hall. Cambridge: Cambridge UP, 1994. 66–88.

Winsor, Mary P. *Starfish, Jellyfish, and the Order of Life.* New Haven: Yale UP, 1976.

Wise, M. Norton, and Crosbie Smith. "Work and Waste: Political Economy and Natural Philosophy in Nineteenth-Century Britain." *History of Science* 27 (1989): 263–301, 391–449; 28 (1990) 221–59.

Wolff, Janet, and John Seed, eds. *The Culture of Capital: Art, Power, and the Nineteenth-Century Middle Class.* Manchester: Manchester UP, 1988.

Thomas Henry Huxley and the Imperial Archive

PATRICK BRANTLINGER

A CENTURY BEFORE TODAY'S PRACTITIONERS OF CULTURAL studies began to privilege sociosymbolic "struggle" and "resistance" as pared-down versions of class conflict, evolutionary theorists such as Darwin, Herbert Spencer, and Thomas Henry Huxley were privileging "the struggle for existence" as the chief aspect of experience universally shared by all life-forms including human societies. As Huxley explained in "Evolution and Ethics" and elsewhere, society had arisen to mitigate the worst types of "struggle" among human individuals and between humanity and other species. But the "struggle for existence" nevertheless continued, albeit in mitigated ways, among individuals within societies and in unmitigated ways between societies through war, conquest, and imperialism. In these circumstances, Huxley contended, the golden rule could only be practiced as an unnatural refusal of what was normal and natural: "Strictly observed, the 'golden rule' involves the negation of law by the refusal to put it in motion against law-breakers; and, as regards the external relations of a polity, it is the refusal to continue the struggle for existence. It can be obeyed, even partially, only under the protection of a society which repudiates it. Without such shelter, the followers of the 'golden rule' may indulge in hopes of heaven, but they must reckon with the certainty that other people will be masters of the earth" (*CE* 9:32).

Here Huxley approximates Nietzsche's repudiation of Christianity and its modern, secular offshoot socialism in favor of something like a universal will to power. Huxley frames his argument in utilitarian terms: pleasure and pain are the universal motivating factors in human experience; "the insatiable hunger for enjoyment" propels "all mankind" and "is one of the essential conditions of success in the war with the state of nature" (*CE* 9:27). But as outcomes of the evolutionary processes of nature, the drive for pleasure (Huxley) and the drive for power (Nietzsche) are perhaps indistinguishable. Whether pleasure or power is the ultimate aim, moreover, imperialism—that is, power over other nations in the struggle for existence—is the natural result of the struggle for existence among societies.

Huxley's social and political opinions have been often and well described; it is not my intention here either to repeat or to critique what William Irvine, Mario di Gregorio, and others have had to say on this topic. Huxley took the "liberal" side in the Governor Eyre controversy and the American Civil War; he was a liberal individualist and utilitarian who, however, also espoused a Burkean "conservatism" about the slow, organic or natural processes of the evolution of social institutions and about the great, natural inequalities among individuals, races, and species in the struggle for existence. Di Gregorio rightly calls Huxley "an open-minded conservative," and Irvine, canvasing Huxley's antisocialist essays, points out that Huxley's essay "The Natural Inequality of Men" reverses Rousseau's *Discourse on the Origins of Inequality.* Huxley regarded "all forms of socialism . . . as utopian illusion rendered dangerous by grave and widespread poverty." Irvine also points out that "in his private letters [Huxley] expressed contempt for Gladstone and despair at Irish Home Rule, which he thought would critically weaken English defenses, virtually dispossess English property owners in Ireland, and launch a poor and ignorant people on a reckless adventure. He seems to have believed in a strong empire as well as a strong government, and sometimes gave cautious and guarded approval to such Conservatives as Salisbury, Chamberlain, and—from an earlier generation—Shaftesbury" (di Gregorio 174; Irvine 334–35, 332).

No doubt much more could be said about Huxley's specific political opinions. My aim, however, is to situate Huxley's general scientific and sociological thinking in the context of late-Victorian imperialism. Quite apart from his specific views on such issues as the American Civil War and Irish Home Rule, Huxley's theories take the history and present-day context of war and empire building as the norm, or in other words as nature's rules applied to international relations. Conquest, colonization, and empire are the natural outcomes of "the ethical process" in its workings among nations. The opposition between ethics and nature in "Evolution and Ethics" is not absolute; Huxley defines ethics in terms of the abeyance of natural processes made possible by society—hence, whatever advances social organization, including conquest and empire, are natural outcomes of the ethical process. Further, the progress of scientific knowledge both promotes and is dependent upon imperialism. The mastery of the laws of nature that is the ultimate goal of Darwinian biology entails an epistemology of conquest. Imperialism in this abstract, ideological sense informs everything that Huxley wrote. First, I explore how Huxley's ideas in his essay "Evolution and Ethics" are informed by imperialism. Next, moving backward in time, I examine how Huxley's ethnological writings, including *Man's Place in Nature,* are also informed by—and written in the service of—imperialism. In the final section, I suggest that Huxley's scientific-evolutionary epistemology must also be regarded as imperialist. Here I should hasten to add that I am not criticizing Huxley for being an imperialist (which would be tantamount to criticizing him for being a Victorian Englishman). I am instead interested in understanding how Huxley's scientific ideas were themselves expressions of an overarching ideological superstructure—the superstructure that Thomas Richards has recently called "the imperial archive."

Ethics and Empire

In writing "Evolution and Ethics," Huxley was constrained by the rules of the Romanes lectures that prohibited discussing either politics or religion. He avoids specific, contemporary political issues; he treats Christian fundamentalism as benighted, but also downplays specific religious issues and challenges, in part by describing, in the second lecture, the ethical traditions of Buddhism and Stoicism while saying little about Christianity. "Yet Ethical Science is, on all sides, so entangled with Religion and Politics, that the lecturer who essays to touch the former without coming into contact with either of the latter, needs all the dexterity of an egg-dancer" (*CE* 9:vi). "Evolution and Ethics" is largely concerned with the progressive development of society in its struggle against nature; it is therefore also a thoroughly political—which is to say, ideological—essay, rooted in the assumption that any historical, sociological change from the simple and primitive to the complex and modern (or Western) is beneficial to the human species in general. Huxley clearly believes, moreover, that whenever possible such so-called progress should be transmitted or taught to those simple, primitive societies capable of achieving greater or lesser degrees of civilization. Savages for him are always Hobbesian: superstitious, unprogressive, self-destructive (9:203; compare di Gregorio 166). That Huxley does not consider his conclusions about savagery or about civilization and its corollary imperialism politically controversial enough to exclude from his lectures seems symptomatic of the extent to which many of his social and political assumptions were generally accepted in the late-Victorian period.

For Huxley, the ethical process in history does not contradict but instead leads to imperialism as the natural outcome of competition among human groups. The ethical process gradually mitigates or even eliminates the struggle for existence within a given society. The "categorical imperative" here (*CE* 9:75), however, is not a secular version of the golden rule as in Kant, but rather the pressure from the struggle for existence upon individuals to band together for mutual protection against other individuals or groups of individuals. Those groups or societies that learn best how to mitigate internal competition among individuals, substituting for such competition cooperation and the "gradual strengthening of the social bond," are "fittest" for survival in the broader struggle—that is, the struggle (competition for scarce resources, warfare) of societies against each other—which Huxley here explicitly identifies with and not against the ethical process (9:35).

Although Huxley deplores "the unfortunate ambiguity of the phrase 'survival of the fittest' " because it equates might with right, or in other words because it confounds what is fittest in the struggle in and against nature with what is morally best (*CE* 9:80), throughout "Evolution and Ethics" he implicitly and sometimes explicitly identifies those societies that succeed in dominating others with progress and civilization, and the dominated with unfitness, barbarism, savagery. Though Huxley does not spell out this conclusion, it follows that those societies capable of constructing empires are more ethical than those that are imperialized. (This conclusion is implicit, of course, in the opposition between savagery and civilization that informs all evolutionary anthropology.) Huxley places his (secular) faith in the gradual improvement of the human condition in that intelligence and science that he everywhere identifies with both civilization and the ethical process. Though all evolutionary processes are ultimately "cyclical" rather than "progressive" (9:49), so that beyond whatever heights of civilization humanity may attain there looms a future retrogression or decline and fall, during the progressive upswing of social evolution he sees "no limit to the extent to which intelligence and will, guided by sound principles of investigation, and organized in common effort, may modify the conditions of existence" for the better. Humanity has the capacity to alter both external and internal nature; "much may be done to change the nature of man himself. The intelligence which has converted the brother of the wolf into the faithful guardian of the flock ought to be able to do something towards curbing the instincts of savagery in civilized men" (9:85).

Viewed through the lens of the ethical process as Huxley understands it, civilization means nothing more nor less than curbing the instincts of savagery—or just curbing savagery. The savage within—not distinguishable from the beast within—must be tamed, repressed. But so must the savage without. In the external struggle among societies, civilization is identical to imperialism, insofar as they both entail the conquest and domestication of savages. In common with other social Darwinists including Darwin, Huxley understands savagery as a state of social organization or disorganization not far removed from Hobbes's state of nature, when life was "nasty, brutish, and short." As in his rejections of egalitarianism, Huxley is thoroughly antagonistic to Rousseauistic versions of primitivism. In his ethnological essays, Huxley displays little interest in cultural differences among savages: there are different degrees of evolutionary progress manifest among different races, but savagery is the name for the negative end of an ideological spectrum or hierarchy at the positive end of which stands civilization. And as the colony-as-garden metaphor in the first lecture of "Evolution and Ethics" makes clear, Huxley understands the civilizing, the ethical, and the imperializing processes of controlling nature and taming savagery as one and the same.

Huxley opens "Evolution and Ethics" with the deceptively simple analogy of society as a garden in which the gardener controls and to large measure excludes "the struggle for existence" by "restrict[ing] multiplication; provid[ing] that each plant shall have sufficient space and nourishment; protect[ing] from frost and drought; and . . . modify[ing] the conditions, in such a manner as to bring about the survival of those forms which most nearly approach the standard of the useful, or the beautiful, which he has in mind" (*CE* 9:14). The garden analogy is less simple than it seems, however, for at least two reasons. First, though Huxley appears to be invoking the biblical associations of Eden and innocence, his garden is not Eden and the gardener is hardly Adam. The equation is instead mainly with modern, western civilization as the most thoroughgoing garden history has yet produced. Second, the gardener becomes specifically an English colonist in Tasmania—a strange substitute for Eden if there ever was one—and then an administrator, or governor, capable of practicing eugenicist breeding and euthanasia upon the population he administers. Huxley presents the English colonist in Tasmania as an ethical sort of gardener; the eugenicist administrator, who carries social Darwinian ideas to an irrational extreme, he presents as a dangerously utopian dictator.

"The process of colonisation presents analogies to the formation of a garden which are highly instructive" (*CE* 9:16). This is clear enough, but what is unclear is why Huxley locates his garden-colony in Tasmania. Not only did imperial Tasmania originate as a penal colony, but in the process its aboriginal population was ruthlessly exterminated. Like Darwin, Huxley considered the Tasmanian aborigines as among the very lowest (or most primitive) of races on the evolutionary totem pole. With the death of the last full-blooded Tasmanian in 1876, however, the Tasmanians became notorious as the clearest example of the total liquidation of a primitive race by civilization (or imperialism) on record. Thus Darwin in *Descent of Man* made the Tasmanians a key exhibit in his consideration of the causes of the extinction of at least some primitive races (Brantlinger; for a contemporary account cited by Darwin as a key authority, see Bonwick). The main causes were disease and genocide, but like most other commentators, Darwin downplays the violence that the English colonists visited upon the hapless aborigines. So does Huxley. Or, rather, Huxley invokes the Tasmanian disaster perhaps because it so dramatically illustrates the antithesis, as he constructs it, between the garden-colony constructed by the English invaders and the state of nature to which the aborigines clung and which the colonists' not-so-innocent gardening effectively eradicated. Huxley does not mention the fate of the aborigines, but he does not have to: their destruction was infamous even as it was occurring. Huxley includes the aborigines implicitly among the old "Flora and Fauna" that the gardener-colonist sweep out of the way of civilization. On landing "in such a country as Tasmania," says Huxley, the colonists "find themselves in the midst of a state of nature. . . . The colonists proceed to put an end to this state of things. . . . They clear away the native vegetation, extirpate or drive out the animal population"—just here, he might have mentioned the extinction of the aborigines (9:16). Instead, Huxley mentions the "native savage" in the next paragraph, as a possible threat to reclaim the colony for the state of nature, should the colonists prove to be "slothful, stupid, and careless." (For Victorian readers, it went without saying that English colonists—even penal colonists—were hardly ever "slothful, stupid, and careless").

Despite his various elaborate taxonomies of races in his ethnological writings, Huxley divides humanity into two great, opposing camps named savagery and civilization. Savages like the Tasmanians are little better than animals; that he simply includes them with the unwanted animal population extirpated by the gardener-colonists should therefore come as no surprise. In his 1888 essay "The Struggle for Existence in Human Society," Huxley writes: "the course shaped by the ethical man—the member of society or citizen—necessarily runs counter to that which the nonethical man—the primitive savage, or man as a mere member of the animal kingdom—tends to adopt. The latter fights out the struggle for existence to the bitter end, like any other animal; the former devotes his best energies to the object of setting limits to the struggle" (*CE* 9:203).

Races in Collision

Huxley's use of Tasmania as an example without directly mentioning the extermination of its aboriginal population is of a piece with what he elsewhere says both about savagery and about the prehistory and history of the races of humanity. From *Man's Place in Nature* forward, Huxley's ethnological writings present a vision of the human past as the struggle for existence among races. Huxley is less prone than many of the race-theorists of the nineteenth century to attribute specific moral, emotional, and intellectual qualities to races, and he also stresses the influence of geographical and climatic factors in the struggle through which the fittest races have survived while apparently unfit ones like the Neanderthals (and like the Tasmanian?) have died out. Huxley also stresses that, though there are apparently great differences between races, with the "lower" ones not far removed from the gorilla and chimpanzee, humanity forms a single species.

As Huxley says in "On the Methods and Results of Ethnology," "ethnology is the science which determines the distinctive characters of the persistent modifications of mankind." The important modifications are morphological, involving bodily structure; these are what differentiate one race from another—or rather, these are the only differences that Huxley believes can be scientifically measured and compared. At the outset of his essay, Huxley insists that the terms "persistent modifications" and "stocks" should be used instead of "varieties," "races" or "species," "because each of these last well-known terms implies, on the part of its employer, a preconceived opinion touching one of those problems, the solution of which is the ultimate object of the science" of ethnology (*CE* 7:209). But Huxley's solution elsewhere is to use "race" while insisting that humanity forms one "species." Anyway, Huxley considers ethnology to be a branch of zoology, "which again is the animal half of BIOLOGY—the science of life and living things" (7:210). One result is that, unlike Edward Burnett Tylor, Sir James Frazer, and many other late-Victorian and early modern anthropologists, Huxley stresses the physical measurements of bones and bodies. Neither language nor artifacts offer reliable evidence, he believes, about the real differences among the human races or about their past: "That two nations use calabashes or shells for drinking-vessels, or that they employ spears, or clubs, or swords and axes of stone and metal . . . cannot be regarded as evidence that these two nations had a common origin, or even that intercommunication ever took place between them" (7:213). Nor can the fact that two nations or races speak the same language prove much: Frenchmen speak French, says Huxley, but so do Haitians. Huxley here, as in later essays such as "The Aryan Question and Prehistoric Man," dismisses the language-based speculations of those philologists like Max Müller who claimed to be able to read the ethnological past and present from the distribution of languages and dialects.

The supposed physical differences among the races provides Huxley with evidence for a taxonomy which, in "On the Methods and Results of

Ethnology," includes eleven distinct racial stocks: Australians, "Negritos" (such as Tasmanians and Papuans), "Amphinesians" (Maoris, Polynesians), American Indians, "Esquimaux," Mongolians, Negroes, Bushmen, "Mincopies" (the inhabitants of the Andaman Islands), and the two great groups that make up most of the population of Europe, North Africa, and much of Eurasia: the blond "Xanthochroi" who may be either "short" or "long-headed" (though in other essays they are described as "long-headed" only) and the dark, "long-headed" "Melanochroi." Huxley remarks that "of the eleven different stocks enumerated, seven have been known for less than 400 years; and of these seven not one possessed a fragment of written history at the time it came into contact with European civilization" (*CE* 7:234, 237–38). This lack or possession of history is, indeed, the main nonmorphological difference that Huxley cites among the eleven stocks. Only three of the eleven stocks have been capable of producing written histories, or—to emphasize the point even farther—have had histories (as opposed to evolutionary prehistories). The Negro stock has been known (to Europeans) for centuries, but is also, from Huxley's standpoint as from Hegel's, historyless. Only the Xanthochroi, the Melanochroi, and—though within limits—the Mongolian stocks have entered history. Huxley says that "archaeological and historical investigations are of great value for all those peoples whose ancient state has differed widely from their present condition, and who have the good or evil fortune to possess a history. But on taking a broad survey of the world, it is astonishing how few nations present either condition. Respecting five-sixths of the persistent modifications of mankind, history and archaeology are absolutely silent" (7:211–12). Like Hegel and many other nineteenth-century thinkers, Huxley identifies history with progress toward civilization (and therefore, with the eventual capacity to compose written histories, even when these are largely mythological). The Xanthochroi and the Melanochroi alone, he claims, are currently civilized and still making progress. It follows that if any of the other racial stocks are to become civilized, then they must do so under the tutelage (i.e., the imperialism) of one or both of the European racial stocks.

While the Mongolians—or that branch of this stock identified as Chinese, at any rate—"have attained a remarkable and apparently indigenous civilization, only surpassed by that of Europe," they have apparently stalled (*CE* 7:229). (In *The Wealth of Nations,* the Chinese were Adam Smith's main example of a "stationary" as opposed to either a progressive or a regressive social condition.) In modern times, therefore, there are only two racial stocks making for progress in civilization. With the Xanthochroi and the Melanochroi, says Huxley, "has originated everything that is highest in science, in art, in law, in politics, and in mechanical inventions. In their hands, at the present moment, lies the order of the social world, and to them its progress is committed" (7:232). Huxley implies, of course, that if the other racial stocks are ever to become progressive or to achieve even a modicum of civilization, then they must take lessons from the two European stocks. This is the same message implicit in the colony-as-garden metaphor at the outset of "Evolution and Ethics," where the actual colony named—Tasmania—suggests the fate of those native savages incapable of gardening or of entering historical time or of becoming civilized: extinction.

In the section on the extinction of primitive races in *Descent of Man,* Darwin provided the most authoritative nineteenth-century examination of what might be called the inevitability-of-extinction thesis: "When civilised nations come into contact with barbarians the struggle is short, except where a deadly climate gives its aid to the native race" (543). Though, like Huxley, humane in his antislavery views, Darwin clearly believes that the extinction of certain primitive peoples is inevitable and ultimately for the best. At times Darwin seems to look forward to the complete triumph of civilized over primitive, or lower, races: "Remember what risk the nations of Europe ran, not so many centuries ago of being overwhelmed by the Turks, and how ridiculous such an idea now is! The more civilized Caucasian races have beaten the Turkish hollow in the struggle for existence. Looking to the world at no very distant date, what an endless number of the lower races will have been eliminated by the higher civilized races throughout the world" (F. Darwin, *Life and Letters* 1:286). Thus for Darwin, as earlier for Robert Knox in *The Races of Men,* the progress of the world dictates not just the peaceful transformation of savagery into its opposite, but its violent liquidation—the triumph of the civilized races over the lower races in the struggle for existence. In this regard, Darwin explicitly agrees with the codiscoverer of natural selection, Alfred Russel Wallace, who in 1864 wrote that the "great law of 'the preservation of favoured races in the struggle for life' . . . leads to the inevitable extinction of all those low and mentally undeveloped populations with which Europeans come in contact" (Wallace cxliv–clxv).

Despite this thesis, Wallace "became a spirited critic of modern war and imperialism" (Crook 57). Also, it seems unlikely that by all "low and mentally undeveloped populations" Wallace means all "primitive" peoples, because by the 1860s it was evident that not all such peoples were withering away upon contact with civilization. Central Africans, for instance, might not avoid being sold into slavery, but many whom that catastrophe befell survived the horrors of the middle passage and of slavery itself in the New World. The decimation of Native American societies contrasted to the successful establishment of African slavery in the New World suggested to some observers both the unfitness of Native Americans and the at least physical fitness of the Africans. Whatever the case, Huxley argues in "Emancipation—Black and White" that slavery should be abolished not because Africans are the equals of Europeans, but because slavery interferes with natural selection; the struggle for existence works to render the great inequalities among the races historically manifest, and "the highest places in the hierarchy of civilisation will assuredly not be within the reach of our dusky cousins" (*CE* 3:67).

It would be inaccurate to suggest that the idea of the inevitable extinction of primitive races was consistently or uniformly applied throughout the British Empire and its history, or that it ever received any official sanction. Indeed, the great abolitionist crusade to end the slave trade and then to eliminate slavery in all British territories, coupled with missionary belief in the convertibility—if not necessarily civilizability—of all peoples, no matter how savage, diffused a humanitarianism that helped to counteract, at least partially, opinions like that expressed by Wallace in 1864. Florence Nightingale declared in the same year: "This question of the fate of aboriginal populations is one closely concerning our national honour" (557). Her view can hardly be said to have prevailed (most definitely not in Tasmania!). Moreover, humanitarianism (both religious and secular) and social Darwinist views were often combined in various ways. Both Huxley and Darwin are major examples of intelligent Victorians who managed to be both humanitarian and abolitionist for social Darwinist reasons—in part because the struggle of race against race should be carried out, as Huxley declared, on "a fair field" with "no favour" (*CE* 3:67).

In contrast to other social Darwinists such as Benjamin Kidd, Karl Pearson, and Darwin's nephew Francis Galton, both Huxley and Darwin took a "mild attitude to the question of race" (di Gregorio 170). This relative mildness is perhaps evident in Huxley's lack of emphasis upon war as an instrument of natural selection, although Paul Crook's alignment of Huxley with Wallace as an opponent of "imperial biology" and therefore as tending toward Wallace's pacifism is misleading (Crook 54–62). In Walter Bagehot's *Physics and Politics* and in other instances of social Darwinism down to Kidd's *Social Evolution,* war is a primary mechanism of history and also of human evolution. This is also true for Huxley, though more implicitly than explicitly. The social Darwinist vision of history is a spectacle of races in collision, with the survival of the fittest (which may not mean morally best, Huxley would hasten to add) as the outcome. As in Bagehot's construction, war is also a primary mechanism of progress and therefore of civilization. And of course, to the fittest go the spoils of war, including empires which, in modern times at least, carry with them "the white man's burden" to try to civilize those perhaps semi-fit savages and barbarians who, though vanquished, do not always vanish. Huxley seems to say little about war in "Evolution and Ethics," but it is subsumed under the general category of struggle. And in the colony-as-garden metaphor, which equates imperialism with the progress of civilization, war is already over: the Tasmanian aborigines have apparently easily been extirpated from the garden along with other unwanted animals and plants (*CE* 9:19).

Despite his relative lack of stress on war, Huxley's reliance on morphology and his skepticism about the reliability of language and of cultural artifacts as historical evidence means that history in his ethnological essays necessarily appears to be a spectacle of races in collision. Communication and economic exchange are beside the point. Huxley's model for the primary interaction of races is conquest and colonization; in his essay "On Some Fixed Points in British Ethnology," the Roman, barbarian, and Norman invasions of Britain are obvious examples. The prime motivator of the great racial migrations and collisions of the past has been economic scarcity (though in "Evolution and Ethics," Huxley recasts economic scarcity in quasi-Nietzschean terms as the desire for increased "enjoyment"), brought on by Malthusian population pressure upon available resources. Writing in "The Aryan Question" about "the Teutonic inroads upon the Empire of Rome," Huxley declares: "Whatever the causes which led to the breaking out of bounds of the blond long-heads, in mass, at particular epochs, the natural increase in numbers of a vigorous and fertile race must always have impelled them to press upon their neighbours" (*CE* 7:287). For Huxley, there is apparently not much more to be said about human history or prehistory unless it sheds light on the evolutionary changes that the physical structures of the different races have undergone. But on this score, he thinks, there is also little to be said, because the main racial stocks, despite much intermingling as in the case of Britain, appear to have been unchanged during most or all of recorded history: "On the whole . . . it is wonderful how little change has been effected by these mutual invasions and intermixtures. As at the present time, so at the dawn of history, the Melanochroi fringed the Atlantic and the Mediterranean; the Xanthochroi occupied most of Central and Eastern Europe, and much of Western and Central Asia; while Mongolians held the extreme east of the Old World. So far as history teaches us, the populations of Europe, Asia and Africa were, twenty centuries ago, just what they are now" (7:238). Compared to evolutionary time, recorded historical time is from this perspective insignificant. During the short span of history, civilization may have begun to allay the struggle for existence within the bounds of certain societies, but that is hardly the case, Huxley thinks, with pre- and antihistorical savagery, and anyway modern intersocietal relations are just more of the same—the struggle continues.

An Imperial Epistemology

As a scientific empiricist and rationalist, Huxley was a child of the Enlightenment. Since Max Horkheimer and Theodor Adorno's *Dialectic of Enlightenment,* much has been written about the transformation of the scientific and democratic expectations of Enlightenment thinkers into their antitheses: technological, bureaucratic, and instrumental rationality on the one hand; totalitarianism on the other. Huxley's primary claims to fame as an exponent of evolutionary theory, as a champion of scientific empiricism and rationalism, and as a writer and teacher of extraordinary rhetorical, polemical, and explanatory power are hardly debatable. At the same time, Huxley brilliantly exemplifies major Victorian assumptions about history, society, and knowledge—assumptions situated midway along the trajectory of the transformation of Enlightenment utopianism (human perfectibility) to postmodernist skepticism, pessimism, or, perhaps, "accomplished nihilism" (Vattimo 19–29).

The key problem for anyone wishing to see Huxley as an exemplar of pure scientific rationality unaffected by ideological factors is the imbrication of all of his ideas with assumptions about the past, about human differences, and about the interactions among human societies and cultures that can only be described as Eurocentric and imperialist. According to his evolutionary ethnology, racial migrations, war, and colonization are the main events in history. The chief modern entry in the historical ledger—an "event" that has just begun to promise a much more complete separation of the ethical process from the natural realm—is the rise of science. But throughout recorded history down to the present, only those fittest races most evidently successful in war and colonization are understood by Huxley to be capable on their own of producing civilization and progress. In some unclear way, science, too, emerges from the struggle for existence; just as it was for Nietzsche and later for Foucault, science is thus a form of power (it seems to have been Bacon who first said that "knowledge is power").

Huxley's scientific epistemology is only partly that of the stereotypic empiricism that takes the humble individual experimenter or observer, emptying his mind of preconceptions, as its starting point (see "On the Method of Zadig," *CE* 4:1–23). Despite its individualism and empiricism, Huxley's concept of the cumulative, progressive essence of scientific knowledge is also an imperialist one, rendered repeatedly in metaphors of conquest and colonization, as in the first paragraph of "The Aryan Question":

> The rapid increase of natural knowledge, which is the chief characteristic of our age, is effected in various ways. The main army of science moves to the conquest of new worlds slowly and surely, nor ever cedes an inch of the territory gained. But the advance is covered and facilitated by the ceaseless activity of clouds of light troops provided with a weapon—always efficient, if not always an arm of precision—the scientific imagination. It is the business of these *enfants perdus* of science to make raids into the realm of ignorance wherever they see, or think they see, a chance; and cheerfully to accept defeat, or it may be annihilation, as the reward of error. (*CE* 7:271)

The passage continues in the same vein. It could, of course, be read as just an instance of Huxley's great rhetorical skill—a vivid but idiosyncratic or isolated analogy. But the military analogy typifies Huxley's combative approach to "the warfare between science and religion." Just as obviously, it also expresses his conception of science's imperializing drive to master and colonize all of nature.

In common with other Victorian ethnologists, including his antagonist James Hunt of the Anthropological Society (see Stocking 247–55), Huxley believed that ethnology was a science capable of providing much useful information to British imperial administrators and colonists. As president of the Ethnological Society in 1869, he "launched what amounted to an ethnological census of the populations of the British possessions" starting with India (di Gregorio 175). Huxley's ethnological goal was the mapping and measurement of all of the races of the world. Only such a total mapping would provide the evidence necessary to taxonomize the races accurately, to understand how they had evolved and interacted through prehistory and history, and to judge their greatly varying capacities for progress and civilization. No doubt Huxley believed that such a mapping of races would also provide the answer to the mystery that absorbed Darwin in the chapter on race in *Descent of Man*—namely, why some or perhaps all primitive races were dying out, apparently on mere contact with civilization. The ethnological map of human races would, of course, be also a hierarchy; it would help in the general process of eradicating savagery, if not the savages themselves.

Huxley always equates new scientific knowledge with progress, and progress with increased civilization. The sciences, moreover, are bound up with—often dependent upon—geographical discovery. The *Beagle* and the *Rattlesnake* followed in the wake of Cook's and many other explorers' voyages, helping to map and taxonomize the flora and fauna of the entire world, including those newly discovered "lowest" races such as the Fuegians, the Australians, and the Tasmanians. For the savage races, the only hope for entering historical time or for becoming at least partially civilized, lies in their being discovered, mapped, conquered, and colonized—though as with the Tasmanians, they may also disappear altogether in the course of "progressive development." Such disappearances, both Darwin and Huxley believe, are only nature taking its course—the course named struggle for existence and survival of the fittest.

Meanwhile science marches on, its progressive development unfolding with every essay that Huxley wrote. This is not to suggest that he considered himself to be a principal scientific discoverer like Darwin; but he everywhere communicates the excitement he experienced with new discoveries and with the relentless banishment of ignorance and superstition.

Huxley not only identified the progress of science unproblematically with social progress, as did most Victorians, he also identified it more generally with evolutionary progress, upward from the apes. Thus in *Man's Place in Nature,* Huxley offers multiple, overlapping narratives of progressive development. There is, first and most obviously, the evolution of man literally upward from the apes (or rather, from the apes' ancestors). Paralleling this evolution but in much foreshortened chronology, there is also the evolution of scientific knowledge, banishing ignorance, as in the opening sentence: "Ancient traditions, when tested by the severe processes of modern investigation, commonly enough fade away into mere dreams" (*CE* 7:1). This second narrative is the intellectual version of eliminating savagery from the world. It is also the recent, most important version of a more general third narrative of progress, "from blind force to conscious intellect and will," which is in turn a recent version of "Nature's great progression, from the formless to the formed—from the inorganic to the organic" (7:151). Every individual organism, moreover, and every species (also every society), undergo a related progressive development from the simple to the complex, from the acorn to the oak tree or the embryo to the adult animal. That organisms also peak, decline, and die, and that all previously recorded or unrecorded empires have declined and fallen, makes all of the processes of nature ultimately cyclical, revolving in "the procession of the great year" as in one of Yeats's gyres (9:85). But usually—as in *Man's Place in Nature*—Huxley emphasizes only "the doctrine of progressive development" (9:208).

In the vanguard of "the cosmic struggle" of nature against itself, and therefore in the vanguard of the ethical process (*CE* 9:35), the European nations and a few of their imperial offshoots (the United States, Canada, New Zealand, Australia) are the most complex, successful examples yet of humanity's ascent out of mere brutality. For Huxley, civilization is the usual name for this vanguard, though an alternative name might be scientific Enlightenment. However named, science for Huxley is that accurate knowledge or consciousness of nature and society that increasingly enables civilized humanity to control its own destiny. And with all their flaws, Britain and its colonial offshoots (even Tasmania) are the latest, most significant manifestations of that imperial and imperializing control. Perhaps after savagery had been finally wiped off the face of the earth and the half-civilized nation-states and empires of the Victorian age had become even more civilized, the time would arrive for a categorically different regime, as far removed as possible from the raw struggle for existence—an ultimate empire of science and reason.

But the late-Victorian fantasy of an imperial archive, uniting all scientific knowledge in an ultimate imperium of truth, arose at just the moment when the very possibility of such encyclopedic coherence was dissolving, fragmented partly by its own Faustian ambition and complexity. Entropy was one late-Victorian name for the incoherence caused by the accumulation of too many facts, too much information (Richards 73–109). And what was true of knowledge was also true of empires like the British, which during the course of Huxley's lifetime reached its apogee and in the 1880s began to disintegrate. Because the struggle for existence was ultimately inescapable, but perhaps also because the Enlightenment dream of total knowledge was beginning to evaporate just as the British Empire was beginning to decline, in "Evolution and Ethics" Huxley declares: "The theory of evolution encourages no millennial anticipations" (9:85).

REFERENCES

Bagehot, Walter. *Physics and Politics, or Thoughts on the Application of the Principles of "Natural Selection" and "Inheritance" to Political Society.* London: H. S. King, 1872.

Bonwick, James. *The Last of the Tasmanians: Or, The Black War of Van Diemen's Land.* London: Sampson, Low, & Marston, 1870.

Brantlinger, Patrick. "'Dying Races': Rationalizing Genocide in the Nineteenth Century." *The Decolonization of Imagination.* Ed. Bhikhu Parekh and Jan Nederveen Pieterse. London: Zed Books, 1995. 43–56.

Crook, Paul. *Darwinism, War, and History: The Debate over the Biology of War from* The Origin of Species *to the First World War.* Cambridge: Cambridge UP, 1994.

Darwin, Charles. *The Origin of Species and the Descent of Man.* N.Y.: Modern Library, n.d.

Darwin, Francis, ed. *The Life and Letters of Charles Darwin.* 2 vols. N.Y.: Basic Books, 1958.

di Gregorio, Mario A. *T. H. Huxley's Place in Natural Science.* New Haven: Yale UP, 1984.

Huxley, Thomas Henry. *Collected Essays.* 9 vols. 1894. N.Y.: Greenwood Press, 1968.

Irvine, William. *Apes, Angels, and Victorians: Darwin, Huxley, and Evolution.* N.Y.: McGraw-Hill, 1972.

Kidd, Benjamin. *Social Evolution.* London: Macmillan, 1894.

Knox, Robert. *The Races of Men: A Fragment.* London: H. Renshaw, 1850.

Nightingale, Florence. "Note on the Aboriginal Races in Australia." *National Association for the Promotion of Social Science Transactions* (1864): 552–58.

Richards, Thomas. *The Imperial Archive: Knowledge and the Fantasy of Empire.* London: Verso, 1993.

Stocking, George. *Victorian Anthropology.* N.Y.: Free Press, 1987.

Vattimo, Gianni. *The End of Modernity.* Baltimore: Johns Hopkins UP, 1985.

Wallace, Alfred Russel. "The Origin of Human Races and the Antiquity of Man Deduced from the Theory of 'Natural Selection.'" *Journal of the Anthropological Society* 2 (1864): clviii–clxxxvii.

The Technical Illustration of Thomas Henry Huxley

GEORGE R. BODMER

> The whole of my life has been spent in trying to give my proper attention to things and to be accurate. . . . You cannot begin this habit too early, and I consider there is nothing of so great a value as the habit of drawing, to secure those two desirable ends.
> —Huxley, "On Science and Art"

IT IS A NATURAL AND LOGICAL STEP TO CONSIDER THE illustration of the scientist Thomas Huxley, who could use the example of a lobster, a piece of chalk, or even the horseshoe shape to explain gracefully the history and principles of biology to his audience. He contributed sketches to his scientific articles and knew well the relationship between the visual and the written, between word and picture. Enthusiastically assuming the visual attitudes of his world, his lectures were advertised as "Profusely Illustrated," and that which he observed throughout his life, he recorded with pen, pencil, and paint. As technology made the publishing of pictures easier and more accurate, especially for scientific illustration, his was an increasingly more visual age, a trend that continues in our own.

In an oft-quoted passage from his taciturn autobiographical sketch, Huxley reflected, "I can hardly find any trace of my father in myself, except an inborn faculty for drawing, which unfortunately, in my case, has never been cultivated, a hot temper, and that amount of tenacity of purpose which unfriendly observers sometimes call obstinacy" (*CE* 1:4). In editing his grandfather's journal and sketches for the account of his scientific voyages of 1846–50, Julian Huxley wrote, "My grandfather had a natural artistic gift, as these sketches will show. As one well-known artist to whom I showed them exclaimed, 'What courage amateurs have! Huxley attempts what most of us would be afraid of tackling, and only the greatest artists could tackle successfully: Yet sometimes he all but pulls it off.' . . . . Huxley was a great scientist and a great man of letters; if he had had time and opportunity, he would have been a great artist as well" (*Diary* 5).

It should be clear from Huxley's writings that he was a close observer of nature. Passages from this early journal (he was twenty-one when the naval research ship HMS *Rattlesnake* left England) demonstrate a strong visual sensitivity in his writing. He may have been trying to dramatize his writing in the manner of a travelogue, but his description of Madeira early in the voyage reads as though he were describing a painting.

> So in this island, a huge monument of some awful volcanic phenomenon—made up of wild peaks with intervening deep gullies and ravines, often carrying within them clear evidence of the ravages of mountain torrents. Nature has so kindly and artistically arranged the various parts: has so softened the variegated brightness of the hills, with here a soft white cloud resting on its summit, here a deep unfathomable-looking valley, and there a patch of vegetation; putting in by way of frame a sky and sea of the bluest blue, and at times as it were in pure wanton sport half decking, half hiding the mountain scenery with rainbows, that the whole produces a soft and harmonious scene on which the eye dwells without novelty. Much of my own pleasure of course arises from mere novelty. Mountain scenery is new to me. (*Diary* 17–18)

He often uses the term "scenery," as in this passage from December 12, 1847, "Cape Upstart is one of the most barren places I have seen, but the rock scenery is grand and rugged" (*Diary* 92). It was a rather recent word in Huxley's time, less than a century old, and still suggested the sense of the scenes in a theatrical production.[1] Carrying his sketchbook with him, Huxley, at least at this point in his thinking, sees nature as a "framed" set to be observed. Another passage, describing Simon's Bay in Malaysia, dating from April 16, 1847, still more sharply shows this view of nature on display:

> The surrounding scenery is grand and rugged, but very arid and sad looking, the vegetation being mainly confined to low dusty-leaved shrubs and heaths which show their beauty only on a near approach. But perhaps to this very circumstance, the absence of any foliage to interfere with the deeper and warmer tints of the soil, may be owing to the gorgeousness of the tints as-

> sumed by the high mountain of the Cape Hanglip side of the bay, when lighted up by the rising of the setting sun. I have never seen anything more beautiful in paintings, where indeed until now I always judged their more scenic effects to be mere painter's license. (*Diary* 34)

Earlier he had used the term "artistically arranged" to describe landscape; his was a voyage to *see* the rest of the world, to catalog the rest of the living things, and by extension also to discover the grand scheme of existence.

On the voyage, Huxley made sketches of the natives and their houses and boats, of the marine animals he had skimmed from the surface of the ocean in his studies, as well as light-hearted drawings of conditions aboard ship for his diary (see fig. 1). At one point he lost his drawing book overboard, and reported a week later,

> I took advantage of the opportunity to review my examination and verify the reproduced lost sketches. Not only did I succeed in doing this but many other new and singular points of structure came out of which I made detailed drawings.
>
> February 25th (1847)
>
> In the morning I made a drawing of the animal of the Ianthina. The figures in Mrs Grey's book are very bad. I also endeavoured to make a dissection of one, but what between the rolling of the ship and the small size and delicacy of the object, I did not succeed at all well. (*Diary* 32)

These entries underscore the fact that drawing was to Huxley an analytic activity, through which he could discover traits of anatomy in those animals he was dissecting. The near-transparent invertebrate sea creatures he was investigating lent themselves to diagram-like pictures. His drawings from these early voyages were used to illustrate his research published in the *Philosophical Transactions of the Royal Society* as well as the chief science officer MacGillivray's *Voyage of the HMS* Rattlesnake, though Huxley was dissatisfied with the reproductions in the latter, which became stylized and softened as they were transferred into woodcuts and lithographs.

Late in life, in his 1882 address to the Liverpool Institute, "On Science and Art in Relation to Education," Huxley clarified his sense of the importance of teaching students to draw:

> I should, in the first place, secure that training of the young in reading and writing, and in the habit of attention and observation, both to that which is

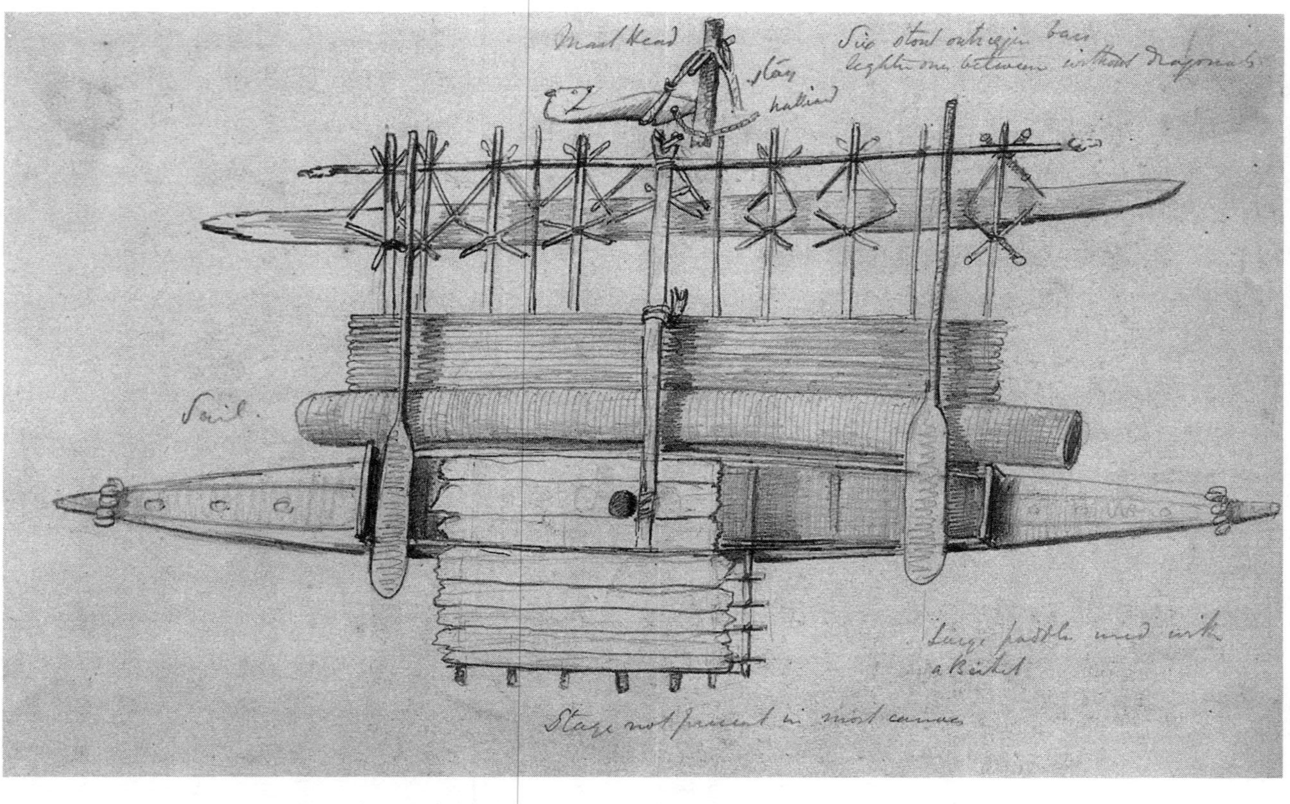

FIGURE 1. Native raft, pencil drawing. By permission of the Archives, Imperial College, London

> told them, and that which they see, which everybody agrees to. But in addition to that, I should make it absolutely necessary for everybody, for a longer or shorter period, to learn to draw. . . . Writing is a form of drawing; therefore if you give the same attention and trouble to drawing as you do to writing, depend upon it, there is nobody who cannot be made to draw, more or less well. . . . I do not think its value can be exaggerated, because it gives you the means of training the young in attention and accuracy, which are the two things in which all mankind are more deficient than in any other mental quality whatever. (*CE* 3:183–84)

Just as his own drawing helped him to analyze his specimens and his dissections, Huxley used his blackboard sketches as a tool in his lectures (see fig. 2). These were recalled years later by his students (at least one noted that Huxley's sketching allowed him to catch up on his notes). Rev. W. J. Sollas, in the supplement to the 1925 issue of *Nature* commemorating the Huxley centennial, wrote, "The diagrams in chalk, drawn from memory on the blackboard, often as a running accompaniment to a description, shared in the same admirable qualities as the spoken words. They were masterly performances, the cod's skull in particular was a triumph. Those who have watched this sketch growing, as bone was added to bone, until this complex structure stood revealed as a whole and in all its parts, will not soon forget the pleasure with which they watched this notable performance" (747).

This sentiment can be found in a number of other reminiscences, such as that of Jeffery Parker: "His blackboard illustrations were always a great feature of his lectures, especially when, to show the relation of two animal types, he would, by a few rapid strokes and smudges, evolve the one into the other before our eyes" (*Life* 2:435). Toward the end of his life, as a governor of Eton College, he advocated that drawing instruction be required in the curriculum, though of course he had had no formal training in it himself (2:3).

The books he published tended to use illustrations by colleagues (e.g., G. B. Howes or W. H. Wesley) or pictures from older texts. His volume *Man's Place in Nature* (1863) is particularly distinctive because as Huxley is describing "Man-Like-Apes" for his presumably skeptical and uncomfortable Victorian audience, he accompanies his words with pictures from the past, when naive artists gave the apes human appearances. For instance,

FIGURE 2. The young, bearded lecturer at work, drawing a gorilla skull. By permission of the Wellcome Institute Library, London

he included a sixteenth-century woodcut of boot-wearing chimps, an eighteenth-century picture of a mannish mandrill, and tool-holding, tail-bearing apemen from Linnaeus, as well as more naturalistic pictures of gorillas and an orangutan. In a book about *Frankenstein,* U. C. Knoepflmacher writes, "In *Man's Place in Nature* Huxley relentlessly brings his civilized readers 'face to face' with 'blurred copies' of themselves, a succession of seemingly alien monsters. And, like Mary Shelley, though for purely scientific reasons, he uses these deformed 'copies' of human beings to force their presumed 'originals' into an act of kinship that will involve a humiliation of their pride" (317).

James P. Paradis is even more struck by the effectiveness of the pictures in *Man's Place,* particularly in a middle section of the book:

> In this chapter Huxley's illustrations take on a symbolic, iconographic significance, something which could never occur in a strictly technical work but which here, given the framework in which Huxley set his study, was nearly inevitable. In endeavoring to establish it in the context of the great systems of human thought, Huxley gave his little treatise symbolic representative status. The most striking example of the symbolism of his illustrations, his scientific iconography, occurs on the first page of the second chapter where photographic reductions of side views of skeletons, beginning with the gibbon, progressing through the orang, chimpanzee and gorilla, and ending with man, are lined up. This simple illustration communicates Huxley's essential vision—a skeletal apocalypse. In its own homely way, it contains an idea as profound as some great work of art, for in these skeletal remains, these artifacts from the scientific bone closet, man stands behind his relations, intimate with nature beyond the wildest dreams of the Romantics. . . .
>
> The second chapter is a visual boneyard, with Huxley making use of comparative illustration to show similarities in nearly every major bone cluster of the primate form. The entire primate order emerges as a progression of variations; the similarities are undeniable. (129–30)

Scientific illustration probably changed more in Huxley's lifetime than in any other period until the recent computer era. At the beginning of the nineteenth century, many book illustrations were wood cuts, as they had been since the beginning of movable type. Printing companies, consequently, kept on hand a file of stock cuts. Sharper-edged pictures could be obtained from etchings or engravings, popular since the middle of the eighteenth century. But since they required that ink be placed under the surface of the plate (in a groove or engraved line), illustrations had to be printed on a separate page from the text. Woodcuts, like type, use a relief process, with the ink placed on the surface of the plate.

One of the most significant events in zoological illustration was the colored engravings of John James Audubon's *Birds of America.* Audubon's project was unique because he tried to portray even the largest birds life-size, producing 435 plates in the so-called double-elephant folio, the largest format published to that date. Though he made his paintings in the 1820s, the two hundred sets of the folios (selling for an impressive one thousand dollars each) were issued between 1827 and 1838 (Peterson and Peterson 4–9). Significantly, 1839 saw the invention of photography in France, which would move the world away from such marvels of hand-drawing. The obvious benefit of this technology ensured that it was soon disseminated throughout most corners of the world.[2]

The improvement of lithography earlier in the century provided for greater detail and tone in drawings. Likewise, the refinement of wood-engraving (in contrast to woodcuts) in England in the 1860s permitted a sharper, more precise line. Lithography, wood engraving, and woodcuts alike required that the original hand drawing be transferred to a block, usually by a craftsman in an engraving firm. While these workers were often highly skilled, the intervention of another hand would often change a picture, in small or large ways. Photography's advent changed the way people viewed objects, and toward the end of the century it also changed printing; henceforward, a plate could be produced photographically directly from the artist's drawing, in effect removing the middle man—essentially the means of printing that we are still using, in offset lithography, for instance.

Though in our day we are enamored of the camera's image, photographs tend best to illustrate single specimens with all the anomalies of an individual, while a drawing can show the general or characteristic aspects of its subject (Knight 67–68). Huxley by and large used drawings rather than photographs even for his latest books, yet his eye must have been shaped by the reality of the process used in reproducing the images he printed. For instance, the drawings of fossils embedded in stone, which accompany "The Crocodilian Remains Found in the Elgin Sandstones with Remarks on the Ignites of Cummingstone" from 1877, seem informed by photographs (*SM* 4:188–247). Nevertheless, Huxley tended to draw from dissected specimens in pencil, stressing the characteristics he was most interested in displaying and rounding out the individual anomalies.

Just as Paradis could describe chapter 2 of *Man's Place* as a "visual boneyard," to read Huxley's scientific articles is to be struck by the many accompanying and illuminating illustrations; it is similarly compelling to be leafing through his papers and come upon his drawings of his life in the South Pacific or his later vacation sketchbooks. Clearly, when he viewed objects in his world, he "saw" in them not only their actual appearance but their history, development, and underlying structure, and as this complex visual response led him to explore and investigate them, he felt the need to describe them in pictures as well as words. His articles are often accompanied by drawings, though he often deferred to a more skillful artist, and as he became busier and important enough to command the support, his own pictures were used less. When he returned in 1850 from the voyages on the *Rattlesnake,* he brought over 180 sheets of drawings back with him, many with multiple pictures (*Life* 1:62); these pictures were the best he produced and he used them throughout his career, just as he would continue to make use of the scientific research he performed in the South Pacific. The four volumes of his *Scientific Memoirs,* edited by Foster and Lankester in 1898, reprint 163 articles, from 1845 to 1894. In addition to textcuts, 128 plates are reprinted, with approximately 38 identified or indicated as Huxley's (25 in vol. 1, 11 in vol. 2, none in vol. 3, 2 in vol. 4).[3] His own pictures adorn two of the last articles to be published in volume 4 of the *Scientific Memoirs,* "The Characters of the Pelvis in the Mammalia, and the Conclusions Respecting the Origin of Mammals which May be Based on Them" (1879; *SM* 4:345–56) and "The Gentians: Notes and Queries" (1888; 4:612–35).

In terms of technique, some of Huxley's pictures, especially his landscapes, are quite beautiful, and he was adept in his use of color and in the painting of watercolor. The paints and materials available to him are similar in quality to those available today.[4] For instance, in a sketchbook of the Pyrenees, dated 1890, pictures are painted mostly in watercolors, but touched in with pencil and thin ink washes. For the white of waterfalls and snow on the mountains, he scraped the paper with a knife, a technique he also used in the *Rattlesnake* pictures (see fig. 3). In the former are two particularly stunning paintings of jellyfish, touched up with color, even though

FIGURE 3. New Zealand waterfall (originally in color).
By permission of the Archives, Imperial College, London

for the kind of scientific publishing Huxley did, only black-and-white pictures were reproduced (see fig. 4). In these, he wasted few lines on the decorative, stressing instead outline and significant anatomical detail. D. R. Newth writes, "It is clear that for Huxley the text figure or the plate was always the servant of the text and he did not let himself be tempted into publishing unnecessary pictures because they were pretty" (Newth and Turlington 72). His shading is adeptly handled to indicate depth, and he was skillful at rendering layers transparent to show underlying structure. By contrast, he never mastered the knack of drawing realistic-looking people; instead they often come out rather stylized or childlike, but of course these rarely figured in his scientific illustrations. Huxley drew in pencil and touched up the sharpness of his drawings with ink, or drew in ink and added the shading with pencil. Despite his earlier unhappiness with the MacGillivray pictures, comparing Huxley's original drafts with his printed pictures shows that he, like other artists of the day, was extremely well

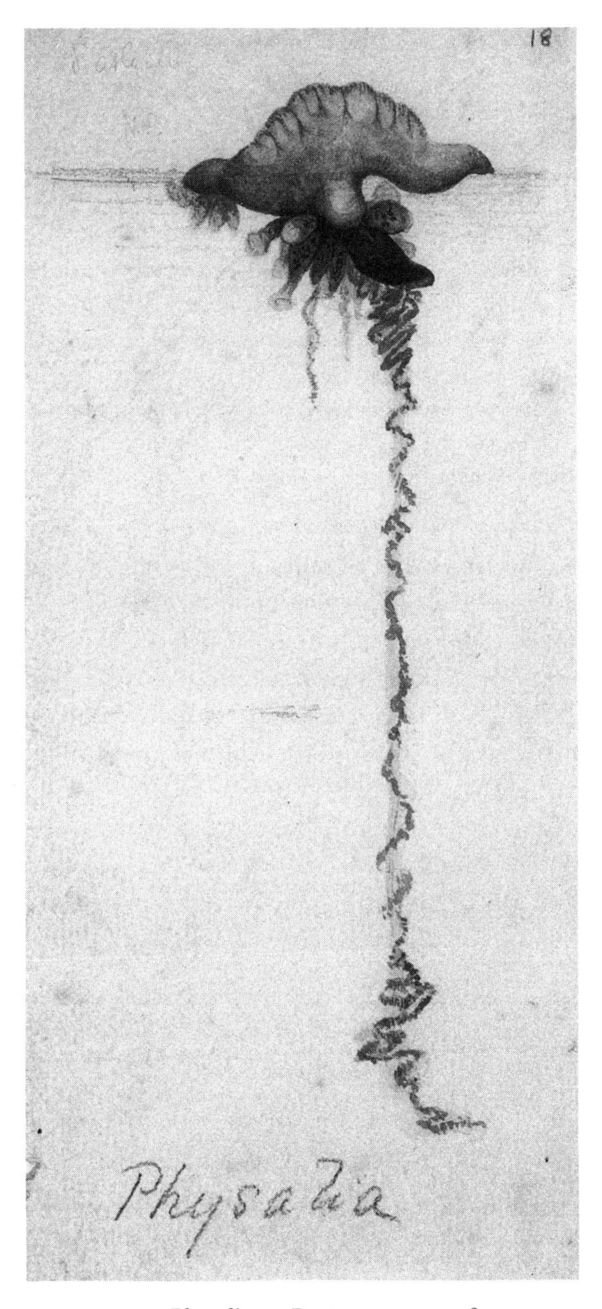

FIGURE 4. Physalia, or Portuguese man-of-war.
By permission of the Archives, Imperial College, London

served by the engravers, who were quite skillful, changing lines, for instance, chiefly for clarity.

In a brief written comment in 1956, E. R. Turlington, an artist of biological drawing for University College, London, found Huxley's drawings

> so skillfully executed and so confident that it is difficult to realize that he was, artistically speaking, an amateur. On the other hand, it must be remembered that he lived in an era when good amateur art was not exceptional, as specialization did not then restrict the work of an individual as it does to-day. In Huxley's drawing, however, there is more to admire than mere technical skill. His work is illuminated by the fascination which the structure of animals held for him and his conviction of the importance of his findings." (Newth and Turlington 74)

A good example of Huxley's adept use of drawing to support his text occurs in his article "Preliminary Note upon the Brain and Skill of *Amphioxus Lanceolatus*," where he offered a tentative but illustrated answer to Agassiz, Haeckel, and Semper, who would reclassify "the singular little fish *Amphioxus Lanceolatus*" out of the fishes "by reason of the supposed absence of renal organs and of any proper skull and brain" (*SM* 4:26–31). By examining and illustrating the nerves of this fish, Huxley argued that *Amphioxus* does have crude versions of the requisite organs to be included in the class of Pisces. Four diagrams are produced on a single page and explained in the lengthy "Explanation of the Figures" at the end (see fig. 5). He emphasized the accuracy and usefulness of these figures:

> A, C, D are diagrammatic, but accurate, representations of the anterior part of the body in *Amphioxus* (A), in an *Ammocoete* 1.6 inch long (C), and in a fully grown *Ammocoete* 5.7 inches long (D). B is a copy of the furthest advanced stage of the young *Petromyzon planeri* six weeks after hatching, as figured by Schultze in his memoir on the development of that fish. The figures are magnified to the same vertical dimension, so as to afford a means of estimating, roughly, the changes in the proportional growth of the various parts of the head of the Lamprey in its progress from the embryonic towards the adult condition" (*SM* 4:30–31).

As printed, the plate bears no title or labels other than key letters, numbers, and abbreviations. The engraving is crisp, sharp, surrounded by ample white space, and especially clear to examine. While it is modern convention

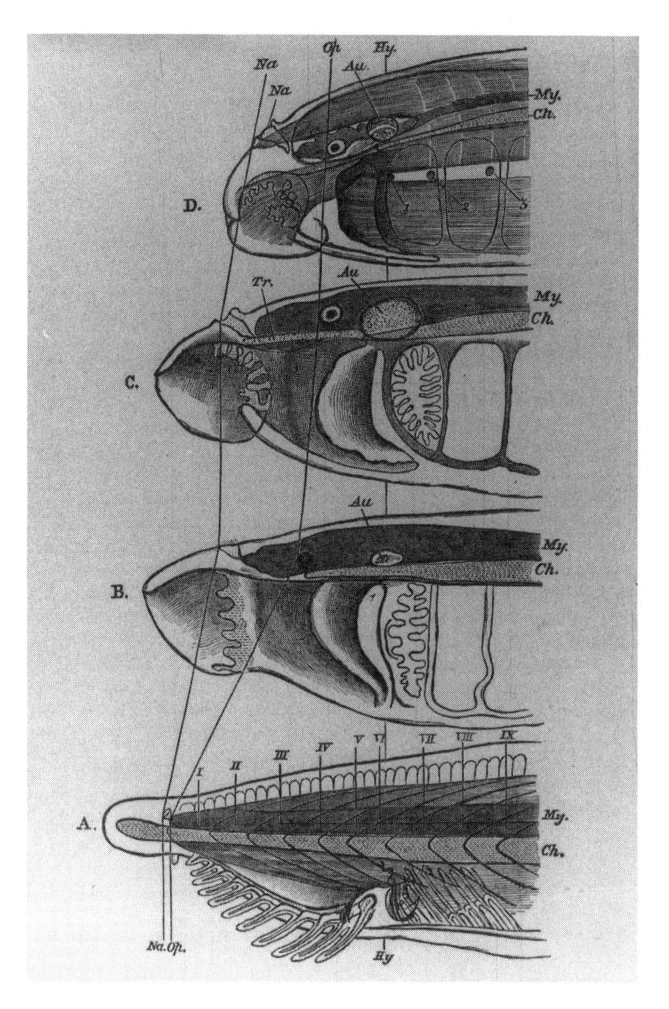

FIGURE 5. Illustration for "Preliminary Note upon the Brain and Skull of *Amphioxus Lanceolatus*," from *Scientific Memoirs*

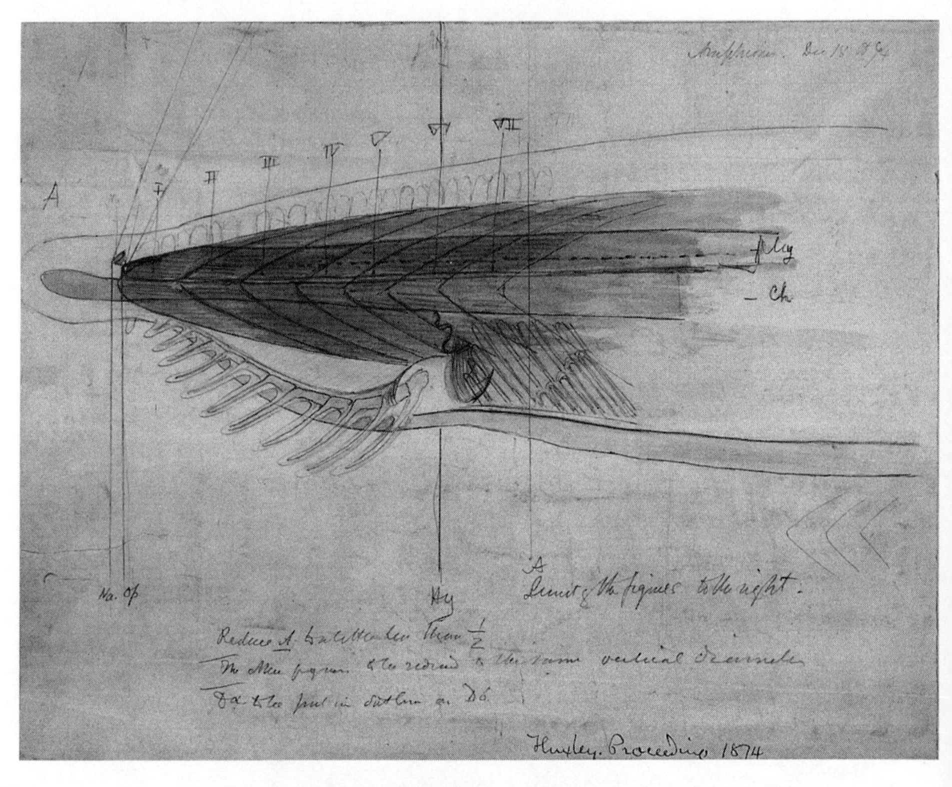

FIGURE 6. Huxley's sketch of *Amphioxus*. By permission of the Archives, Imperial College, London

to place illustrations at that point in the text where the pertinent subject is discussed, publishing practices in the nineteenth century often printed the pictures in separate plates, as is done here. On the other hand, it must be said that Huxley frequently intermingled words and pictures elsewhere, especially in his textbooks, but here the four fish cross-sections are lined up for comparison. The plate bears no assignation of artist or engraver, but the drawings exist in the Huxley archives.

Those original drawings are large, three times as large as their printed versions, slightly cruder than most of Huxley's other works, made with pencil and shading on squared drafting paper (see fig. 6). The use of the personal pronoun in the article—"I have found nothing representing an auditory organ in this position" (*SM* 4:30)—suggests that at least some of these drawings were made from specimens, "drawn from life." One, included for comparison, was copied from the printed article of another scientist.

As was generally the case, Huxley was well served by his engraver, who reproduced line for line his cut-away of organs, cilia, and irregular surfaces. Because of his particular interests and studies, many of the animals Huxley sketched tended to be diagrammatic anyway, and he was adept at rendering different layers in transparent anatomy through shading. In the case of *Amphioxus,* the engraver replaced Huxley's scrawled labels with print type, sharpened the scientist's sketched lines, and stylized his shading as finely dotted and parallel lines, which the medium requires. A steel engraving uses lines and cannot produce a solid black; if this were a lithograph, for example, the shading would be graded gray. The sketches on three large sheets in the original bear instructions to the printer in Huxley's handwriting for scale and alignment; for instance, he indicates the squaring off of the specimens on the right margin with a ruled line, "Limit of the figures to the right." Clearly the look and arrangement of the graphic on the page were determined by the author.

Contemporary rhetorical theory in the area of graphics provides a basis for the evaluation of scientific illustration. Sam Dragga, in "Evaluating Pictorial Illustrations," uses questions such as these to judge the effectiveness of pictures:

> Syntactic unity: Do the various markings constitute a single picture?
>
> Semantic unity: Are the characteristics of the pictured object distinctive enough to identify the object?
>
> Pragmatic unity: Is the accompanying contextual information sufficient to comprehend the pictured object?
>
> Syntactic location: Is value and texture used effectively to designate depth or distance of the pictured object?
>
> Semantic emphasis: Is the pictured object intrinsically interesting?
>
> Semantic text parallels: Is the picture explicitly linked to accompanying verbal explanation? (52)

Such an approach attempts to consider the reader-viewer and to establish the connection between text and picture in terms of the rhetorical effect.

The four fish Huxley has reproduced are cleanly rendered and draw the viewer to them. He was addressing, in the *Proceedings of the Royal Society of London,* an audience that might have been expected to have some knowledge of the attendant debate and that would have brought a prior interest to the pictures. The pictures are noted in Huxley's prose text, keyed by the captioned letters, and discussed in great technical detail in the final note. In fact, "Explanation of the Figures" (a device he often used) allowed him to shift some of the more technical details out of the discursive article. The plate is only marked with keys and could not stand independently without close recourse to the article and the explanation. Likewise, the scale is visually misleading without it, since one fish at 1.6 inches and another at 5.7 inches are drawn the same size. Nevertheless, the significance of this illustration is comparative; the nasal sacs (marked *Na*), the eyes *(Op)*, and the region of the auditory sac and hyoidean arch *(Hy)* are delineated by vertical lines to show how the *Amphioxus* corresponds anatomically to the *Ammocoetes* and the *Petromyzon planeri* (lamprey), thereby visually making Huxley's point. It is as if Huxley were offering the dissected specimens for the reader's perusal, but marked and made comparable in size to stress the pertinent correspondences. The plate is a highly effective illustration with a clear sense of what a picture can add to a written text and the way a reader would view it.[5]

In nineteenth-century England, drawing was a common component of a schoolboy's education, and by the time Huxley made his voyages to the South Pacific, it was a pastime that contributed to his scientific investigations as well as helped him fill the hours. Even though he was recording the flora and fauna and the local people, it is clear that he enjoyed painting his watercolors and developed an eye for the visual, that is, rendering life in two dimensions. Later, when his keen visual sense would merely have helped him in directing the artists and printers who illustrated his texts, he nevertheless continued to use his own drawings in his words. He skillfully used pictures as a rhetorical device in books like *Man's Place in Nature,* as a teaching tool in his textbooks and lectures, and as an extension of his text in his scientific articles. Seeing and drawing were two more analytical tools with which to understand and to explain his world. Roland Barthes was describing this inclusiveness when he wrote of Jules Verne, Huxley's contemporary: "His tendency is exactly that of an eighteenth-century encyclopaedist or of a Dutch painter: the world is finite, the world is full of numerable and contiguous objects. The artist can have no other task than to make catalogues, inventories, and to watch out for small unfilled corners in order to conjure up there, in close ranks, creations and the instruments of man" (65). Evolving the implications of Darwin's theory, traveling to the antipodes, chasing the odd ends of taxonomy, all these bespeak a sense of the numerable and contiguous world. Whether it was Huxley's art or his tool, drawing was a way of disciplining and extending his sight, uncovering the corners of the world and the hidden parts of nature.

NOTES

1. This visual sense of description is not uncommon in books contemporary to Huxley. For instance, in the original book about the voyages of the HMS *Rattlesnake,* the expedition's chief naturalist, John MacGillivray, writes, "The harbour presented a busy scene from our anchorage. The water was alive with small craft of every description. . . . The sea breeze died away, and was succeeded by a sultry calm; after a short interval, the grateful land wind, laden with sweet odours, advanced as a dark line slowly stealing along the surface of the water, and the deep boom of the evening gun echoing from hill to hill may be said appropriately to have closed the scene" (18). Likewise, Charles Darwin, in *The Voyage of the* Beagle, writes, "One day, two of the officers and myself rode to Ribeira Grande, a village a few miles eastward of Porto Praya. Until we reached the valley of St. Martin, the country presented its usual dull brown appearance; but here, a very small rill of water produces a most refreshing margin of luxuriant vegetation. In the course of an hour we arrived at Ribeira Grande, and were surprised at the sight of a large ruined fort and cathedral. This little town, before its harbour was filled up, was the principal place in the island: it now presents a melancholy, but very picturesque appearance" (3).

2. For example, John L. Stephens made an expedition to eastern Mexico in 1841 to write his historically significant *Incidents of Travel in Yucatan,* which reintroduced the Western world to some of the great sites of Mayan culture. In doing so, his group set up a photographic portrait studio in the Yucatan capital of Merida and carried the daguerreotype equipment into the jungle to help the artist Catherwood, who made more than 120 engravings for the printed volumes.

3. One of the problems in researching Huxley's drawings and their transference to a medium that could be reproduced in his writings, that is, comparing a drawing to its printed version, is that although Huxley obviously valued his drawings, he was probably not scrupulous in valuing each individual picture. Therefore, while many of his investigative drawings exist in the Huxley Archives in the Imperial College, London, when a drawing was forwarded to the craftsman who would make

an engraving or lithograph of it for publishing, the drawing, kept or destroyed by the engraver, or kept by the printer for comparison, often did not find its way back to Huxley's files.

Likewise, it is not always easy to determine who produced which illustration for Huxley's articles. The printed pictures are sometimes marked "T.H.H. del.," meaning "T. H. Huxley *delineavit*," or "Huxley drew it." "Ad. Nat. Delt. T.H.H." suggests Huxley made the sketch from nature. Credit to the printer, lithographer, or engraver may be assigned as "G. Jarman sc.," "Jos. Dinkel, lith. Ad. Nat.," "W. West, imp.," or "A. Hammond, lith."

4. Reba Fishman Snyder notes, "Prepared, dry-cake watercolors were widely available after about 1780. Moist colors, which usually contained gum arabic, glycerine, and sugars, were stickier than dry cakes but not really wet. Watercolors in tubes were marketed in 1846 by Windsor & Newton" (67).

5. Huxley followed up this article in 1876 with his discussion of the lamprey and the tadpole skull in "On the Nature of the Craniofacial Apparatus of *Petromyzon*," in *Journal of Anatomy and Physiology*, (*SM* 4:128–44).

Lancelets, small, fishlike, and pointed on both ends, are now classified as invertebrates, as cephalochordates. Cecil Starr and Ralph Taggart, *Biology: The Unity and Diversity of Life*, 5th ed., Belmont, Calif.: Wadsworth, 1989, p. 677.

REFERENCES

Barthes, Roland. "The *Nautilus* and the Drunken Boat." *Mythologies*. N.Y.: Farrar, Straus & Giroux, 1972.

Blinderman, Charles S. "Thomas Henry Huxley." *Scientific Monthly* 84 (1957): 171–82.

Darwin, Charles. *The Voyage of the* Beagle. Annotated by Leonard Engel. N.Y.: Doubleday, 1962.

Dragga, Sam. "Evaluating Pictorial Illustrations." *Technical Communication Quarterly* 1.2 (spring 1992): 47–62.

Huxley, Leonard. *Life and Letters of Thomas H. Huxley*. 2 vols. N.Y.: Appleton, 1901.

Huxley, Thomas Henry. *Collected Essays*. 9 vols. 1894. N.Y.: Greenwood Press, 1968.

———. *T. H. Huxley's Diary of the Voyage of HMS* Rattlesnake. Ed. Julian Huxley. London: Chatto & Windus, 1935.

———. *The Scientific Memoirs of Thomas Henry Huxley*. Ed. Michael Foster and E. Ray Lankester. 4 vols. plus supplement. London: Macmillan, 1898–1903.

Knight, David. *Zoological Illustration: An Essay towards a History of Printed Zoological Pictures.* Folkestone, Kent, Eng.: Dawson, 1977.

Knoepflmacher, U. C. "'Face to Face': Of Man-Apes, Monsters, and Readers." *The Endurance of Frankenstein.* Ed. George Levine and U. C. Knoepflmacher. Berkeley: U of California P, 1982. 317–24.

MacGillivray, John. *Narrative of the Voyage of HMS* Rattlesnake (Australiana facsimile editions no. 118). 1852. Adelaide: Libraries Board of South Australia, 1967.

Newth, D. R., and E. R. Turlington. "The Drawings of T. H. Huxley." *Medical Biology Illustrated* 6 (1956): 71–76.

Paradis, James G. *T. H. Huxley:* Man's Place in Nature. Lincoln: U of Nebraska P, 1978.

Peterson, Roger Tory, and Virginia Marie Peterson. *Audubon's* Birds of America. N.Y.: Abbeville, 1991.

Snyder, Reba Fishman. "Complexity in Creation: A Detailed Look at the Watercolors for *The Birds of America.*" *John James Audubon: The Watercolors for* The Birds of America. Ed. Annette Blaugrund and Theodore F. Stebbins Jr. N.Y.: New York Historical Society, 1993.

Sollas, W. J. "The Master." *Nature* 115 *Supplement* (May 9, 1925): 747–48.

Thomas Henry Huxley, H. G. Wells, and the Method of Zadig

BRUCE SOMMERVILLE AND
MICHAEL SHORTLAND

T. H. HUXLEY'S WORK HAS NEVER LACKED PRAISE, BUT THAT praise has seldom lacked qualification. In *Apes, Angels, and Victorians,* an outdated but still much-read account of Darwin and Huxley, William Irvine evokes Huxley as a man whose writing is powerful but nevertheless "arid" and whose mind was "antiseptically free from speculation" (10–11, 24). Spinning one of the many fancies in his book, Irvine invites us to consider Huxley reading his daily newspapers. There he would be "seated bolt upright in his laboratory, with his microscope on one side and his dissecting knives on the other." Drawing quickly to his moral, and bringing his criticism to a point, Irvine concludes that Huxley "allowed himself no prejudices, no sentimentalities, no illusions. He sometimes faced facts so courageously that they bent over backwards" (184–85).

Irvine's portrait burlesques the view that Huxley's stand against the bishops was as unyielding as the fossiliferous strata and his defense of evolution as dry as the bones of paleontology. And indeed, this is precisely the self-image Huxley strives to cultivate in his lectures and essays where he speaks as "a worshipper of the severe truthfulness of science" ("Scientific Aspects" 653). He repeatedly insists on the mental discipline and ruthless honesty required of the scientific investigator. In "Agnosticism" Huxley states that the demands of scientific truth can be met by the rigorous application of only a single principle. It is a principle as old as Socrates, the foundation of the Reformation, the canon of Descartes, the very axiom of modern science.

Positively stated, it instructs: "In matters of the intellect, follow your reason as far as it will take you, without regard to any other consideration. And negatively: In matters of the intellect do not pretend that conclusions are certain which are not demonstrated or demonstrable" (*CE* 5:246). Although the original agnostic, Huxley is conscious of how far he falls short of this ideal. But to accept his self-made image at face value is to fall victim to the very rhetoric that is Huxley's forte, the rhetoric by which he takes the high moral ground from the bishops in his role as secular theologian removing obstacles in the way of science (Jensen 125–42). His hot-flushed rhetoric, indeed, tends to suspend disbelief and carry us to his demonstrable, speculation-free conclusions. We must therefore be aware of the pitfalls Huxley's rhetorical strategies pose for us as interpreters of his work and strive to see what he *does* as well as what he says.[1]

The received view of H. G. Wells is quite different from that of his mentor Huxley. It is, as a matter of fact, almost the reverse. Praise Wells's writing certainly attracts, for its wide-ranging energy and generous eclecticism, but this is qualified by the criticism that the corpus as a whole is overly imaginative, even romantic, and lacks verisimilitude. Interestingly, Wells's and Huxley's supposed strengths and weaknesses are such as to make the former's work far more congenial to late twentieth-century readers. Wells's books are, one might say, difficult to put down, Huxley's easy not to pick up. Bernard Bergonzi, writing about such works as Wells's *Time Machine, Invisible Man,* and *War of the Worlds,* finds them captivating but adds that "despite their air of scientific plausibility, [they] are much more works of pure imagination. They are, in short, *fantasies,* and the emphasis should be on 'romance' rather than 'scientific' " (39).

It is not our wish to present a detailed account of contemporary estimates of either Huxley or Wells, nor of the vagaries of their reputations over the past century. We do not think that the often heated denunciations of either man and his work call for our reproving rally to their defense. Our purpose is to show, then search, the extensive common ground that exists between Huxley and Wells, and then to explore the quasi-fictional technique (which we term "conjection") that each deploys at the center of his own project. If Wells and Huxley had not suffered to receive such divergent readings from critics, the connection between them would perhaps be thought obvious, even banal. Wells, after all, spent some time under the direct tuition of Huxley at the Normal School of Science in 1884, just before Huxley's retirement. The impact on him, and indeed on fellow students, was enormous (Wells, "Huxley"). In the years that followed, Wells adopted an interpretation of Darwinian biology, a philosophical outlook, and a literary style very close to Huxley's. Wells would be described simply as following in his master's footsteps. We might even picture him reading his newspapers "seated bolt upright."

While we believe the comparison between Huxley and Wells is illuminating, this essay is principally concerned with Huxley, with his work and with some solidly encrusted Huxleyan myths. Our work, like this essay collection as a whole, is, we think, timely, for recent movements in the literary magma have begun to soften and crack the Huxley monolith. Criticisms of Huxley have recently been attenuated, and more interestingly, Huxley's essays have begun to be viewed as achieving a synthesis of the scientific and the literary by their appeal to our sense of awe, mystery, and beauty and in their use of rhetoric, metaphor, personification, humor, and poetic symbolism.[2] We hope to be able to extend such approaches, and to do so by taking as our major reference points Huxley's important essays "The Progress of Science" (1887; *CE* 1:42–129) and "On the Method of Zadig" (1880; *CE* 4:1–23). The first informs us that Huxley is strongly anti-Baconian, aware of the theory-dependent nature of observation, inclined to the invention of speculative hypotheses, and alive to the limitations of the truth claims of science. Huxley's essay on Zadig's method of prophecy authorizes the invention of hypotheses in scientific and literary practice. Using a quasi-fictional technique of scientific hypothesizing, Huxley pursues evolutionary theory into the remotest corners of time and space. At the point where imaginative creativity is highest, the scientific substrate enters the literary receptacle to create a highly conjectural form of literature. It is in this literature that Wells and Huxley find common ground.

Huxley on Hypothesis, Observation, and Truth

In "The Progress of Science" Huxley advances an anti-Baconian philosophy of science. He describes Bacon's attempt to summarize the past of physical science and provide a method to guide its future as "a magnificent failure." The view that a formulaic kind of method or industry will lead to progress in science is the greatest of delusions, for it makes no allowance for "motherwit," for the play of genius, or for hypothesizing (*CE* 1:46–47). Huxley observes: "To anyone who knows the business of investigation practically, Bacon's notion of establishing a company of investigators to work for 'fruits,' as if the pursuit of knowledge were a kind of mining operation and only required well-directed picks and shovels, seems very strange" (1:57).

Huxley's main complaint is that Bacon specifically condemns the invention of hypotheses on the basis of incomplete inductions. The practical and historical reality is quite otherwise, argues Huxley. Indeed, the progress of science has been effected by men such as Galileo, Harvey, Boyle, and Newton, "who would have done their work just as well if neither Bacon nor Descartes had ever propounded their views respecting the manner in which scientific investigation should be pursued" (*CE* 1:48–49). Huxley's rejection of Bacon's method could not be more emphatic than when he describes Bacon as "a man of great endowments, who was so singularly devoid of scientific insight that he could not understand the value of the work already achieved by the true instaurators of physical science" (1:46). What Huxley terms the "anticipation of nature" by the invention of hypotheses, he claims to have been "a most efficient, indeed an indispensable, instrument of scientific progress" (1:47).

By the "anticipation of nature" Huxley means educated guesswork, which involves going beyond established observations. A scientific theory is not developed by induction from a collection of facts. A more subtle process is involved, of which an important part is the framing of verifiable hypotheses. Often based on slender assumptions, these hypotheses do not necessarily arise from observational data and induction but frequently precede them. "Any one who has studied the history of science," Huxley writes, "knows that almost every great step therein has been made by the 'anticipation of Nature,' that is, by the invention of hypotheses, which, though verifiable, often had very little foundation to start with" (*CE* 1:62). The process of scientific discovery, then, depends on the invention of initial working hypotheses, often of a speculative nature.

Huxley's use of the phrase "invention of hypotheses" implies the fabrication of an idea, a supposition, or conjecture, only partly derived from existing knowledge, which further investigation may verify or contradict. It also implies a veneration for the scientific imagination. Huxley conveys this most effectively elsewhere by a metaphor that compares the body of scientific workers to an army. The army's advance is facilitated by "clouds of light troops," armed with the weapon of scientific imagination, whose task is to "make raids into the realm of ignorance" (*CE* 7:271). These light troops are often annihilated or forced to retreat to the main body of science, where they regroup to make fresh skirmishes. As Donald Watt notes, this metaphor implies that Huxley sees the scientific imagination as being rash, daring, and courageous (34). Indeed, the connotations of movement, exploration, adventure, and risk illustrate how speculative Huxley considered scientific advance to be. Moreover, it hints at how speculative Huxley himself was prepared to be in his own writing.

Three main phases in the development of a scientific theory are distinguished by Huxley in "The Progress of Science." These are (1) the determination of the phenomena of nature through observation and experiment, (2) the determination of the constant relations between these phenomena, which are expressed as rules or laws, and (3) the explication of these laws by deduction under the most general laws of physical science. The invention of hypotheses plays its part in stages two and three. The three stages, however, are an idealization of actual practice: no branch of science has ever strictly followed this series since "observation, experiment, and speculation have gone hand in hand" (*CE* 1:64–65). Again, Huxley acknowledges that speculation may precede, follow or accompany observation and experiment. He even carries this view to the extreme of approving of Kepler as "the wildest of guessers" (1:62).

Huxley's philosophy of science is not Baconian; it incorporates a much more sophisticated understanding of the role of speculative but verifiable hypotheses. However, his comparison of the body of science to an army seems to indicate a belief that science has a solid and inexorably advancing core. An examination of Huxley's interpretation of the history of science will clarify the extent to which he considers science to be a cumulative process.

The Historical "Progress" of Science

Astronomy provides an excellent example of the process of the invention, the employment and the eventual rejection of erroneous hypotheses. The geocentric model of the solar system, with its circular orbits, eccentrics, and epicycles, gave way to the Copernican heliocentric system, which, following Kepler, adopted elliptical orbits. Even though the elliptical hypothesis takes no account of the immensely complicated curve traced in space by the center of gravity of a planet, it is an approximation suitable for ordinary purposes. Newton's corpuscular theory of light, now defunct, and the current hypothesis of the existence of an ether, are further examples of the provisional nature of scientific knowledge. "It sounds paradoxical to say," writes Huxley, "that the attainment of scientific truth has been effected, to a great extent, by the help of scientific errors." Equally paradoxical, we might add, is Huxley's reference to "scientific truth," since it would seem to imply belief in an absolute, objective truth. For Huxley, truth is in fact a relative concept. The errors that typify scientific progress are caused by the limits of our faculties, "while, even within those limits, we cannot be certain that any observation is absolutely exact and exhaustive" (*CE* 1:63). So, any given generalization may be true at one time but untrue at a later time, when practical developments may enhance our powers of observation. What is true at one period of scientific history may not be true in another.

Does Huxley consider that successive hypotheses have different degrees of truth that approach an absolute truth? He would appear to do so when he writes: "Or, to put the matter in another way, a doctrine which is untrue absolutely, may, to a very great extent, be susceptible of an interpretation in accordance with the truth." The assumption that the planets have circular orbits was true enough at a time in history when observations were cruder than now. The circular orbit hypothesis was sufficient to accommodate those observations. But the state of observational astronomy in Kepler's time supported a new truth, namely that of elliptical orbits. Today, it is a "matter of fact" that the center of gravity of a planet follows a complicated undulating line (*CE* 1:63–64). Huxley seems to be implying that continual improvements in our powers of observation result in the correction of previous hypotheses, yielding a series of generalizations that approach an absolute truth.

However, Huxley's final word in this section of the text leaves little doubt as to his view: he believes that the idea of absolute truth is another working hypothesis, which provides a basis for a *symbolic* interpretation of nature. "It may be fairly doubted whether any generalisation, or hypothesis, based upon physical data is absolutely true, in the sense that a mathematical proposition is so; but, if its errors can become apparent only outside the limits of practicable observation, it may be just as usefully adopted for one of the symbols of that algebra by which we interpret Nature, as if it were absolutely true" (*CE* 1:64). The existence of an absolute truth, for Huxley, is just a useful assumption that assists in our interpretation of nature, and then only within the limits of the observational capacities of the time. In considering Huxley's philosophy, therefore, we should bear in mind that his statements about truth are qualified by this very important caveat.

Huxley spells out his assumptions as a practitioner of physical science. One postulate is the objective existence of a material world, composed of an extended, impenetrable, mobile substance. The universality of the law of causation is a second postulate from which derives the view that the state of the universe at any given moment is the result of its state at any preceding moment. A third postulate is that the "laws of Nature," which define the relationship between phenomena, are "true for all time," by which Huxley means that they operate uniformly throughout time (*CE* 1:60–61).

Huxley summarizes his view of the progress of science, based on his understanding of the history of science: "The progress of physical science, since the revival of learning, is largely due to the fact that men have gradually learned to lay aside the consideration of unverifiable hypotheses; to guide observation and experiment by verifiable hypotheses; and to consider the latter, not as ideal truths, the real entities of an intelligible world behind phenomena, but as a symbolical language, by the aid of which Nature can be interpreted in terms apprehensible by our intellects" (*CE* 1:65).

We could hardly ask of Huxley a more concise statement of his position. Scientific method involves the continual invention of new hypotheses having the potential to correct or replace the old. New hypotheses are subject to verification by observations and more sophisticated experiments. The concept of absolute truth enters Huxley's philosophy only as a convenient assumption, a working hypothesis, a term in "the algebra of science," which helps give form and direction to periods of scientific development. His discussion of the major changes that scientific theories undergo through history, such as the Copernican and Keplerian revisions, shows an informed concern with the historical processes of scientific change. His basic postulates are metaphysical. In short, Huxley's philosophy of science shows a surprising modernity and sophistication that is not Baconian, positivist, nor crudely realist.

Writing on the relations of science and literature, George Levine has interpreted contemporary philosophical studies of science as moving scientific statements closer to a literary mode of interpretation. The literary relation emerges because, "like fiction, like poetry, science, on this account, achieves its status by virtue of its 'coherence' rather than its correspondence to external reality" (17). While not proposing a literary status for Huxley, our account of his philosophy of science permits a greater freedom in the exploration of his essays as a form of speculative writing.

Ed Block has noted the modernity of Huxley's idea of scientific method and its importance as an imaginative springboard for his essays. "Each essay," writes Block, "is an attempt to extend the applicability of the hypothesis by means of rigorous inductive and deductive reasoning, imaginatively linked by complex comparisons of process and morphology to the objects of earlier research" (369). In effect, this means that Huxley draws out the implications of evolutionary hypotheses in a logically rigorous way. The imaginative achievement lies in the mental translation of the hypothesis beyond the limits of existing observations.

The Method of Zadig

This extension of hypotheses is often highly speculative and constitutes a form of prophecy. The prophetic activity of the scientist is explicated by Huxley in his lecture entitled "On the Method of Zadig."

In this lecture Huxley defines prophecy as a form of "outspeaking" in which the seer seeks to impart new knowledge by extending a knowledge of existing causes into new spheres of time and space. Huxley retells Voltaire's story of a Babylonian philosopher, Zadig, a man learned in both natural philosophy and metaphysics. Zadig's main qualities are wisdom, integrity, and great powers of observation and ratiocination. These latter qualities are displayed in the tale of "The Dog and the Horse" in which Zadig deduces the appearance of the Queen's dog and the King's horse from marks and footprints they have left over the land.

Huxley's admiration for Voltaire's philosopher is obvious enough. Zadig not only brings intelligence and acute observation to the problems of nature but also uses his wisdom to diffuse conflict between religious sects, dissolve religious superstition, and extend freedom of belief and action. Huxley's attacks on the superstitions of theological dogma have much in common with Zadig's program.

Huxley considers Zadig's ratiocinative method to be the scientific method. Scientific prophecy may be prospective, retrospective, or spatial; it may include divination forward in time, backward in time, and of events in distant places. It is applied predictively in fields such as astronomy and retrospectively in the disciplines of archaeology, geology, and paleontology. These latter "historical or palaetiological" sciences, writes Huxley, "strive towards the reconstruction in human imagination of events which have vanished and ceased to be" (*CE* 4:6–12). As an example, Huxley cites Cuvier's triumphant prediction that the lower part of a particular fossil embedded in rock would show the pelvic bones of a marsupial, an inference drawn from the exposed jaw and teeth (4:18–19).

The fundamental axiom on which the prophetic method rests is the constancy of the order of nature, which Huxley considers to be the common foundation of all scientific thought (*CE* 4:12). It is also one of Huxley's axioms, as he shows in "Lectures on Evolution," a series of three lectures he gave in New York during his 1876 American tour. Huxley plunges into the remote past to gather fossil evidence to test three hypotheses concerning the history of nature. The first two hypotheses—that of the indefinite existence of the earth and its organisms in the present state, and that of special creation in six days—are not supported by the succession of forms revealed in the stratified rocks. On the other hand, demonstrative evidence of evolution is found in many places, such as in fossil remains from the Western Territories of America. The ancestors of the *Equidae* (horses, asses, and zebras) as they are traced back through geological history show an increasing separation of the limbs of the forearm and foreleg, an increasing number of digits, and modifications of the molar teeth, a history that "is exactly and precisely that which could have been predicted from a knowledge of the principles of evolution" (4:131–32).

A central feature of these lectures is the combination of broad mental sweep and detail, which correspond to the two main elements of the method of Zadig, namely, imaginative reconstruction on a foundation of existing knowledge. The temporal journey is aided by Huxley's device of referring the mind of his audience to that of an imaginary spectator overseeing the evolution of life. As if to cap his demonstration, Huxley gives a prophecy of his own. Using the latest paleontological investigations by O. C. Marsh, Huxley predicts the existence of an even earlier ancestor of the horse, which would have four complete toes on each foot with additional rudimentary first and fifth digits in the front and hind feet respectively. This fossil form, *Eohippus,* was soon discovered (*CE* 4:129–32; see Rudwick 252–54). This trilogy of lectures constitutes an evolutionary tour d'horizon of remote ages on the basis of existing knowledge. It is matched in Huxley's work perhaps only by his 1868 lecture "On a Piece of Chalk," which Tess Cosslett has examined in relation to the method of Zadig (28–29).

If the dead events of the past can be made to live by the prophetic method, so can the immanent events of the future. What course would evolution take on the return of a glacial epoch? Natural selection would favor the lower organisms over the higher: Cryptogamic vegetation (ferns, mosses, lichens), for example, over Phanerogamic vegetation (flowering plants), Crustaceans over Insects, Cetaceans and Seals over Primates (*CE* 2:91). Huxley also ponders such an eventuality in reference to the inevitable heat-death of the earth predicted by cosmologists. If the "physical philosophers" are correct in their view that the sun and the earth are cooling, then "the time must come when evolution will mean adaptation to a universal winter, and all forms of life will die out, except such low and simple organisms as the Diatom of the arctic and antarctic ice and the Protococcus of the red snow" (9:199).

The prophetic vision need not only be displaced in time but may also be displaced in space. Transferring the vantage point away from earth helps dispel the anthropocentrism that puts man at the center of creation. This is particularly useful when Huxley wishes to consider man as an animal with close evolutionary affinity to the higher apes, as in *Man's Place in Nature:* "let us imagine ourselves scientific Saturnians, if you will, fairly acquainted with such animals as now inhabit the Earth, and employed in discussing the relations they bear to a new and singular 'erect and featherless biped,' which some enterprising traveller, overcoming the difficulties of space and gravitation, has brought from that distant planet for our inspection, well preserved, may be, in a cask of rum" (*CE* 7:95).

A perspective of almost total omniscience is created by Huxley in the context of "Supernaturalism" and religion where he suggests the possibility of higher, even omnipotent, forms of life: "Without stepping beyond the analogy of that which is known, it is easy to people the cosmos with entities, in ascending scale, until we reach something practically indistinguishable from omnipotence, omnipresence, and omniscience. If our intelligence can, in some matters, surely reproduce the past of thousands of years ago and anticipate the future, thousands of years hence, it is clearly within the limits of possibility that some greater intellect, even of the same order, may be able to mirror the whole past and the whole future" (*CE* 5:39).

This quite astonishing statement by Huxley carries the warrant that his prophecy adopts "the most rigidly scientific point of view" (*CE* 5:39). Although the passage bears some rhetorical freight, being part of a Huxleyan tirade against religion, it is significant that Huxley is willing to stick his scientific neck out so far. The stress between hypothesis and veracity is underscored when we compare Huxley's speculative enterprises with his very cautious support for Darwin's principle of natural selection. Although Huxley considers natural selection to be a better hypothesis of species transformation than any of its predecessors, he acknowledges "it is quite another matter to affirm absolutely either the truth or falsehood of Mr. Darwin's views at the present stage of the inquiry" (2:20). The tension inherent in these extreme stances of caution and extraterrestrial extravagance seems to confirm Peter Morton's comment that the emotional pressure of speculation in Victorian science was very high (11). It also provides a way to help us understand how Huxley's work so powerfully stirred a writer of the stamp of H. G. Wells.

Wells's 1891 essay "Zoological Retrogression" uses the data of geology, paleontology, and embryology to tell the story of man's terrestrial ancestry in the upper Silurian period. The reader is invited "to imagine the visit of some bodiless Linnaeus to this world" who attempts a classification of fauna. This disembodied researcher would scarcely suspect that the inheritance of the earth was vested in an emerging species of mud-fish, swimming in the pluvial waters or caked in mud on the river banks. Wells then foreshortens in time the evolutionary stages by which the skull, skeleton, and swimming-bladder became adapted to terrestrial life.

The main theme of Wells's essay, however, is evolutionary *retrogression.* The developmental stages of the Ascidians, or sea-squirts, tell of an organism that evolved from a relatively complex, active, free-swimming form into a simple, sessile form, with an associated degeneration of organs. Evolution, warns Wells, does not always take an upward path. To this extent "Zoological Retrogression" constitutes an attack on anthropocentrism and evolutionary optimism. The temporal journey allows us to follow the evolutionary fates of several organisms through the possibilities of progression, or degeneration, or extinction. "There is, therefore," Wells concludes, "no guarantee in scientific knowledge of man's permanence or permanent ascendancy" ("Zoological Retrogression" 253).

William Thomson's estimates of the duration of the sun's life, along with works on cosmology by Sir Robert Ball and Emerson Reynolds, prompted Wells to consider the remoter past and distant future of earth.[3] "Another Basis for Life" describes states where extremes of temperature and differences in chemical composition would prevent the formation of protoplasm and the oxidation of chemical elements vital to terrestrial life. However, at the high temperatures of the primordial earth, compounds of silicon and aluminum, which are now inert oxidized matter, "may conceivably have presented cycles of complicated syntheses, decompositions, and oxidations essentially parallel to those that underlie our own vital phenomena." The chemical analogies between silicon and carbon enumerated by Reynolds led Wells to consider the startling possibility that if "the chemical accompaniments of life were rehearsed long ago and at far higher temperatures by elements now inert, it is not such a very long step from this to the supposition that vital, sub-conscious, and conscious developments may have accompanied such a rehearsal" (677).

"Another Basis for Life" also takes the reader to the far end of the earth's history. Drawing on the cosmologists' predictions of the heat-death of the solar system, Wells anticipates a time when the carbon-based organic matter of our age will become another geological stratum in a frozen world: "In that remote future we may anticipate that the former basis of life will form under the chilly atmosphere a lifeless covering to the dead earth, and that if we could travel forward in time we should find above the present rocks layers of ice, solid ammonium carbonate, solid carbon dioxide, a large series of compounds of carbon with its former contributors to the vital process, once quasi-fluid and active, but now solid and inert" (677).

As remarkable as these speculations may seem, Wells here merely draws on the same method explicated by Huxley in his essay on Zadig: the extension of a knowledge of existing causes into remote areas of time and space. Wells simply opens the curtains of time more widely.

From such a wide temporal span to an omniscient view the step is comparatively small. This is apparent from Wells's first major work, *The Time Machine.* First published serially in the *New Review* in 1895, the opening chapter, "The Inventor," discusses the concept of a "rigid" four-dimensional universe. "Perceiving all the present, an omniscient observer would likewise perceive all the past and all the inevitable future at the same time. Indeed, present and past and future would be without meaning to such an observer: he would always perceive exactly the same thing. He would see, as it were, a Rigid Universe filling space and time—a Universe in which things were always the same. He would see one sole unchanging series of cause and effect to-day and to-morrow and always" (100). This passage achieves an effect similar to Huxley's speculation concerning an intelligence capable of knowing the entire past and the entire future. Both writers portray a deterministic universe where the forces of nature dictate cosmic and biological evolution.[4]

Both writers also have in common a quasi-fictional technique. To illustrate how this is reflected in their literary practice, we explore a particularly suggestive observation by Block that Huxley uses a very modern form of imaginative construct that anticipates J. B. S. Haldane's concept of "possible worlds." Huxley's description of two philosophical death-watch beetles living in a clock and attempting to divine its workings, and his discussion of the limits to consciousness of a "rational" piano, are Block's examples. Huxley's analogical, spatial, and hypothetical methods are seen by Block as quasi-fictional techniques that "punctuate the already richly various structure of individual essays and point to a fundamentally imaginative mode of perception and thought which puts the stamp of originality on Huxley's work" (384).

This quasi-fictional technique is closely allied to the method of scientific prophecy we have been examining, and greatly strengthens Huxley's hand in a socioscientific debate of the late nineteenth century. This debate evoked Huxley's essay "The Struggle for Existence in Human Society" (1888; *CE* 9:195–234), his "Evolution and Ethics" lecture of 1893 (*CE* 9:46–116), and his subsequent "Evolution and Ethics: Prolegomena" (1894; *CE* 9:1–45), as well as Wells's first book, *The Time Machine.* Huxley's use of quasi-fictional methods and the development of this technique by Wells form part of a significant episode in the history of late nineteenth-century biology.

Huxley, Spencer, and the Time Traveller

Huxley's insistence, in the late 1880s and early 1890s, on the amoral nature of the cosmic process is a response to the "optimistic evolutionism" popular at the time, in which evolutionary forces were considered by the educated public and some scientists to be inherently progressive.

Herbert Spencer, Alfred Russel Wallace, Henry Drummond, and Winwood Reade, and literary artists such as George Meredith all coupled the main tenets of Darwinian theory to progressive social doctrines or visions. As early as the 1860s, despite its negation of inherently progressive or teleological factors in nature, Darwinism was considered by these writers to be robust enough to be yoked to optimistic evolutionism (Morton 53–70). A doctrine of Social Darwinism is expressed by Herbert Spencer: "All these [social] evils which afflict us, and seem to the uninitiated the obvious consequences of this or that removable cause, are unavoidable attendants on the adaptation now in progress. Humanity is being pressed against the inexorable necessities of its new position—is being moulded into harmony with them, and has to bear the resulting unhappiness as best it can" (*Man versus State* 68).

Huxley's response to these views is presented in a series of microcosms (or "possible worlds"), which offer three distinct but related arguments:

1. The cosmic process of evolution is *non-moral;* that is, has no inherently moral or immoral features and therefore cannot be considered "progressive."

2. Whatever social perfection may be achieved will be threatened by a return to struggle because of the pressure of an increasing population on limited resources.

3. The struggle for existence in modern civilization has for practical purposes ceased and an ethical process has supplanted it.

The first argument prefaces both "The Struggle for Existence in Human Society" and "Evolution and Ethics." In the former, Huxley condemns the "optimistic dogma" that the evolutionary state of nature is "the best of all possible worlds" as being "little better than a libel upon possibility" (*CE* 9:196). The evidence of benevolence in evolutionary progress must be balanced against the evidence of malevolence; the tendency toward the consummation of pleasure must be balanced by that of the production of pain. If the skill displayed by a wolf in tracking a deer, for example, can be considered a positive feature of nature, it must be weighed against the pain the wolf inflicts on the deer. The study of nature can lead to only one conclusion: that nature is neither moral nor immoral but non-moral, "a materialized logical process, accompanied by pleasures and pains, the incidence of which, in the majority of cases, has not the slightest reference to moral desert" (9:202).

In "Evolution and Ethics" Huxley emphasizes the evolutionary legacy of pain and suffering that accompanies the human animal in the struggle for existence. Success in the struggle is due not only to our sociability, curiosity, and imitativeness, but also to our cunning and "ruthless and ferocious destructiveness." These "ape and tiger" qualities have become defects for contemporary humans because their intrusion into civilized life "adds pains and griefs, innumerable and immeasurably great" to those already imposed on man in the natural state (*CE* 9:51–52).

The second argument addressing the application of evolutionary theory to society occurs in "The Struggle for Existence in Human Society" and the "Prolegomena," where Huxley invents quasi-fictional microcosms to illustrate the Malthusian problem of population pressure on limited resources. Huxley uses the fiction of a perfect society, Atlantis, to imagine a world where the production of food exactly meets the wants of the population, and where the struggle for existence has been abolished. "In that happy land," writes Huxley, "the natural man would have been finally put down by the ethical man," and the millennium would have finally set in (*CE* 9:206). But this perfect state may be only temporary; any economic disturbance, such as a shortage of food caused by the pressure of population, would readmit the forces of nature and reintroduce struggle. "The Atlantis society might have been heaven upon earth. . . . Yet somebody must starve. Reckless Istar, non-moral Nature, would have riven the ethical fabric" (9:207).

The next, much more extensive, microcosm is developed in the "Prolegomena," where Huxley extends the Atlantis analogy. He tells a story in which human intervention in the state of nature creates a perfect order, or state of art, secured by the inventions of modern science. Huxley imagines a group of English colonists establishing a bridgehead against nature from which the cosmic process is excluded, order established, and artificial selection imposed upon plants, animals, and humans. "With every step of this progress in civilization, the colonists would become more and more independent of the state of nature; more and more, their lives would be conditioned by a state of art" (*CE* 9:19).

An omnipotent Administrator is set over this world to artificially select, according to ideals of beauty and utility, those organisms, human and nonhuman, that will share this earthly paradise. However, this garden of Eden will have its serpent: the instinct of reproduction. Overpopulation would threaten the artificial order so carefully established. The Administrator would be forced to destroy the superfluous population or permit the reintroduction of the struggle for existence. The population pressure on finite resources, therefore, will tend either to despotism or struggle, neither of which is acceptable to a civilized society (*CE* 9:20–23). In each of the microcosms put forward by Huxley the artificial order ultimately yields to the cosmic process and struggle resumes. No comfort or cause for optimism can be derived from the theory of evolution; the cosmic process has little to offer civilized humanity.

Huxley's alternative to optimistic evolutionism—his third argument—is expressed in "Evolution and Ethics," where he proposes that improvement in the human condition may be aided by cultivating an existing ethical process in society in opposition to the cosmic process of evolution in nature. "In place of ruthless self-assertion," Huxley states, the ethical process "demands self-restraint; in place of thrusting aside, or treading down, all competitors, it requires that the individual shall not merely respect, but shall help his fellows; its influence is directed, not so much to the survival of the fittest, as to the fitting of as many as possible to survive" (*CE* 9:82).

The contrasting of the state of nature to a state of art flags an increased emphasis on the possibilities of an artificial environment created by man through the practical effects of science. This view is best seen in "The Progress of Science," where Huxley speaks of "a new Nature" created by mechanical artifice, the employment of new chemicals in manufacture, and the artificial breeding of plants and animals. The practical outcomes of material science are transforming Europe and America and protecting the population from pestilence and famine. They are the foundation of wealth, comfort, and security, and "conduce to physical and moral well-being" (*CE* 1:51) Moreover, inventions in transport and telegraphic communications are reducing, or "levelling up," inequalities between individuals, classes, and nations (1:108).

In contrast to the romantic and a priori arguments of the evolutionary social idealists, Huxley's "transformational optimism" lays increasing emphasis on civil or artificial history rather than natural history as a basis for ethics and morality (Paradis 21–22, 37). The microcosm Huxley develops in the "Prolegomena" succinctly conveys his view that under the aegis of science civilization has banished the cosmic process and replaced it with an ethical process. "I have endeavoured to show that, when the ethical process has advanced so far as to secure every member of the society in the possession of the means of existence, the struggle for existence, as between man and man, within that society is, *ipso facto,* at an end. And, as it is undeniable that the most highly civilized societies have substantially reached this position, it follows that, so far as they are concerned, the struggle for existence can play no important part within them" (*CE* 9:35–36). It is in this ethical process, improving the "State of Art of an organized polity" in opposition to the State of Nature, that Huxley finds our salvation (9:44–45).

This third argument of Huxley's, that the struggle for existence in society has ended, seems at first glance inconsistent with his second argument that population pressure is invariably a cause of struggle. Huxley appears on the one hand to be employing evolutionary theory to support his Malthusian argument on population, while denying its applicability elsewhere. Michael Helfand has addressed this problem by proposing that Huxley is attempting to withhold the scientific weapon of evolution from his political rivals, such as Spencer, Wallace, and Henry George, while at the same time using it to support a Liberal imperialist political policy then under attack from these social scientists (Helfand). Nevertheless, the relationship Huxley establishes between the cessation of struggle in society and the role of population increase in natural selection is better understood as a warning: the struggle for existence has ceased, but could reemerge if the population increases beyond the agricultural resources of the nation. This warning is clearly sounded in "The Struggle for Existence in Human Society" (*CE* 9:209–10).

Wells, Huxley, and Degeneration

Wells's essays and books in the 1890s continue this socioscientific debate about ethics of evolution, as Huxley developed it, even deploying techniques recognizably similar to those of Huxley. Mark Hillegas has pointed out that most of Wells's early works are designed to "jolt the English-speaking world out of its complacency" by an imaginative presentation of Huxley's "cosmic pessimism" (656–57). Wells began his attack on optimistic evolutionism in his piece "Zoological Retrogression." Many other short, speculative pieces, such as "On Extinction" (1893), "The Extinction of Man" (1894), and "The Rate of Change in Species" (1894), all seek to undermine Victorian attitudes of security and self-satisfaction *(Early Writings).*

In *The Time Machine* Wells develops the microcosm of a Golden Age somewhat similar to Huxley's Atlantis. The world of the year 802,701 depicts the same opposition of social perfection yielding to a struggle over limited resources. At first Wells's Time Traveller believes the struggle for existence between rich and poor, the Overworld and the Underworld, has indeed led to a millennium, a Golden Age: "Once, life and property must have reached almost absolute safety. The rich had been assured of his wealth and comfort, the toiler assured of his life and work. No doubt in that perfect world there had been no unemployed problem, no social question left unsolved. And a great quiet had followed" (130).

The Time Traveller's theory of a Golden Age is an illusion, for the idyllic existence of the childlike Overworld people covers a grim reality. The social perfection the two races once enjoyed has given way first to speciation and then to a Darwinian struggle for existence for limited resources. The shortage of food causes the Underground species to prey on its degenerating Overworld competitor. "Mother Necessity," concludes the Time Traveller, "who had been staved off for a few thousand years, came back" (131).

Wells's microcosm reverses the optimistic arguments of the social evolutionists by depicting, with cool logic and satirical overtones, his view of a world where the doctrine of evolutionary necessity has triumphed. "Man had been content to live in ease and delight upon the labours of his fellow-man, had taken Necessity as his watchword and excuse," says the Time Traveller, "and in the fullness of time Necessity had come home to him" (105–6).

However, in criticizing the view that evolution can be applied prescriptively to human society Wells gives Huxley only qualified support, for he presents a paradox, shown by the Time Traveller's theory of the degeneration of the Overworlders: "I thought of the physical slightness of the people, their lack of intelligence, and those big abundant ruins, and it strengthened my belief in a perfect conquest of Nature. For after the battle comes Quiet. Humanity had been strong, energetic, and intelligent, and had used all its abundant vitality to alter the conditions under which it lived. And now came the reaction of the altered conditions" (53).

The utopian conditions under which the Overworlders live has eliminated the premium on intelligence and physical strength so important for survival in the natural world, and the biological degeneration of these traits has inevitably followed. Wells pointed out the theme of degeneration in *The Time Machine* in a letter to Huxley dated May 1895: "I am sending you a little book that I fancy may be of interest to you. The central idea—of degeneration following security—was the outcome of a certain amount of biological study" (HP 28:233–34).

Although degeneration is associated with a wide range of biological and social issues in post-Darwinian England (Greenslade), the degeneration of nonadaptive organs as an outcome of natural selection is most clearly spelt out by E. Ray Lankester, in his 1880 book *Degeneration: A Chapter in Darwinism.* Many examples are given of organisms, such as various parasites, barnacles, and Ascidians, which have become adapted to less-varied and less-complex conditions of existence and have thus evolved from a complex ancestor to a simpler descendant. Such less-complex conditions may accompany adaptation to parasitism, immobility, vegetative nutrition, or a reduction in size of an organism. The human species, too, may be subject to degeneration under civilized conditions, and Lankester warns, "we have to fear lest the prejudices, pre-occupations, and dogmatism of modern civilization should in any way lead to the atrophy and loss of the valuable mental qualities inherited by our young forms from primeval man" (61).

The use of degeneration by Wells is also informed by the Spencer-Romanes-Weismann controversy over the mechanism of degeneration, carried on in the *Contemporary Review* during 1893 and 1894. Spencer holds that the diminution in the size and efficiency of organs occurs by disuse, which through the inheritance of acquired characteristics is then passed on from one generation to the next. In reply, both August Weismann and G. J. Romanes consider what Romanes aptly terms "Cessation of Selection" to be the primary cause of degeneration (Spencer, "Weismann's Theories" 758–59; Weismann 332; Romanes 611–12).

Degeneration, then, is a possible, perhaps even probable, outcome of the amelioration of the conditions of existence implied by the "levelling up" of Huxley's new nature or the evolution toward perfection envisaged by Spencer. It matters not whether the millennium is attained by Huxley's cultivation of the State of Art of an organized polity, or by Spencer's eventual triumph of the fittest, for in either case the longer-term result is cessation of selection, degeneration, and the loss of the mental and physical traits that distinguish us from the brutes.

Huxley shows an awareness of the importance of degeneration in evolutionary theory (*CE* 9:4, 6, 80–81, 88). He even uses it as a supplementary weapon in his armory, admonishing, "And, again, it is an error to imagine that evolution signifies a constant tendency to increased perfection. That process undoubtedly involves a constant remodelling of the organism in adaptation to new conditions; but it depends on the nature of those conditions whether the direction of the modifications effected shall be upward or downward. Retrogressive is as practicable as progressive metamorphosis" (*CE* 9:199).

But this reference to degeneration concerns adaptation to secular cooling, not to the artificial conditions of civilization. The problem of degeneration following security is something Huxley does not address in "Evolution and Ethics" and associated essays—a peculiar omission considering the topicality of degeneration. Why Huxley declined to discuss the evolutionary outcome of "the fitting of as many as possible to survive" is an open question. Perhaps his preoccupation in these essays with the immediate environment and short-term future reflects a belief that, given the slowness of evolutionary change, human nature can be considered practically fixed (Paradis 32). Nevertheless, the interplay between evolutionary theory and sociopolitical concerns in these later essays seems to have caused Huxley to fall away somewhat from his usual scientific rigor.

Nowhere near as reticent, Wells the student overreaches Huxley the master in his use of the method of Zadig by adhering to a biological determinism that Huxley fails to sustain. By strictly applying Darwinian natural selection to modern conditions of existence, Wells points out a weakness in the arguments of both Huxley and Spencer. Unfortunately, we have no way of knowing Huxley's reaction to Wells's critique as it is unlikely that he read it—Huxley was ill in May 1895 and died the following month.

Neither Huxley nor Wells, however, had any illusions about the longer-term prospects for the planet. The comparatively short life-span Thomson had assigned to the sun presaged the extinction of all life in a few tens of millions of years.[5] Although some noble work may be done in the millennia remaining, Huxley is resigned to the ultimate resumption of the cosmic process over our domain: "The theory of evolution encourages no millennial anticipations. If, for millions of years, our globe has taken the upward road, yet, some time, the summit will be reached and the downward route will be commenced. The most daring imagination will hardly venture upon the suggestion that the power and the intelligence of man can ever arrest the procession of the great year" (*CE* 9:85)

To witness the end, Wells moves his Time Traveller forward 30 million years. Tidal drag has now caused one side of the earth to face an enlarged and cooling western sun. The atmosphere is thin and cold. A deep twilight broods over the earth and snow falls continuously. The Time Traveller observes an eclipse that threatens to extinguish the already diffuse light:

> The darkness grew apace; a cold wind began to blow in freshening gusts from the east, and the showering white flakes in the air increased in number. From the edge of the sea came a ripple and whisper. Beyond these lifeless sounds the world was silent. Silent? It would be hard to convey the stillness of it. All the sounds of man, the bleating of sheep, the cries of birds, the hum of insects, the stir that makes the background of our lives—all that was over. . . . The breeze rose to a moaning wind. I saw the black central shadow of the eclipse sweeping towards me. In another moment the pale stars alone were visible. All else was rayless obscurity." (140–41)

The cosmological predictions of physicists such as Thomson had great intellectual and emotional significance for evolutionary writers such as Huxley and Wells. Whereas Wells expresses the heat-death of the solar system in quite moving prose, Huxley had a few years earlier gone one step further by expressing it poetically. Although ostensibly an obituary for Tennyson, the last few lines of Huxley's poem may just as well be called "Ode to Thermodynamics":

> Not thine [mourner] to kneel beside the grassy mound
> While dies the western glow; and all around
> Is silence: and the shadows closer creep
> And whisper softly: All must fall asleep.
>
> ("Westminster" 832)

THE PRINCIPAL CHARACTERISTICS of Huxley's work that we have been examining may now be summarized. A scientific method of prophecy, having its warrant in Huxley's philosophy of science and explicated in his essay "On the Method of Zadig," informs the invention of hypotheses that extend evolutionary theory beyond the sphere of immediate knowledge. The construction in imagination of events distant in space and time has a quasi-fictional quality.

How may this aspect of Huxley's essays be characterized? Huxley considers the method of Zadig to be essential to the science of paleontology as practiced by Cuvier. The method merely reflects the rapidly growing worldview that informs the uniformitarianism of Charles Lyell's *Principles of Geology* and a modern conception of evolutionary history that Charles Darwin displays in *Origin of Species.* Lyell and Darwin provide a type of historical narrative in which a knowledge of the laws of nature, operating uniformly throughout time, permits the mental construction of the past from the fragmentary geological record. The method is a scientific form of historical construction. Thus evolutionary theory involves an attention to time, cause, and effect that yields a form of hypothetical history. In Darwin's work, for example, it is the historical or "proto-historical" element that calls upon the imagination as he studies not only existing species but the transformation of species through time (Beer 98).

The ratiocination that underlies the enterprises of geology and paleontology, along with its importance for literature, is discussed by Lawrence Frank, who shows how the historical consciousness created by Lyell and

Darwin is also found in the work of Conan Doyle. The reconstruction of a fragmentary text demanding decipherment is essentially the method of Zadig. The fictive elements in the method would hardly have been lost on nineteenth-century naturalists such as Darwin, and were not lost on Huxley, either in his philosophy of science or in his actual practice as scientific essayist.

The ratiocinative character of geology and evolutionary biology is also emphasized by Wells when he states his views on how science should be communicated to the public. Considering that the educated public comes to scientific books for "problems to exercise their minds upon," Wells proposes that the scientific writer should be guided by the same principles of construction that underlie the writing of Poe and Conan Doyle, in which a problem is presented and then the solution pieced together ("Popularising Science" 301).

In *The Time Machine* Wells's Time Traveller constructs a historical narrative from fragmentary evidence. Through a succession of hypotheses the Time Traveller rebuilds the evolutionary history of the race over the time through which he has traveled. He applies the hypothesis of natural selection to his observation of the physical characteristics of the two races, their behavior, and their conditions of existence. From the point of view of the Time Traveller, these hypotheses are a historical reconstruction; from our point of view, Wells is constructing a hypothetical future history based on a knowledge of existing laws and observations.

A formal statement of his belief that the future would yield to this kind of scientifico-historical analysis is given by Wells in "The Discovery of the Future," his 1902 address at the Royal Institution. In the same way that an inductive past has been built by the analysis during the nineteenth century of geological strata, fossils, and the laws of organic development, so an "inductive future" may be known: "By seeking for operating causes instead of for fossils and by criticising them as persistently and thoroughly as the geological record has been criticised, it may be possible to throw a searchlight of inference forward instead of backward" (329).

Just as Huxley's approach to science in his essays is subtle, imaginative, and quasi-fictional, so Wells's *The Time Machine* may be considered a scientific essay, derived from his science journalism and employing the method of Zadig: a contribution to the evolutionary debates of the day with Huxley among its audience.

The personae of Huxley as a "bolt upright" model of Victorian rectitude with a mind set in the terra firma of scientific truth and of Wells as a pseudoscientific fantasist with his head in the clouds easily obscure their extensive common ground. In particular, the view of Huxley's writing as dry and speculation-free is erroneous. On the contrary, the "reconstruction in imagination" of scientific hypotheses lends itself to a freedom of thought and expression fully exploited by Huxley.

We need not see Huxley simply as a scientist partaking of literary techniques, nor Wells as just a literary artist employing scientific data, but as two thinkers employing a common method. Consequently, we need not feel obliged to classify their writing within the disciplines of historical narrative, or scientific theory, or literary fiction. The expansive literary product that results from the use of the method of Zadig in practiced hands displays an imaginative kinesis and logical precision that each category alone fails to encompass. The tendency to investigate the relations of science and literature in terms of the influence of one discipline on the other, of crosscurrents flowing across boundaries, fails to recognize that such boundaries are neither fixed nor given, but are cultural artifacts (Christie and Shuttleworth 3).

Here an attempt has been made to emphasize the *process* involved rather than the product that results from it, to emphasize method over presentation, and intellection over style. Since the method involves a quasi-fictional mode of scientific conjecture, we propose "conjection" to capture the essence of the process. It is quite possible, indeed likely, that contemporaries of Huxley and Wells, such as Grant Allen, use some form of conjection in their imaginative constructions of the past and future (see Morton 137–42). But we think it that it is Huxley and Wells who best display the power of conjection. In developing the exact projection of existing scientific theory beyond the limits of observation, they illustrate both the reach and power of the method of Zadig.

NOTES

1. Huxley's rhetorical strategies are discussed in Block; Morton 42–43; and Jensen.

2. See Stanley; Blinderman; Gardner; Watt; and Cosslett 1–38.

3. Thomson's highest estimate for the length of time the sun has illuminated the earth is something less than about 100 million years, with about the same predicted

for the future (1:375). His lowest estimates are about 20 million years of past heat with about 5 or 6 million years of heat left (1:397).

4. The omniscient view described by Wells and Huxley may have as its source Laplace's *Théorie analytique des probabilités* (1812–20): "A mind which could, in a given instant, know all the forces which act in nature and the respective positions of the bodies which make it up—if, in addition, it were vast enough to be able to submit these givens to differential analysis—would incorporate in the same formula the movements of the largest bodies in the universe and those of the lightest atom: nothing would be uncertain for it, and the future, as well as the past, would be present before its eyes" (qtd. in Bell 183).

5. In 1869 Huxley and Thomson clashed over these estimates in addresses to the Geological Societies of London and Glasgow, respectively; see Thomson 2:10–64; Huxley's reply *CE* 8:305–39; and Thomson's rejoinder 2:73–113.

REFERENCES

Beer, Gillian. *Darwin's Plots: Evolutionary Narrative in Darwin, George Eliot, and Nineteenth-Century Fiction.* London: Routledge & Kegan Paul, 1983.

Bell, David F. "Balzac with Laplace: Remarks on the Status of Chance in Balzacian Narrative." *One Culture: Essays in Science and Literature.* Ed. George Levine. Madison: U of Wisconsin P, 1987. 180–99.

Bergonzi, Bernard, ed. "*The Time Machine:* An Ironic Myth." *H. G. Wells: A Collection of Critical Essays.* Englewood Cliffs: Prentice Hall, 1976. 39–55.

Blinderman, Charles S. "Semantic Aspects of T. H. Huxley's Literary Style." *Journal of Communication* 12 (1962): 171–78.

———. "T. H. Huxley's Theory of Aesthetics: Unity in Diversity." *Journal of Aesthetics and Art Criticism* 21 (1962): 49–55.

Block, Ed, Jr. "T. H. Huxley's Rhetoric and the Popularization of Victorian Scientific Ideas, 1854–1874." *Victorian Studies* 29 (1986): 363–86.

Christie, John, and Sally Shuttleworth. "Introduction: Between Literature and Science." *Nature Transfigured: Science and Literature, 1700–1900.* Manchester: Manchester UP, 1989. 1–12.

Cosslett, Tess. *The "Scientific Movement" and Victorian Literature.* Sussex: Harvester Press, 1982.

Frank, Lawrence. "Reading the Gravel Page: Lyell, Darwin, and Conan Doyle." *Nineteenth-Century Literature* 44 (1989): 364–87.

Gardner, Joseph H. "A Huxley Essay as 'Poem.'" *Victorian Studies* 4 (1970): 177–91.

Greenslade, William. *Degeneration, Culture, and the Novel, 1880–1940.* Cambridge: Cambridge UP, 1994.

Helfand, Michael S. "T. H. Huxley's 'Evolution and Ethics': The Politics of Evolution and the Evolution of Politics." *Victorian Studies* 20 (1977): 159–77.

Hillegas, Mark R. "Cosmic Pessimism in H. G. Wells's Scientific Romances." *Papers of the Michigan Academy of Science, Arts, and Letters* 46 (1961): 655–63.

Huxley, Thomas Henry. *Collected Essays.* 9 vols. 1894. N.Y.: Greenwood Press, 1968.

———. "The Scientific Aspects of Positivism." *Fortnightly Review* 5 n.s. (1869): 653–70.

———. "Westminster Abbey: October 12, 1892." *Nineteenth Century* 32 (1892): 831–32.

Huxley Papers. Imperial College of Science, Technology, and Medicine Archives, London.

Irvine, William. *Apes, Angels, and Victorians: Darwin, Huxley, and Evolution.* N.Y.: McGraw-Hill, 1972.

Jensen, J. Vernon. *Thomas Henry Huxley: Communicating for Science.* Newark: U of Delaware P, 1991.

Lankester, E. Ray. *Degeneration: A Chapter in Darwinism.* London: Macmillan, 1880.

Levine, George. "One Culture: Science and Literature." *One Culture: Essays in Science and Literature.* Madison: U of Wisconsin P, 1987. 3–32.

Morton, Peter. *The Vital Science: Biology and the Literary Imagination, 1860–1900.* London: George Allen & Unwin, 1984.

Paradis, James. "'Evolution and Ethics' in its Victorian Context." *Evolution and Ethics: T. H. Huxley's "Evolution and Ethics," with New Essays on Its Victorian and Sociobiological Context.* Ed. James Paradis and George C. Williams. Princeton: Princeton UP, 1989. 3–55.

Romanes, George J. "A Note on Panmixia." *Contemporary Review* 64 (1893): 611–12.

Rudwick, Martin J. S. *The Meaning of Fossils: Episodes in the History of Paleontology.* London: MacDonald, 1972.

Spencer, Herbert. *The Man versus the State.* London: Williams & Norgate, 1885.

———. "Professor Weismann's Theories." *Contemporary Review* 63 (1893): 743–60.

Stanley, Oma. "T. H. Huxley's Treatment of 'Nature.'" *Journal of the History of Ideas* 28 (1957): 120–27.

Thomson, William. *Popular Lectures and Addresses.* 3 vols. London: Macmillan, 1891–94.

Watt, Donald. "Soul-Facts: Humor and Figure in T. H. Huxley." *Prose Studies* 1 (1978): 30–40.

Weismann, August. "The All-Sufficiency of Natural Selection: A Reply to Herbert Spencer, 1." *Contemporary Review* 64 (1893): 309–38.

[Wells, H. G.]. "Another Basis for Life." *Saturday Review* 78 (1894): 676–77.

Wells, H. G. "The Discovery of the Future." *Nature* 65 (1902): 326–31.

———. *Early Writings in Science and Science Fiction.* Ed. Robert M. Philmus and David Y. Hughes. Berkeley: U of California P, 1975.

———. "Huxley." *Royal College of Science Magazine* 13 (1901): 209–11.

———. "Popularising Science." *Nature* 50 (1894): 300–301.

———. *The Time Machine: An Invention.* London: Heinemann, 1895.

———. "The Time Machine. 1. The Inventor." *New Review* 12 (1895): 98–112.

———. "Zoological Retrogression." *Gentleman's Magazine* 271 (1891): 246–53.

"Fighting Even with Death"

Balfour, Scientific Naturalism, and Thomas Henry Huxley's Final Battle

BERNARD LIGHTMAN

THROUGHOUT THE SPRING AND EARLY SUMMER OF 1895, AN elderly Huxley was brought out each day to the veranda of his home at Eastbourne, so he could enjoy the unusual "endless sunshine" (*Life* 2:425). According to his son Leonard, who visited him from the twenty-second to the twenty-fourth of June, the tanned Huxley looked well in spite of all he had been through that year. In early March, in the midst of rushing to complete a book review for John Knowles, editor of the *Nineteenth Century,* Huxley had been struck down by influenza. He wrote to Knowles on March 8, apologizing for the delay and criticizing his doctor's diagnosis. "Doctor says it's a mild type," Huxley declared. "I find coughing continuously for fourteen hours or so a queer kind of mildness" (2:423–24). Eleven days later he warned Knowles not to expect the review to be done by the end of March, adding that he was now fighting a severe case of bronchitis.

By May, after a long and painful struggle against his illness, Huxley had recovered from both the influenza and bronchitis. He was even strong enough to walk outside the house. But his work for Knowles remained incomplete, for it became apparent that Huxley's heart and kidneys had been affected by his sickness. And so, Huxley just sat each day on the veranda, trying to regain his full strength. What was Huxley thinking about as he sat in the sun? Knowledgeable as he was about medicine and biology, Huxley may have wondered why, if the doctors were right about the damage to his kidneys, he wasn't already in a coma (*Life* 2:425). Or did he worry about the review for Knowles, which he had tried unsuccessfully to finish before his illness made it impossible to continue?

The subject of Huxley's review was *The Foundations of Belief*, published in February 1895. The author was none other than Arthur J. Balfour, one of the most important political figures of the day and a future Tory prime minister. In *Foundations* Balfour launched what amounted to a direct attack on Huxley's concept of "scientific naturalism." Scholars have paid scant attention to the clash between the middle-class, liberal, agnostic scientist and the aristocratic, conservative, Christian politician. Biographers of Balfour, though devoting a few paragraphs to *Foundations*, ignore Huxley's response (see Dugdale; Egremont). Even biographers such as Young and Mackay, who propose to explore Balfour's intellectual life in more detail, fail to mention Huxley's attack on *Foundations*. Huxley's biographers have dealt, albeit briefly, with the debate between Huxley and Balfour. Subscribing to the notion that Huxley needed controversy to rouse his spirits late in life, Bibby refers to *Foundations* as a "fillip" that allowed Huxley to struggle against sickness (87). In their evaluation of the battle between two great minds, Ashforth and Irvine both award the victory to Huxley. Though Ashforth admits that Huxley left many of Balfour's questions unanswered, "his superior understanding of the question and his precise, forceful use of language gave him the best of the exchange" (151). Irvine judges that after Huxley had finished examining Balfour's arguments in *Foundations*, "hardly a word or an idea was left standing" (357). By contrast, in what is the most sustained analysis of the controversy, Peterson uses Huxley's response to Balfour (referred to by Peterson as one of Huxley's "swan-songs") as a springboard for probing the weaknesses within "scientific naturalism" (300–14).

Huxley's attack on Balfour, his final battle, deserves further scholarly treatment for several important reasons. First, as Peterson suggests, Balfour was particularly adept at exposing the problems inherent within Huxley's scientific naturalism (302–8). Second, Huxley's response to Balfour contains his most mature thoughts on the nature of scientific naturalism, arrived at after years of intellectual activity, countless debates with Christian opponents, and under the pressure of imminent death. Third, Huxley's attempt to meet Balfour's penetrating attack illuminates the rhetorical strategies that Huxley used successfully in previous battles with defenders of the Christian faith.

Huxley's strategy in the past centered on controlling the meaning of the key terms of the debate, and sometimes inventing new terms over which he could exercise complete control, like "agnosticism" and "scientific naturalism," which defined both his position and the enemy's. Huxley may very well have found that these strategies were ineffective against Balfour, who had the benefit of learning from the mistakes of Huxley's past foes and who could draw strength from the emerging hostility toward the pretensions of dogmatic scientific naturalists like Huxley. Eager to give an old warrior the glory of winning his final battle, Ashforth and Irvine overlooked the serious challenge that Balfour represented to Huxley's position as the leader of scientific naturalism and its chief interpreter of the social, political, and religious meaning of scientific thought.

Scientific naturalism was the English version of the cult of science in vogue throughout Europe during the second half of the nineteenth century. Since science was perceived during that period as providing the most legitimate path to certain truth, whoever could lay claim to speak on behalf of science could present themselves as authoritative intellectual leaders who knew how to interpret the larger significance of modern science. Scientific naturalists put forward new interpretations of man, nature, and society derived from the theories, methods, and categories of empirical science. This cluster of ideas and attitudes was naturalistic in the sense that it would permit no recourse to causes not empirically observable in nature, and therefore scientific naturalists were notoriously critical of spiritualism and other beliefs that allowed for the existence of the supernatural. Scientific naturalists claimed to be scientific because their picture of nature drew on three major mid-nineteenth-century theories: the atomic theory of matter, the conservation of energy, and evolution. The ideology of scientific naturalism became the apologetic tool of the Victorian middle class in its attempts to generate a new Weltanschauung appropriate for a competitive, urban, and industrial world. Besides Huxley, the leaders of scientific naturalism included John Tyndall, Herbert Spencer, William Kingdom Clifford, Frederic Harrison, G. H. Lewes, Edward Tylor, John Lubbock, E. Ray Lankester, Henry Maudsley, Leslie Stephen, Grant Allen, and Edward Clodd (see Turner, *Between Science and Religion* ch. 2).

Balfour was well placed to launch an effective attack on Huxley and the cult of scientific naturalism. Born in 1848, he was the eldest son in a family with enormous political influence and wide social connections. His godfather was the Duke of Wellington. Balfour's father, James Maitland Balfour, was a Scottish lord; his mother, Lady Blanche, a devout Christian, belonged to the distinguished Cecil family of Tory aristocrats. The landowning family was so wealthy that upon receiving his inheritance at the age of twenty-one, Balfour became one of the richest young men in the country. He was educated at Eton and then in 1866 went on to Trinity College, Cambridge, where he met Henry Sidgwick, classicist and philosopher. Sidgwick later became one of his closest friends and his brother-in-law. Together they shared an interest in going beyond the empiricist confines of the worldview of scientific naturalism, and they played a key role in the founding of the Society of Psychical Research in 1882 (Egremont 28, 25).

Though Balfour was intensely interested in the major intellectual issues of the day, especially the relationship between science and religion, he was drawn increasingly into the world of politics by his mother's brother, Lord Salisbury, who carefully molded his nephew into the future Tory prime minister. It was Salisbury who suggested that Balfour run for Parliamentary office in 1874 as Conservative M.P. for Hertford, and who, as Tory leader, offered his nephew the position of secretary of state for Scotland in 1886 and the same position for Ireland the following year. Balfour's effective handling of the explosive Irish situation won him the confidence of the Tory party and led to his appointment as leader of the House of Commons in 1891.[1]

Though Salisbury concentrated on developing his nephew's political potential, he did not extinguish Balfour's fascination with the world of ideas. In fact, Balfour saw a clear connection between his philosophical interests and his political activities. In the same year that he became leader of the House, Balfour delivered an address at Glasgow University, "A Fragment of Progress," where he argued that through the state, politicians can provide the necessary conditions, such as security and freedom, under which positive advances can be accomplished, but that politicians themselves can effect only the tiniest bit of progressive change. Later, in 1908, Balfour spelled out the full implications of his position. Progress is not brought about by the state, by politicians, or by the majority in a democratic society, whose tendency is to control or check movement. Only the above-average minority, the intellectual elite, can initiate and realize meaningful change through the force of their ideas (*Essays Speculative* 50). Balfour attempted throughout his life to combine the role of intellectual with that of political leader. Scholars have tended to neglect the former for the sake of the latter. But Balfour's intellectual activity was important in its own right, so much so, as Root points out, that he "would deserve the historian's attention even had he never become such a remarkable statesman" (121).

Though Balfour's contribution to the world of ideas is significant, it cannot be studied in isolation from the social and political context; his philosophical and political concerns were intertwined (see Jacyna 11). Throughout the eighties and nineties, Tories perceived distressing signs of deep social discord in the strength of socialism, in the rise of unbelief, and in the turmoil in Ireland. Ireland was the most obvious threat. In his position as chief secretary to Ireland, Balfour had seen the problem firsthand. To Balfour and Salisbury the cause of the disorder was not so much the result of Irish nationalism as of the simmering class war between landlords and tenants, which, they feared, could easily spread to the mainland. In England the 1880s were marked by high unemployment, urban poverty, strikes, and working-class unrest. Socialism, a creed that, in the eyes of conservatives, stressed the antagonism between various parts of society, merely exacerbated the problem, while bourgeois individualism offered little or no solution to the prevailing distress. In addition, the rise of various forms of unbelief, accompanied by materialism, was also seen as divisive by conservatives, for they presented a challenge to the authority of the Anglican Church, an integral part of the Tory establishment.

In opposition to these disruptive social, political, and intellectual developments, the Conservative Party increasingly adopted the role of the protector of order, the upholder of unity, and the guardian of spiritual values (Jacyna 20). Balfour's steady rise to political eminence coincided with this reorientation of conservatism, and in his political activities he played an important role in the efforts of his fellow Tories to prevent political revolution and social upheaval. The theme of unity and authority provided the guiding thread in his philosophy, as it did for a number of his contemporaries. Even Balfour's interest in psychical research can be seen in this light. The attempt by upper-class Cambridge intellectuals such as Sidgwick, Frederick Myers, Balfour, and other leading members of the Society for Psychical Research to provide scientific legitimacy for the existence of a spiritual world continuous with and superior to the material world echoes the emphasis on the ether as an all-unifying medium by Cambridge physicists who were closely connected with the elite of conservative politics. In both cases ineffable, unseen essences underlying the diversity of material nature were interpreted as indications of a higher unity, or as Wynne puts it, as symbolic of "the enduring organic ties of society," which "acted as the 'natural law' witness to a higher spiritual or political law, to be revealed to the ignorant masses by an upper-class intellectual elite" (180).[2]

Balfour's *Foundations of Belief* can also be read as representative of the conservative response to the problem of social disorder. Like Jacyna, Irvine characterized it as a very political work, "a kind of prolonged speech from the Opposition benches attacking the fashionable scientific-utilitarian universe and urging the Tory universe in its stead" (355). The main target of the book was what Balfour called "Naturalism," whose "leading doctrines," he declared, "are that we may know 'phenomena' and the laws by which they are connected, but nothing more" (*Foundations* 6). Naturalism was seen by conservatives as promoting a materialist challenge to the Anglican establishment because it denied the activity of God in nature. Huxley, along with Herbert Spencer and John Stuart Mill, was singled out by Balfour as a prominent naturalist to whom his criticisms applied (124).

Balfour divided *Foundations* into four parts. In the first part, "Some Consequences of Belief," Balfour examined the negative impact of naturalism on our cherished concepts of morality, aesthetics, and human reason. "If naturalism be true," Balfour declared, "or, rather, if it be the whole truth, then is morality but a bare catalogue of utilitarian precepts; beauty but the chance occasion of a passing pleasure; reason but the dim passage from one set of unthinking habits to another. All that gives dignity to life, all that gives value to effort, shrinks and fades under the pitiless glare of a creed like this" (77). To highlight the impoverished quality of the naturalistic creed, Balfour outlined its rather barren five-point catechism and compared it point by point to what he called "the current teaching," in the conclusion to part one (83–86).

In part two, "Some Reasons for Belief," Balfour considered the possibility that naturalism is true even though its practical tendencies are intolerable. But he found the philosophic basis of naturalism to be faulty; indeed, the primary presuppositions of science cannot be justified if we are bound by naturalism's insistence that only empirical methods are valid. Balfour also examined the transcendental idealism of thinkers such as T. H. Green as an escape from the difficulties of naturalism, but rejected it largely on theological grounds.

The third part of *Foundations,* "Some Causes of Belief," introduced the key conservative notion of authority. Balfour admitted that up to this point the results of his study had all been negative: naturalism is practically insufficient and speculatively incoherent, and idealism is not a viable alternative. Balfour's plan in this section was to draw the reader's attention to the positive role of tradition or authority as the chief source of all our beliefs, and to argue that this is a desirable state of affairs. Rather than classifying the clash between opposing forces of new thoughts and old as a conflict between science and religion, Balfour saw the battle as taking place between reason and authority. Though it was fashionable to see reason as more properly molding the convictions of humankind and to applaud the victories of reason in its fight with authority, Balfour pointed out that not all reason is right reason, just as not all authority is legitimate. Furthermore, Balfour believed that if we only relied on reason to determine our beliefs the result would be chaos. Reason was a nihilistic force that divided and disintegrated both thought and society (237). Fortunately, the unconscious work of authority, which, Balfour maintained, was far more influential than the conscious work of reason, provided us with our beliefs about religion, ethics, and politics. It was upon authority, rather than reason, that the "deep foundations of social life" were laid. Balfour even went so far as to suggest that authority supplied us with the "essential elements in the premises of science." Inverting the customary celebration of reason by evolutionists as that which distinguishes us from the animals, Balfour asserted that what elevated us above "brute creation" was "our capacity for influencing and being influenced through the action of Authority" (238).

Entitled "Suggestions Towards a Provisional Philosophy," part four offered a tentative "unification of all beliefs into an ordered whole compacted into one coherent structure under the stress of reason" (241). Reason did have a role to play in Balfour's provisional philosophy, but rather than conceive of reason as totally independent of authority, he aimed at embedding it within a larger system shaped by the power of authority. Having argued in part three that authority is a legitimate source of belief, Balfour constructed a thoughtful reconciliation between science and religion, which drew upon a traditional framework of belief. He maintained that scientific, ethical, and theological beliefs can be brought into a more coherent whole "if we consider them in a Theistic setting, than if we consider them in a Naturalistic one" (344). Naturalism, which assumed that our powers of reasoning were produced during evolution by physiological processes designed to satisfy fear and hunger, cannot tell us why our powers are adequate to supply us with a proper metaphysics for grounding science and religion. In contrast, Theism did not culminate in a position that portrayed human reason as weak. If we presupposed that the world was the "work of a rational Being, who made *it* intelligible, and at the same time made *us,* in however feeble a fashion, able to understand it," then we could have confidence in our powers of reasoning (309).

What distinguished Balfour's contribution from other refutations of scientific naturalism appearing during the eighties and nineties was his philosophical critique of its epistemological foundations (Jacyna 22). The sections of his work dealing with naturalism's theory of knowledge were the most convincing and the most philosophically sophisticated. Though he was following the lead of his old teacher and brother-in-law Henry Sidgwick in centering on the incoherence of the empiricist theory of knowledge, Balfour was able to apply this type of critique far more effectively and more specifically to scientific naturalism. Balfour's whole approach in this area of his thought allowed him to present a brilliant demolition of Huxley's conception of science, as well as its ideological implications, without at the same time undermining the scientific enterprise itself. However, the positive side of Balfour's work, wherein he attempted to build a conservative ideology that is not in conflict with the results of modern science, was not nearly as strong.

In part two of *Foundations,* Balfour focused on the reliance of naturalism's "empirical theory of knowledge" on sensations as giving us immediate experience of natural objects. According to Balfour this narrow view of experience provided defenders of naturalism with a very shaky foundation for real knowledge. How, Balfour asked, can we move from sensations to the perception of independent material objects? Just as the science of physiology tells us that colors are not in nature, but are only secondary qualities, it also tells us that every element in every perception is an affectation of our sensibility. "We are directly cognizant of nothing but the mental results of cerebral changes: all else is a matter of inference," Balfour declares in a very Humean fashion (113). Supporters of naturalism have been wrong to suppose that we have immediate experience of some independent things. We are asked to believe in an inconsistent, "hybrid world," which is half primary qualities of matter and half secondary qualities, but by what means are we to distinguish between the primary and the secondary if we are limited to sensations as the final court of appeal (122)? It is not possible to treat sensations in such a way as to "squeeze out of them trustworthy knowledge of the permanent and independent material universe" (119).

Or, Balfour continues, what are we to make of the law of causation, another principle grounding science according to upholders of naturalism? Is not this also another "general principle," which is "presupposed in any appeal to general experience" and which therefore cannot "be proved by it" (128)? How can we move from our very narrow experience, based on sensation, to a principle claiming universal jurisdiction "to the utmost verge both of time and space" (129)? Balfour concluded that we cannot. The failure of naturalism to provide a coherent defense of its primary principles is, for Balfour, a philosophical scandal, especially since those devoted to its expansion try to dictate terms of surrender to every other system of belief (135).

Balfour made it clear, however, that his critique of naturalism is not meant to annihilate science. He carefully disengaged naturalism, which is "nothing more than the assertion that empirical methods are valid, and that no others are so," from science itself (134). The leaders of naturalism illegitimately tried to "associate naturalism and science in a kind of joint supremacy over the thoughts and consciences of mankind" (136). To Balfour, naturalism was but a "poor relation" of science that had forced itself into the "retinue of science" and now claimed to "represent her authority and to speak with her voice" (135). Despite the inability of naturalism to satisfy human needs and human reason, its influence continued to increase. But science existed before naturalism, and he was confident it would survive the decline and fall of naturalism (134). Balfour's insistence on referring to "scientific naturalism" as "naturalism" throughout *Foundations* encapsulated his underlying plan of denying to scientific naturalism any intrinsic connection to science. Distinguishing naturalism from science was absolutely crucial, for it allowed him to claim that advocates of naturalism, such as Huxley, did not speak with the authority of science behind them. It also permitted Balfour to maintain that science and religion were not in conflict. "The differences between naturalism and theology are," Balfour admitted, "irreconcilable, since naturalism is by definition the negation of all theology." But true science has no quarrel with theology or religion, for all science says is that matters such as the existence of God are beyond its jurisdiction and must be tried "in other courts, and before judges administering different laws" (301).

Balfour was now free to conceive of a form of science that did not contain the ideological baggage accompanying naturalism. Instead of being used as an instrument of political radicalism, science could be exploited by conservatives looking for resources to justify existing social forms (Jacyna 33).[3] Balfour would have encountered this very different vision of science and its broader social meaning in the work of those Cambridge physicists and intellectuals who interpreted science as providing support for religious belief and notions of social unity.

Unlike other Christian critics of science, such as Wace or Gladstone whom Huxley had engaged in controversy, Balfour was no stranger to the contemporary scientific scene. Even within his family circle he was surrounded by individuals who were actively engaged in scientific pursuits and who kept him in touch with some of the leading scientific men. John Strutt, later Lord Rayleigh, a Nobel Prize–winning chemist, married one of Balfour's sisters in 1871. Balfour's brother Frank, tragically killed at the age of thirty-one in a climbing accident in the Swiss Alps in 1882, had been a promising young biologist. Cambridge had created a chair for him in animal morphology. Ironically, Huxley was so impressed with Frank Balfour that he regarded him as someone who would carry on Huxley's work after he retired (*Life* 2:37; Peterson 217). But there was an even more personal relationship between Frank Balfour and his brother's future foe. To Huxley, Frank was almost like a son, while he reminded Mrs. Huxley of the young Huxley in his *Rattlesnake* days (Peterson 216; *Life* 2:40–41). When Huxley heard of Frank's death he was so distraught that for three days he was totally incapacitated and almost unable to eat or sleep (*Life* 2:40). There is every reason to believe that Balfour's interest in reconciling science in its full integrity with religion was genuine.[4]

Considering that it contained some very abstruse philosophical arguments, *Foundations of Belief* sold well. According to the publishers, they sold 8,926 copies between February and June 1895, and 2,179 copies during the following twelve months (Peterson 301). A lively controversy took place, which included positive responses from the Unitarian James Martineau and a number of Roman Catholic writers, as well as the negative reviews of unbelievers such as Herbert Spencer (Root 128–30). One of the earliest reviews appeared in the *Times* (London) on February 8, 1895. Hailing the book as "one of the chief contributions to philosophy made for many years in England," the anonymous author compared Balfour to Pascal, Bishop Butler, and J. H. Newman. He lavished praise on the wit and beauty of the writing and the success of the overall argument. Then, in a mocking tone, the reviewer remarked that "there will be consternation in the little Bethels of 'crude realism' and Agnosticism, and among those who will resent this attack with ardour proportioned to the ignorance of its nature" ("Mr. Balfour").

Earlier that day, Huxley had received a telegram from Knowles, the editor of the *Nineteenth Century*, requesting that he review *Foundations*. He replied to Knowles with hesitation, but after reading the *Times* (London) piece he sent Knowles a second telegram agreeing to do the review (*Life* 2:421–22). Knowles responded the following day, sending the book, and expressing his gratitude to Huxley for tackling Balfour. He was doubly thankful, both as an editor eager to enlist the pen of one of the great controversialists of the day and as an individual with a personal interest in seeing a spiritualist like Balfour get his due: "It seems to me high time that some smashing blow should be struck against spook-speculation—in all its forms. Of course I do not *know* that this book is part of that literature for I have not read it—and the newspaper notices are mostly chaotic. But as the author is the President of the Spook society . . . I feel much disposed to misdoubt him as a philosophic guide—however greatly I value him as a politician" (HP 20:176). Having secured Huxley's agreement to review the book, Knowles now attempted to rouse his combative spirit. He knew that Huxley hated table-rappers with a vengeance. But here was an important politician who was also the president of the Society for Psychical Research. On February 13 Knowles emphasized again the necessity for Huxley to challenge someone like Balfour who was an influential political figure. Knowles wrote that "in his position of *political* authority the poor wondering British

Public will (if not clearly and plainly disabused) follow him—according to its custom—as a philosophic and speculative authority also to its still greater and greater confusion of brain" (HP 20:177). Knowles provided Huxley with an almost irresistible target, while implying that Huxley should feel some sense of responsibility to the public to answer Balfour.

As a leader of scientific naturalism, Huxley had many times in the past taken on defenders of the Tory-Anglican establishment. Turner has explored the social and political significance of scientific naturalism and has shown that intellectual controversies during the Victorian period often reflected the collision between established and emerging intellectual elites vying for popular cultural preeminence. Throughout the nineteenth century the older, established intellectual elite, who defended the interests of the aristocracy and the Anglican church, fought to preserve their social and intellectual positions. Scientific naturalism served the interests of sections of the new professional middle class and provided a rationale for their leaders to wrest cultural and social control from the clergy (Turner, *Contesting Cultural Authority*). An aristocratic, Tory politician, educated at Cambridge, a devout Christian believer with an interest in spiritualism to boot, Balfour represented almost everything Huxley despised. As they attained positions of prestige and succeeded in professionalizing science during the sixties, seventies, and eighties, however, scientific naturalists had moved away from their earlier, youthful radicalism of the forties and fifties (Jacyna 28). By the nineties, the scientific establishment and the Tory-Anglican establishment had much in common. The elderly Huxley would have shared Balfour's views on the dangers of socialism and Irish nationalism. Huxley's affection for Balfour's late brother, Frank, would also have held him back from an all-out attack. Knowles knew that he must provoke the aged Huxley to take action.

Egged on by Knowles, Huxley immediately went to work on *Foundations*. By February 18, he sent off the larger part of the piece to Knowles with the promise that the rest would follow in two or three days. "I am rather pleased with the thing myself," Huxley joked, "so it is probably not so very good." The next day Huxley wrote to Knowles that he would have the rest of the article the following day, but Huxley was unable to keep his promise. On February 20, Huxley was again having problems with completing the review, even though he had put in seven hours on the project.

The piece had grown too long for one paper and Huxley requested that he be allowed to divide it into two separate articles. Ominously, Huxley informed Knowles of the outbreak of an influenza epidemic in Eastbourne and expressed fears that he would contract the illness. If that happened he would have to "go to bed and stop there . . . until I am killed or cured" (*Life* 2:422). Knowles replied to Huxley on February 22 that he was "only too glad to have two articles instead of one." He also suggested that Huxley title the piece "Mr. Balfour's Attack on Agnosticism," which would be more "rousing" than the title of Balfour's book (HP 20: 178).

The following morning Huxley received the proofs of the first article, quickly reviewed them, and sent them off early in the afternoon. After promising that the second installment would be "no worse," Huxley told Knowles that his estimation of Balfour as a thinker "sinks lower and lower, the further I go. God help the people who think his book an important contribution to thought" (*Life* 2:423). The next day Knowles acknowledged the extra effort Huxley had put into polishing off the proofs with such promptness: "It was indeed noble of you to work with such splendid fury and effect" (HP 20:179). Huxley went back to writing the rest of the second part of the review, but the danger of contracting influenza made him feel as if he were in a race against time. Writing Knowles on March 1 that his wife had come down with the influenza, Huxley fretted that "chances are I shall follow suit—though I don't mean to if I can help it." Just in case he did fall sick, Huxley told Knowles that he had been working hard on part two of his piece and hoped to have the manuscript, "even if rather in the rough," ready by the following day. "If the demon seizes me I may at any rate hope to have stuff enough left for correcting proofs" (HP 20:188).

Huxley completed the manuscript sometime in early March. Just after he sent it off to Knowles, his worst fears came to pass when the influenza hit him. On March 8 Huxley received the proofs for part two but was too sick to look at them (*Life* 2:423). In the meantime, part one had appeared in the March issue of the *Nineteenth Century*. On March 18 Knowles sent Huxley a cheque for forty-two pounds in payment for part one, informing him that he would need the corrected proofs for part two in the next four or five days in order for it to appear in next month's issue. If Huxley wasn't well enough to do the work, Knowles offered to put a note in the *Nineteenth Century* that the influenza had prevented him from finishing it (HP 20:183). Huxley replied the next day, saying that he could not summon the strength to check the proofs (*Life* 2:424). He never completed this task before his death. The rough draft of "Mr. Balfour's Attack on Agnosticism 2" was not published until 1932, as an appendix to Houston Peterson's *Huxley: Prophet of Science.*

A close examination of Huxley's two-part "Mr. Balfour's Attack on Agnosticism" reveals the problems that Huxley encountered in trying to respond effectively to Balfour. Before even reading *Foundations,* and on the basis of perusing the review in the *Times* (London) and whatever he picked up from friends, Huxley had intended to build his strategy around exploiting the double-edged sword of skepticism by turning the critical, destructive elements in *Foundations* against Balfour himself. On February 8 Huxley wrote to Knowles that the review of *Foundations* would be worth doing in order to prove that "Balfour is an agnostic after my own heart" (*Life* 2:422). Huxley had had great success in the past demonstrating that the arguments of Christian skeptics, who emphasized faith by undermining the validity of reason, could be twisted around to support agnosticism. In "Agnosticism and Christianity" (1889; *CE* 5:309–65), where he argued that the Bible, as a historical document, can hardly be considered an authority for accepting the reality of miracles, Huxley's ally was J. H. Newman (see Lightman, *Origins* 115). Newman took the position that faith should determine the true Christian's belief in miracles, but Huxley merely borrowed elements from Newman's attack on the inadequacy of external evidence. H. L. Mansel, the High Church author of *The Limits of Religious Thought* (1859), was appropriated by Huxley in a similar fashion. Though Mansel's aim was to defend the doctrine of biblical infallibility by showing that the inherently limited human mind is incapable of criticizing a communication from an infinite God, Huxley was able to fasten upon Mansel's notion of limits of knowledge to uphold the validity of agnosticism (see Lightman, *Origins* 6–10, 32–67, 106–13).

Huxley anticipated that Balfour's skepticism could be turned against Balfour's Christianity. However, Huxley found that he could not treat Balfour in the same way as he had other Christian skeptics such as Newman or Mansel. Whereas previous Christian skeptics undermined the authority of human reason in religious matters in order to drive the reader toward the authority of the Bible, or the Church, or faith, Balfour's skepticism was aimed solely at destroying the power of scientific naturalism. Once he read

Foundations, Huxley had to rethink his strategy. He fell back on another strategy he had used many times in the past: controlling the meaning of the key terms of the controversy. Huxley knew all about the power of words. During his career he coined new terms, such as "agnosticism," and was forced to abandon old ones, such as "Darwinism," when they no longer served his purpose (Lightman, *Origins* 10–13; Moore 353–408). Controlling the meaning of key terms involved contesting the meaning of his opponent's terms, inventing new terms, and juxtaposing important terms in such a way that the reader was confronted by a stark choice between two opposing positions. But this strategy, so successful in the past, was not well designed to meet the new challenge that Balfour represented. Balfour's astute critique of scientific naturalism was unlike anything Huxley had had to confront in the past, for it effectively called into question the crucial assumptions grounding Huxley's conception of scientific naturalism, agnosticism, and the relationship between science and religion. Working against Huxley as well was the fact that the controversy took place at a time when scientific naturalism was being questioned by important intellectuals throughout Europe.

Huxley began the first part of "Mr. Balfour's Attack on Agnosticism" by laying out for the reader what was at stake in this controversy through the juxtaposition of two opposed terms. Nothing less than the "future of our civilisation" depended on the result of the contest between "Science and Ecclesiasticism" (530). Placing himself on the side of "Science" and Balfour with "Ecclesiasticism," Huxley expressed satisfaction in finding that one of "our political chiefs" who was "sure to go higher" in "official rank" and "in the estimation of his countrymen" was aware of the important consequences flowing from "an antagonism in the world of thought of a much sharper and more serious kind than has ever yet existed" (531). This was an old theme for Huxley. He often referred to a war between agnosticism, or science, and ecclesiasticism, and identified the upholders of the latter as the champions of a particular form of dogmatic theology (Lightman, *Origins* 132). From the start, Huxley attempted to place Balfour with those who resisted the growth of science.

Huxley then introduced a parallel pair of opposing terms, "Naturalism" and "Demômism." Balfour's reticence to name the opposite of "Naturalism," other than the "current teaching," led Huxley to coin the term "Demômism," which he derived from a Greek verb that means to "talk popularly" (531). Though Huxley offered a linguistic justification for the new term "Demômism," he obviously wished to capitalize on the unpleasant sound of the word, which is too close to "demonism" for the selection to be accidental. In Huxley's examination of *Foundations* he subjects the two terms to a systematic analysis in order to show that Balfour's understanding of both is flawed.

It was absolutely crucial for Huxley to retain control over the meaning of "naturalism," having himself coined the term "scientific naturalism" in 1892 in the prologue to *Controverted Questions* (*CE* 5:38). Here Huxley had conceived of scientific naturalism as an intellectual movement that had its origins in Descartes and the Renaissance, and as a term that was antithetical to "supernaturalism" (Paradis 178–80). He discussed both the destructive, or critical, side of scientific naturalism, as well as the constructive character of its nineteenth-century incarnation (5:41). Balfour's attempt to determine the meaning of "naturalism" was a serious threat to Huxley's lifelong work. Turner has pointed to the importance of Balfour's book and the ensuing controversy for "giving currency to *naturalism* as the term under which scientific writing and philosophy became subsumed" (*Between Science and Religion* 11). Balfour's aim was to attach a negative meaning to the term and to disengage it from science proper.

Balfour was not alone in his assault on "naturalism." Whereas the fortunes of scientific naturalism had been on the rise since the middle of the nineteenth century, the 1890s was characterized by a "revolt against Positivism." Eminent thinkers from around the Western world, including the American William James, Henri Bergson and Henri Poincaré of France, the German Edmund Husserl, and James Ward of England, all contributed to a critique of naturalism that began to gather strength by the end of the century (see Turner, *Between Science and Religion* 228). During a time when scientific naturalism was under attack it was particularly important for Huxley to respond effectively to Balfour. In contesting Balfour's views on the meaning and validity of scientific naturalism, Huxley was fighting for his life. Everything he had worked for—professionalizing science, establishing the authority of the scientific elite, even his own authority as spokesperson for the true meaning of science—was at risk.

On February 17, a period when Huxley was in the middle of writing the first part of his response to Balfour, he lunched with friends and spoke about his reactions to *Foundations.* As reported by the Roman Catholic Wilfrid Ward, Huxley's neighbor at Eastbourne, the need for retaining control of the meaning of scientific naturalism was already fixed in his mind: "Mr. Balfour ought to have acquainted himself with the opinions of those he attacks. . . . An attack on us by some one who understood our position would do all of us good—myself included. But Mr. Balfour has acted like the French in 1870: he has gone to war without any ordnance maps, and without having surveyed the scene of the campaign. No human being holds the opinions he speaks of as 'naturalism.' He is a good debater. He knows the value of a word. The word 'Naturalism' has a bad sound and unpleasant associations. It would tell against us in the House of Commons, and so it will with his readers" (*Life* 2:419). Huxley recognized that Balfour's formidable debating skills, learned and honed in the political arena, were just as dangerous when applied to the politics of science. By focusing on the word "Naturalism," and instilling it with negative connotations, Balfour had struck a telling blow. However, Huxley thought he could defeat his enemy if he convinced readers that Balfour's naturalism did not really exist. Challenging Balfour's definition of the term became a central feature of Huxley's strategy.

Just after introducing the opposing terms "Naturalism" and "Demômism," at the beginning of "Mr. Balfour's Attack on Agnosticism," Huxley launches into a discussion of the meaning of "Naturalism." It is a "well-known and perfectly understood technical term of philosophy," Huxley declares, "and applies to all systems of speculation from which the supernatural is excluded." According to Huxley, naturalism therefore had nothing to do with the specific doctrines of materialism, idealism, determinism, or libertinism, but was compatible with any of those doctrines. A supporter of naturalism could be an empiricist, a believer in innate ideas, a Platonist, an Epicurean, a pantheist, or an atheistic Buddhist. But in his *Foundations of Belief,* Balfour, Huxley pointed out, attempted to redefine naturalism as materialistic and deterministic. "Serious errors of connotation may arise from this novel use of old language," Huxley warned (533). By using the term "Naturalism" in a sense different from that employed by the rest of the world, Balfour merely clouded the issues. "I am afraid it must be admitted," Huxley declared, "that the brilliancy which hovers over the pages of the

Foundations of Belief is sometimes so vague and shifty that, like a hostile search-light, it often spreads confusion where it professes to illuminate" (537). More importantly, if Balfour's opponents did not hold to naturalism as defined by him, then his critique failed to reach its real target (532).

Huxley followed up on this approach in "Mr. Balfour's Attack on Agnosticism 2," the majority of which is devoted to a close examination of the conclusion to part one of *Foundations of Belief,* where Balfour laid out the five point catechisms of both naturalism and the "current teaching" (*Foundations* 83–86). Here Huxley also resorted to a strategy he had used in the past. In his essay "Agnosticism" (1889; *CE* 5:209–62), Huxley ridiculed self-proclaimed agnostic Samuel Laing for offering an agnostic creed composed of eight articles. Objecting that agnosticism had been "furnished with a set of 'articles' fewer, but not less rigid, and certainly not less consistent than the thirty-nine," Huxley insisted that agnosticism was a method, not a creed to be pinned down in some dogmatic, pseudo-Christian formula (*CE* 5:245–47; see also Lightman, *Origins* 143–44). In his response to Balfour, Huxley adopted an ironic tone as he questioned how appropriate it was to supply naturalism with a catechism. Furthermore, Huxley explicitly questioned the accuracy of Balfour's depiction of naturalism, attempting to cleanse it of the negative connotations that clung to it in *Foundations* ("Balfour's Attack, 2" 316–25). By the end of part two of his response to Balfour, Huxley had fulfilled his promise, announced at the beginning of part one, of proving that "if 'Naturalism,' as defined and catechetically represented by him, is a body of doctrine which nobody holds . . . the champion of Demômism is doing battle with the air" (532).

But Huxley was not just concerned with reclaiming control over the meaning of naturalism. He also contested Balfour's depiction of agnosticism. Not all scientific naturalists were agnostics. But all Victorian agnostics were scientific naturalists, and Huxley had often used his agnosticism as a powerful epistemological tool in support of the scientific naturalists' goal of displacing the Anglican clergy from their position as intellectual elite. The central notion of Huxley's agnosticism, staying within the limits of knowledge, helped scientific naturalists to define the boundaries of proper science and to attack those within and outside the profession who were determined to go beyond these boundaries by bringing improper theological concepts into science (Lightman, *Origins* 92). Since Huxley had coined the word "agnosticism" in 1869, it evoked in him a particularly strong proprietary feeling (see Lightman, *Origins* 10–13). He had even written a series of articles in 1889 to reestablish his rightful ownership of its meaning (see Lightman's *Origins* 136–44 and "Ideology"). Balfour's negative depiction of agnosticism in *Foundations* led Huxley to reassert his claim to determine what it really meant.

In part one of his response to Balfour, Huxley attempted to establish the "highly respectable lineage" of agnosticism, which he found in the works of High Church Anglican Henry Mansel and eminent Scottish philosopher Sir William Hamilton, and which also could be "traced back for centuries." Though he modestly affirmed that he did not invent the idea, Huxley noted that because he was responsible for christening agnosticism he was qualified to speak authoritatively on its real meaning. Huxley's purpose here was to convince his readers that *Foundations* presented a misleading portrait of agnosticism, allowing Balfour to associate it with all sorts of unsavory doctrines such as positivism, empiricism, materialism, and atheism. "I object to making Agnosticism the scapegoat," Huxley insisted, "on whose head the philosophic sins of the companions with whom it is improperly associated may be conveniently piled up" (533–34).

Huxley spent several precious pages attempting to distinguish his position from other scientific naturalists. In fact, Huxley told Knowles that he accepted the task of reviewing *Foundations,* in part, because it gave him the opportunity to distance himself from Herbert Spencer (*Life* 2:422). Indeed, Huxley's account of Hamilton's philosophy stresses that since God is not a subject of knowledge, His existence cannot be affirmed with certain knowledge—an implicit critique of Spencer's position that we can assert the existence of a shadowy deity he referred to as the "Unknowable" (535; see also Lightman, *Origins* 85–86). At the beginning of part two of his response to Balfour, Huxley protested being lumped together with Spencer, and he explicitly rejected Spencer's a priori method and his form of the doctrine of evolution (316). As for English positivism, which Balfour had included within the ranks of scientific naturalism, Huxley refused to defend it from Balfour's criticisms and cavalierly left it "to take care of itself" ("Balfour's Attack" 540).

Huxley's defense of scientific naturalism and his attempt to distinguish his agnosticism from other scientific naturalists like Spencer revealed a weakness in his strategy that ultimately subverted his attempt to respond effectively to Balfour. While Huxley positioned himself as the spokesperson for scientific naturalism, he simultaneously denied responsibility for the beliefs of any scientific naturalists but himself. Though he claimed that Balfour had set up a straw man in his false depiction of naturalism, Huxley's desire to control the meaning of the terms "naturalism" and "agnosticism" led him to give up fellow scientific naturalists like Spencer and his followers, and the positivists, who constituted a large faction within scientific naturalism, to the tender mercies of Balfour's critique. Huxley may have reasserted his authority as spokesperson for scientific naturalism and agnosticism, but he also insisted that he spoke for himself only.

Did scientific naturalism necessarily speak with the authority of science behind it? Balfour's attempt to disengage naturalism from science through a critique of its epistemological basis was particularly difficult for Huxley to counter. Huxley chose not to respond directly.[5] In the past, Huxley had offered epistemological proofs that justified basing science on axioms such as the uniformity of nature or the existence of an external natural world (see Lightman, *Origins* 161–76). But in 1895 he avoided this approach and his resounding silence on this issue is curious, given that Balfour's distinctive contribution to the literature refuting scientific naturalism during the eighties and nineties was his philosophical critique of its epistemological premises. Perhaps Huxley felt that epistemological arguments ultimately were not very convincing. It may be that the repetition of the same old arguments was not sufficient in the 1890s when the scientific worldview had come under attack and Huxley could not depend on an audience that would unquestionably accept his assumptions. In any case, Huxley tried two strategies, neither of which was very convincing, aimed at putting the science back into naturalism.

First, he denied that the science of scientific naturalism was limited to empirical experience of material phenomena. He argued that scientific naturalism and science also included the realm of "mental phenomena" or mind ("Balfour's Attack" 539). Though Huxley may have scored some points in showing that Balfour was ignorant of the development of new sciences such as psychology, he neither acquitted scientific naturalism of materialism nor of the lack of philosophical justification for its axioms. For here Huxley merely treated mind as if it were another type of phenomenon akin to the material and therefore open to scientific investigation. Furthermore, the admission of the reality of mind did not speak to the universal validity of the notion of cause.

Second, Huxley tried to show that Balfour could not be trusted as a spokesperson for science because he really didn't understand it. Wilfrid Ward recollected that during the lunch on February 17, 1895, Huxley had nothing but praise for the scientific abilities of Balfour's brother Francis. "'I used to say there was science in the blood,'" but Arthur's new book, Huxley added, smiling, "'shows I was wrong'" (*Life* 2:420). At one point in the second part of his review of *Foundations,* Huxley asserted that Balfour did not understand how evolution worked. "I am afraid," Huxley sadly observed, "Mr Balfour has hardly given the *Origin of Species* the attention which that great little book really needs if one is to understand its teachings." How could a politician-philosopher, unacquainted with one of the most important scientific works of the century, presume to speak on behalf of science and, worse yet, claim that scientific naturalists like Huxley, who were professional scientists, did not speak with the full authority of science behind them? With mock politeness, Huxley asked Balfour, if "I may presume to know what" a "'purely scientific point of view'" is ("Balfour's Attack, 2" 321). Ashforth has argued that "Huxley did good work in demonstrating how imprecise and unscientific Balfour's language was," which implies that Huxley succeeded in demonstrating that Balfour was an uninformed amateur unworthy of speaking for science (151). Yet the fact that Balfour became the president of the British Association for the Advancement of Science in 1904, reveals that Huxley failed to convince his contemporaries that Balfour lacked all scientific credibility.

Not content with challenging Balfour's right to speak for science, Huxley questioned the legitimacy of Balfour's religious authority. Huxley could not treat Balfour as he had other Christian opponents such as Gladstone or Wace. Balfour had insisted that his *Foundations,* subtitled *Being Notes Introductory to the Study of Theology,* was not intended to present a system of theology. Rather, it was meant to recommend "a certain attitude of mind" or a "particular way of looking at the World-problems," which, prior to a study of theology, would make that study fruitful (4). As a result, Balfour did not offer Christian dogmatic beliefs that Huxley could use as convenient targets for his agnosticism. So Huxley hit upon a different strategy: he attempted to show that Balfour's catechism for Demômism was not a faithful representation of "current teaching." The source for the five articles of this catechism, Huxley argued, was to be found in "pre-Christian Greek philosophy," not in Christianity ("Balfour's Attack, 2" 325). Balfour's preliminary notes to a study of theology were pagan. Balfour not only misunderstood the naturalism he attacked, he also did not grasp "the current teaching" he defended.

If Balfour could not speak for either science or religion, then Huxley thought he was spared the task of examining the reconciliation of science and religion put forward in *Foundations.* Huxley may have avoided the issue in order to draw attention away from a comparison of Balfour's reconciliation with his own. For such a comparison would have revealed the brilliance of Balfour's overall strategy in *Foundations* to undermine scientific naturalism. Like other agnostics, Huxley often argued that, if rightly conceived, there was no war between science and religion. While religion belonged to the realm of feeling (including emotion, imagination, inwardness, and symbol), science was part of the distinct realm of the intellect (including fact and knowledge). If science and religion, whose authorities were sovereign within their own spheres of interest, did not seek to go beyond their own realms, they could never come into conflict. However theology, which was to be distinguished from religion, attempted to embody feelings in concrete facts and was potentially in conflict with science (see Lightman, *Origins* 128–34). If any war existed, it was only between science and theology. Huxley held to this position right up to the end of his life (Barton 261–87).[6] In a letter dated 1894 he wrote that "most people mix up 'Religion' with Theology and conceive that the essence of religion is the worship of some theological hypostasis or other" (HP 12:342).

But in *The Foundations of Belief,* Balfour questioned the notion of separate spheres of authority for science and religion. He objected that the "two spheres" approach fragmented our world and left us with no recourse for judging between science and religion when boundary disputes arose. "And the two regions of knowledge lie side by side," Balfour observed, "contiguous but not connected, like empires of different race and language, which own no common jurisdiction nor hold any intercourse with each other, except along a disputed and wavering frontier where no superior power exists to settle their quarrels or to determine their respective limits." Balfour labeled this approach to the problem a "patchwork scheme of belief" (194). Huxley's resolution, which depicted science and religion in peaceful harmony, where no disputes arise and no superior power is needed, seemed naive in light of Balfour's analysis.

There is, however, a deeper irony to be found in Balfour's and Huxley's contrasting methods of reconciling science and religion. Huxley separated religion from theology in order to reconcile science and religion. This strategy allowed him to pose as a spokesperson for science and for religion, even when he attacked the doctrines of Christian theology. Balfour imitated the strategy behind Huxley's separation of true religion from theology in order to reconcile science and religion. But instead of dividing religion into religion and theology, Balfour divided science into authentic science and naturalism, which permitted him to attack naturalism without disturbing the harmonious relationship between science and religion.[7] In effect, Balfour used Huxley's basic strategy to turn the tables on scientific naturalism. If Huxley had drawn attention to Balfour's strategy, his own might have seemed more transparently opportunistic. At the very least, he could hardly criticize Balfour for adopting the very same strategy that had served him so well for so long. On this point, as on others, Balfour had proven himself to be a formidable foe. When Huxley agreed to take on the review of *Foundations,* at the urging of Knowles, he seriously underestimated the challenge Balfour would present to his rhetorical skills and to his lifelong defense of scientific naturalism.

What was Huxley thinking about, day after day, as he sat in the sun during the spring and early summer of 1895? Did he recall his life-and-death struggle with the influenza, which had resulted in serious damage to his heart and kidneys, so serious that he could slip into a coma at any time? Did his mind turn to the unrevised proofs to the second part of his review of *Foundations*—a life-and-death struggle of another kind. Had the two become inextricably connected in his mind? Shortly after his son Leonard had visited on June 24, Huxley's condition began to deteriorate. On June 26 Huxley wrote to his old friend Hooker that he was plagued by bouts of nausea and vomiting, which sometimes lasted for hours. A slight improvement over the next two days was followed by signs of heart failure on the morning of the twenty-ninth. Anesthetics were administered to relieve the excruciating pain. By midafternoon Huxley was gone, dead at the age of seventy. He was buried at Finchley on July 4 beside his brother George and son Noel (*Life* 2:426).

After waiting a decent period of time for the grieving widow to recover, Knowles wrote to Mrs. Huxley on August 16, asking if the unrevised proofs of Huxley's second piece on Balfour could be published uncorrected. "It is so very fine and good as it stands," Knowles pleaded, "that under any circumstances it seems wrong not to give it to the world." Moreover, Knowles pointed out, "as his last unfinished labour it would have a pathos as well as a power which would carry it home in a thousand quarters where it would do great good" (HP 20:184). Knowles contacted Mrs. Huxley again four days later, assuring her that he had a copy of the proofs and that many readers wanted to see the second part of the essay (20:186).

When Mrs. Huxley remained unconvinced, Knowles did not give up; on October 11, he wrote a long letter to Mrs. Huxley in a final effort to obtain her approval. In his letter Knowles revealed Mrs. Huxley's main reason for stubbornly refusing to give way to his request. In his final days Huxley had given explicit instructions to both his wife and his doctor that the publication of the piece "be postponed until he got better enough to give it final polishing." Knowles argued that if Huxley had realized he would not recover and if he had known that Mrs. Huxley would question whether the conclusion of the article should be published without revisions, "I cannot help thinking he would have answered—'Oh publish it for it is a good bit of work and my last—but say I left it uncorrected and might have modified the form of it—tho' not the substance" (HP 20:188).

Was it merely Huxley's perfectionist desire to give the piece a final polish that led him to instruct, on his deathbed, that the second part of the Balfour piece not be published? Or did he realize, in his heart of hearts, that he had failed to respond adequately to Balfour? Certainly friends like Knowles and Hooker had spoken highly of the first part. Hooker wrote on March 27 that Huxley had "probed" Balfour's "weak places with effect" (HP 3:425). Huxley told Knowles that he was pleased with the first part of the piece and that the second installment would be just as good. Several times during the period that he was working on the review, he made disparaging comments about Balfour's abilities, implying that it was easy to demolish such a foe (*Life* 2:422–23). Yet the light-hearted barbs at Balfour's expense should be contrasted to Wilfrid Ward's report that Huxley was "very angry with the book," an observation Leonard Huxley leaves out of his account of his father's lunch with Ward (Ward 352). And not all of Huxley's friends were impressed with Huxley's response to Balfour. Even legal writer and Oxford professor of jurisprudence Frederick Pollock (1845–1937), who counseled Mrs. Huxley to publish the second part of the review, was not enthusiastic about the first published piece. "I must confess," Pollock told Mrs. Huxley, the first article "rather disappointed me as being a needlessly long preliminary skirmish" (HP 24:160).

Certainly Balfour himself was not perturbed by Huxley's attack. Early in the month Huxley died (June 1895) Balfour had been invited by L. J. Maxse, editor of the *National Review,* to write a reply to those who had assailed *Foundations of Belief.* In a letter dated June 4, 1895, Balfour answered that he did not intend to respond, explaining that it would be difficult to deal with the diverse viewpoints of his critics within the limits of a single article. He added that if he devoted an article to individual writers, "such as Huxley or Spencer, I shall probably feel myself bound to send it to the Review in which their criticism originally appeared" (Balfour to Maxse). Balfour evidently never felt the need to write a separate response to Huxley's attack.

Balfour turned out to be far more difficult an adversary than Huxley had anticipated. Huxley's attempts to control the meaning of the key terms of the controversy—words such as "naturalism," "agnosticism," "science," and even "religion"—did not reduce the impact of Balfour's perceptive arguments. Huxley found himself in the fight of his life, for he had to defend his authority as a spokesperson for science and the validity of both scientific naturalism and agnosticism. He moved heaven and earth to complete the piece, but ultimately could not defeat the grim reaper, a foe even more relentless than Balfour. In one of his letters pleading with Mrs. Huxley to allow the publication of Huxley's last essay, Knowles tried to change her mind with a moving description of her husband's efforts to complete the essay. It was an act that seemed to Knowles "as grand as it is pathetic." Battling old age and the influenza in order to respond to a serious challenge from Balfour "had cost him such immense and courageous effort—fighting even with death to accomplish it" (HP 20:188). Did Knowles mean here that Huxley heroically held back death in order to complete his task? Or did he believe that Huxley hastened his own demise by taking on Balfour at a time when he needed all his strength to combat a deadly sickness? In any case, there is also a more profound meaning to Knowles' remark: in a sense, Balfour's *Foundations of Belief* symbolized movements of thought that threatened the hegemony of scientific naturalism. Huxley, who spent his entire life clearing the way for the triumph of scientific naturalism in Victorian society, could not help but feel obliged to battle Balfour to the death.

NOTES

I express my gratitude to John Greene and Sydney Eisen, whose comments on early drafts of this paper were helpful in the revision process, and to Richard Hayes, whose discussion of Balfour in a letter to me led to the idea for this essay. Thanks also to the Imperial College of Science, Technology, and Medicine Archives for permission to quote from the Huxley Papers, and to the West Sussex Record Office and the Maxse family for allowing me to refer to one of their Balfour manuscript letters. This essay was researched and written while I held a Research Grant from the Social Science and Humanities Research Council of Canada.

1. Balfour later became prime minister from 1902 to 1905 and resigned the Conservative leadership in 1911. He served in various important political positions, including first lord of the admiralty and then foreign secretary during World War I, and lord president of the council from 1919 to 1922 and from 1925 to 1929.

2. McGeachie argues that late-Victorian organicism and its language of complexity are far more complex than Wynne or Jacyna realize. According to McGeachie, organicism was not just conservative in nature but was also drawn upon by others such as the Victorian New Liberals (211–52).

3. Jacyna sees in Balfour's attitude toward science an inexplicable ambiguity, for in certain contexts it was to be condemned while in others it was an aid to the Conservative cause. I argue that no dichotomy exists in Balfour's thought, for the distinction between flawed naturalism and praiseworthy science was clear in his mind.

4. Later, in the 1920s, Balfour was able to use his political clout to pursue one of his long-standing passions, the government support of scientific research. See Egremont, *Balfour* 332.

5. In the Huxley Papers kept at the Imperial College in London, there exists a manuscript marked "Another incomplete essay on Balfour and Agnosticism" (see HP 47:108–44). Though resembling in sections the other two essays written by Huxley on Balfour's *Foundations,* this seems to be an entirely different essay, in

which Huxley spends more time on the epistemological dimension of the debate. Huxley appears to have been unhappy with this draft and elected not to use it. It is unclear as to when in 1895 Huxley wrote this piece, for there are no dates on the manuscript and he never mentions a discarded draft in his correspondence with Knowles or others. But its existence is indicative of the difficulties Huxley faced when he tried to respond effectively to Balfour's critique of scientific naturalism.

6. Barton was the first to explore in any depth Huxley's complex conception of the relationship between science, religion, and theology.

7. In a forthcoming article, John C. Greene explores how Huxley and Balfour's opposing views of the relationship between science, religion, and theology led to their differing assessments of the origins of the Victorian crisis of faith.

REFERENCES

Ashforth, Albert. *Thomas Henry Huxley*. N.Y.: Twayne, 1969.

Balfour, Arthur James. *Essays and Addresses*. Freeport, N.Y.: Books for Libraries Press, 1972.

———. *Essays Speculative and Political*. N.Y.: George H. Doran, 1921.

———. *The Foundations of Belief: Being Notes Introductory to the Study of Theology*. London: Longmans, Green, 1895.

———. Manuscript letter to L. J. Maxse, June 4, 1895. Maxse 443 f704. West Sussex Record Office.

Barton, Ruth. "Evolution: The Whitworth Gun in Huxley's War for the Liberation of Science from Theology." *The Wider Domain for Evolutionary Thought*. Ed. D. Oldroyd and I. Langham. Dordrecht: D. Reidel, 1983. 261–87.

Bibby, Cyril. *T. H. Huxley: Scientist, Humanist, and Educator*. London: Watts, 1959.

Dugdale, Blanche E. C. *Arthur James Balfour: First Earl of Balfour, K.G., O.M., F.R.S.* London: Hutchinson, 1936.

Egremont, Max. *Balfour: A Life of Arthur James Balfour*. London: Collins, 1980.

Greene, John C. "Darwin, Huxley, and Balfour and the Victorian Crisis of Faith." *General Linguistics*. Forthcoming in vol. 35.

Huxley, Leonard. *Life and Letters of Thomas H. Huxley*. 2 vols. London: Macmillan, 1901.

Huxley, Thomas Henry. *Collected Essays*. 9 vols. 1894. N.Y.: Greenwood Press, 1968.

———. "Mr. Balfour's Attack on Agnosticism." *Nineteenth Century* 37 (March 1895): 527–40.

———. "Mr. Balfour's Attack on Agnosticism 2." *Huxley: Prophet of Science.* By Houston Peterson. London: Longmans, Green, 1932. 300–14.

Huxley Papers. Imperial College of Science, Technology, and Medicine Archives, London.

Irvine, William. *Apes, Angels, and Victorians: Darwin, Huxley, and Evolution.* N.Y.: McGraw-Hill, 1972.

Jacyna, L. S. "Science and Social Order in the Thought of A. Balfour." *Isis* 71 (1980): 11–34.

Lightman, Bernard. "Ideology, Evolution, and Late-Victorian Agnostic Popularizers." *History, Humanity, and Evolution: Essays for John C. Greene.* Ed. James R. Moore. Cambridge: Cambridge UP, 1989. 285–309.

———. *The Origins of Agnosticism: Victorian Unbelief and the Limits of Knowledge.* Baltimore: Johns Hopkins UP, 1987.

Mackay, Ruddeck F. *Balfour: Intellectual Statesman.* Oxford: Oxford UP, 1985.

McGeachie, James. "Organicism, Culture, and Ideology in Late Victorian Britain: The Uses of Complexity." *Approaches to Organic Form: Permutations in Science and Culture.* Ed. Frederick Burwick. Dordrecht: D. Reidel, 1987. 211–52.

Moore, James. "Deconstructing Darwinism: The Politics of Evolution in the 1860s." *Journal of the History of Biology* 24.3 (fall 1991): 353–408.

Paradis, James G. *T. H. Huxley:* Man's Place in Nature. Lincoln: U of Nebraska P, 1978.

Peterson, Houston. *Huxley: Prophet of Science.* London: Longmans, Green, 1932.

Root, John David. "The Philosophical and Religious Thought of Arthur James Balfour (1848–1930)." *Journal of British Studies* 19.2 (spring 1980): 120–41.

Turner, Frank Miller. *Between Science and Religion: The Reaction to Scientific Naturalism in Late Victorian England.* New Haven: Yale UP, 1974.

———. *Contesting Cultural Authority: Essays in Victorian Intellectual Life.* Cambridge: Cambridge UP, 1993.

Ward, Maisie. *The Wilfrid Wards and the Transition.* London: Sheed & Ward, 1934.

Wynne, Brian. "Physics and Psychics: Science, Symbolic Action, and Social Control in Late Victorian England." *The Natural Order.* Ed. B. Barnes and S. Shapin. London: Sage, 1979. 167–84.

Young, Kenneth. *Arthur James Balfour: The Happy Life of the Politician, Prime Minister, Statesman, and Philosopher, 1848–1930.* London: G. Bell & Sons, 1963.

CONTRIBUTORS

ALAN P. BARR is professor of English at Indiana University Northwest. His writings about Victorian literature include *Victorian Stage Pulpiteer: Bernard Shaw's Crusade* and articles for *Victorian Literature and Culture* and the *Shaw Review.*

GEORGE R. BODMER is professor of English at Indiana University Northwest. His articles on the relationship between illustration and text have appeared in such journals as the *New Orleans Review,* the *Journal of Popular Culture,* and the *Children's Literature Association Quarterly.*

PETER J. BOWLER is professor of history and philosophy of science in the Department of Social Anthropology at the Queen's University of Belfast, Northern Ireland. His numerous books on evolutionary science include *Darwin: The Man and His Influence; The Eclipse of Darwinism; Evolution: The History of an Idea;* and *The Non-Darwinian Revolution.*

PATRICK BRANTLINGER is professor of English at Indiana University Bloomington and previous editor of *Victorian Studies.* His books on Victorian literature and culture include *Rule of Darkness: British Literature and Imperialism, 1830–1914; The Spirit of Reform: British Literature and Politics, 1832–1867;* and *Fictions of State: Culture and Credit in Britain, 1694–1994.*

MARIO A. DI GREGORIO is associate professor of history of science at the University of L'Aquila, Italy, often dividing his time between L'Aquila and Darwin College, Cambridge; he is also a fellow of the Linnean Society of London. In addition to his study *T. H. Huxley's Place in Natural Science,* he has edited *Darwin's Marginalia* and is working on a scientific biography of Ernst Haeckel.

DAVID KNIGHT is professor of philosophy at the University of Durham, England, and president of the British Society for the History of Science. His books on the history and philosophy of science include *The Age of Science; The Scientific World-View in the Nineteenth Century; Humphry Davy: Science and Power;* and *Zoological Illustration.*

BERNARD LIGHTMAN is professor of humanities and associate dean of arts at York University, Toronto. He is the author of *Origins of Agnosticism* and the coeditor of *Victorian Faith in Crisis* and *Victorian Science and Religion.*

SHERRIE L. LYONS recently finished her Ph.D. in the history of science at the University of Chicago and is now assistant professor at Daemon College, Amherst, New York. She has published a series of articles on Huxley and the history of science.

JOHN R. REED is distinguished professor of English at Wayne State University. Among his books on Victorian literature and culture are *Perception and Design in Tennyson's* Idylls of the King; *Decadent Style; Victorian Will; Victorian Conventions;* and most recently, *Dickens and Thackeray: Punishment and Forgiveness.*

ROBERT G. B. REID is professor of biology at the University of Victoria, British Columbia. Besides articles on biology and the history of biology, he has published *Evolutionary Theory: The Unfinished Synthesis* and is now completing a further study of evolutionary theory.

MICHAEL RUSE is professor of philosophy and zoology at the University of Guelph, Canada. His studies that deal most closely with evolutionary science are *The Darwinian Paradigm; The Darwinian Revolution; Darwinism Defended; But Is It Science? The Philosophical Question in the Creation/Evolution Controversy;* and *Monad to Man: The Concept of Progress in Evolutionary Biology,* which particularly argues the influence of Huxley on evolutionary thinking's subsequent history.

JOEL S. SCHWARTZ is associate professor of biology at the College of Staten Island of the City University of New York. About half of his dozen or so articles appear in the *Journal of the History of Biology* and are about nineteenth-century evolutionary science.

MICHAEL SHORTLAND is associate professor in the history of science at the University of Sydney, Australia. He has authored, coauthored, or edited five books that deal with teaching science and the history of science; he has also written a number of articles on evolutionary science.

BRUCE SOMMERVILLE recently completed his master's degree at the University of Sydney, Australia, with a thesis on H. G. Wells and T. H. Huxley. He is now working as an independent scholar on the history of nineteenth-century English psychology.

PAUL WHITE recently completed his doctorate in the history of science at the University of Chicago. He is now living in England pursuing his research interests in the culture of scientific exploration in Victorian England.